Advanced Industrial Electronics
Second Edition

Other McGraw–Hill titles by Noel Morris:

Control Engineering (Third Edition)
Industrial Electronics (Second Edition)
Digital Electronics for Works Electricians
Logic Circuits (Third Edition)

Advanced Industrial Electronics

Second Edition

NOEL M. MORRIS
Principal Lecturer
North Staffordshire Polytechnic

McGRAW–HILL Book Company (UK) Limited

London · New York · St Louis · San Francisco · Auckland
Bogota · Guatemala · Hamburg · Johannesburg · Lisbon · Madrid ·
Mexico · Montreal · New Delhi · Panama · Paris · San Juan
Saõ Paulo · Singapore · Sydney · Tokyo · Toronto

Published by
McGRAW–HILL Book Company (UK) Limited
MAIDENHEAD · BERKSHIRE · ENGLAND

British Library Cataloging in Publication Data

Morris, Noel M.
  Advanced industrial electronics.–2nd ed.
  1. Electronic apparatus and appliances
  I. Title II. Morris, Noel M. Industrial electronics
  621.381   TK7870

  ISBN 0-07-084694-4

Library of Congress Cataloging in Publication Data

Morris, Noel Malcolm.
  Advanced industrial electronics.
  Includes index.
  1. Industrial electronics. I. Title.
  TK7881.M65 1984   621.381   83-25538
  ISBN 0-07-084694-4

12345 IM 8654

Printed and bound in Malta by Interprint Limited.

Contents

Preface to the Second Edition

Dynamic developments in the field of Electronics have led to new demands not only on engineers but also on College curricula. This edition of *Advanced Industrial Electronics*, together with its sister book *Industrial Electronics*, provides broad coverage both of traditional and new technology. The eleven chapters in the first edition have been thoroughly revised, new material has been added and dated material removed.

The book has been enlarged by the addition of a chapter on microprocessor-based systems together with an appendix containing details of the instruction set of the Intel 8085 CPU. The 8085 CPU is one of the most popular microprocessors in use today, and for this reason it has been selected for illustration in this book. Its instruction set not only contains all the instructions of the 8080 CPU but also contains all the important instructions of the Z80 family of CPUs. The chapter on microprocessor-based systems includes details of the architecture, addressing modes, subroutine operation, and interrupt operation of a typical microprocessor. The chapter also contains a description of a typical programmed input/output port, and goes on to study examples of its use including digital-to-analogue and analogue-to-digital convertors.

Integrated circuits have led to a 'systems' approach to electronics and to digital techniques. To this extent it is an economic proposition, even in small systems, to build networks consisting completely of integrated circuits. Before the electronics engineer can reach this stage, however, he must appreciate the basic design principles of amplifier circuits together with their characteristics and performance specifications. It is with this in mind that the book contains the elements of circuit design principles, together with the necessary information which leads to an understanding of system design.

The book covers the field of advanced electronics, which can broadly be classified into: basic devices, linear and feedback amplifiers, oscillators, digital electronics, high power electronics, and control systems. To provide a unified approach to the subject, devices described in the first chapter are introduced into later chapters as they arise naturally. Thus, in the chapter on logic circuits we find bipolar and field-effect transistor circuits described, including complementary MOS integrated circuits. In order that readers may test their grasp of the subject, a selected list of problems (with solutions) is provided at the end of each chapter.

The analysis of electronic circuits is vital to the understanding of the operation of circuits, and great care has been taken in this book to select the most suitable analytical techniques. As a result, the book should be of wide interest and will be suited to students following a wide range of disciplines and courses, including degree courses and higher TEC Certificate and Diploma courses in Electrical, Electronic, and Telecommunication subjects.

I would like to acknowledge with gratitude the contributions made during the preparation of the book by colleagues and students. The electronics industry at large has also made significant contributions to the material in the book, and I would like to record my thanks for the assistance received; in particular, I would like to record appreciation in this edition of the book to the Intel Corporation and R. S. Components Ltd.

I am greatly indebted to the work put in by my wife on the manuscript, and for all her help during the writing of the book.

Finally, I hope you will enjoy reading this book, which is dedicated to the continuing development of electronics.

NOEL M. MORRIS

1. Active devices

1.1 Basic technology

Perhaps the most useful group of materials used in modern electronics lies in group IV of the periodic table, which includes silicon (Si) and germanium (Ge). In a perfect crystal, the atomic structure of silicon or germanium is in the form of a regular array. At certain electronic energy levels the interaction between the electric field within the crystal and electrons causes the electrons to be reflected so that none are found at these energy levels. This gives rise to the concept of a *forbidden energy band* within the structure, shown in Fig. 1.1. The *valence energy band* is the highest electronic energy level within the semi-conductor at which one would

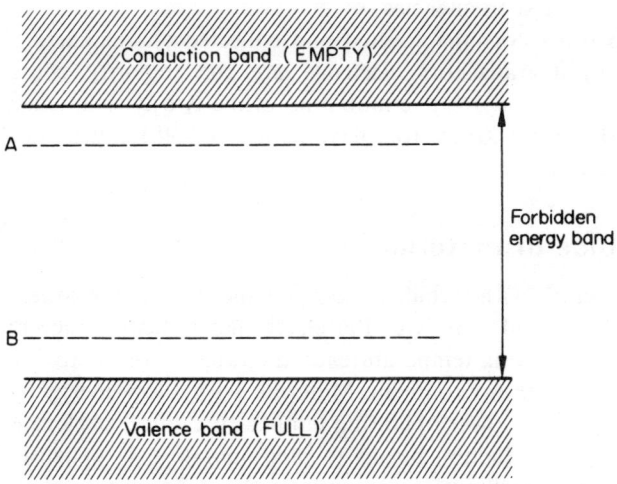

Fig. 1.1   Energy band structure of semiconductors.

expect to find electrons, and electrons must be raised to the empty *conduction band* before conduction can take place.

Electronic energy is measured in *electron-volts* (eV), and the forbidden energy gap in the most useful of the semiconductors is about 1 eV. In germanium it is 0.7 eV and in silicon 1.1 eV. *Intrinsic conduction* takes place in semiconductors when the electrons in the valence energy band have sufficient energy to cross the forbidden energy gap. The introduction of a controlled amount of impurities (about 1 part in 10^8) gives rise to additional available bands in the band gap. In the case of n-type semiconductors, a 'full' level is introduced close to the conduction band, shown at A in Fig. 1.1. In p-type semiconductors, an 'empty' level is introduced close to the valence band, shown at B in Fig. 1.1. As a result, electrons can enter the conduction band in n-type material and leave the valence band in p-type material using less energy (typically 0.02 to 0.07 eV) than is required to cross the forbidden energy gap. This gives rise to *extrinsic conduction*, which is the principal mechanism for bringing about current flow in semiconductor devices.

When an electron leaves a given energy level it leaves behind it a *hole* or positive charge carrier. Since electrons are generally found at higher energy levels than are holes, then the *mobility* (ability to move in an electric field) of the electrons is higher than the mobility of holes. In germanium, the ratio of electron-to-hole mobility is about 2:1 and in silicon about 3:1, with the electron mobility in germanium being about 2.8 times that of silicon. From this it follows that germanium transistors are potentially better for high frequency applications than silicon devices. However, as shown later, the technological developments in the manufacture of silicon devices have outstripped those in germanium devices so that silicon devices can also be used at the highest frequencies.

In p-n-p transistors the emitted and collected charge carriers are holes, while in n-p-n transistors electrons are the majority charge carriers. Consequent upon the relative mobilities of the two types of charge carriers, it follows that n-p-n transistors have higher cut-off frequencies than otherwise similar p-n-p devices.

1.2 Choice of materials

Certain materials with forbidden energy band gaps of the order of 1 eV are well suited to the demands of the electronics industry since they lead to devices with operating temperatures in the range $-100°C$ to $+200°C$, such as are found in our environment. Materials with a low energy band gap are better suited to the lower operating temperature since, at elevated temperatures, intrinsic conduction begins to predominate over extrinsic (impurity) conduction. For instance, germanium with its band gap of 0.7 eV has a maximum operating temperature (typically 85°C) which is high enough to allow it to be used for the manufacture of power transistors. Silicon, with its band gap of 1.1 eV enables transistors to be manufactured which can operate at elevated temperatures (typically up to 200°C).

The material gallium arsenide (GaAs) has a band gap of 1.53 eV and is not a particularly efficient emitting material at room temperature, but it has other valuable properties which make it more suited to specialized branches of engineering, including its use in microwave work. Devices manufactured in GaAs and similar materials include IMPATT [An acronym derived from IMPact ionization and Avalanche Transit Time] diodes and Gunn effect devices (see section 1.12) giving outputs both above and below the X-band (5.2–10.9 GHz).

Gallium phosphide (GaP) has a band gap of 2.26 eV at room temperature and finds its main contribution to electronics in the manufacture of *light-emitting diodes*. The range of values of photon energy visible to the human eye lies between about 1.8 eV and 3.1 eV, so that the light colour radiated by GaP devices depends on device fabrication. Gallium phosphide and gallium arsenide phosphide (GaAsP) devices are used in the construction of several forms of light-emitting diode, the most popular providing red or green displays, which work at low voltage (5 V being typical) with efficiencies between 0.5 and 7 per cent, depending on the colour radiated. These are widely used in alphanumeric displays for panel and instrument meters and other forms of indicator.

The substance indium antimonide, with an energy gap of 0.2 eV does not appear to be of much use in semiconductor devices, yet it is found to have applications where magnetic fields are involved. In the so-called *magnetic diodes*, the resistance of the diode is a function of the external magnetic field and applications include contactless push-buttons, tachogenerators, altimeters, etc.

1.3 Bipolar junction transistors (BJT)

The great majority of BJTs are produced by the *planar technique* in which many thousands of components can be manufactured on a single *slice* of semiconductor material of diameter 2.5 to 7.5 cm (1 in to 3 in). A section through an n-p-n *planar epitaxial* BJT is shown in Fig. 1.2. Construction commences with a heavily doped (n$^+$) substrate of semiconductor material which provides a low resistivity collector region. The thickness of the substrate is, typically, 150 to 200 μm which is about the thickness of two pages of this book. On top of the substrate a lightly doped epitaxial layer is grown, having a depth of about 5 or 6 μm. In Fig. 1.2 the thickness of the epitaxial layer is exaggerated to illustrate the remainder of the construction. The lightly doped *n*-type epitaxial layer forms the high resistance region of the collector which also gives a high collector breakdown voltage.

The resulting semiconductor *chip* is processed further to have p-type and

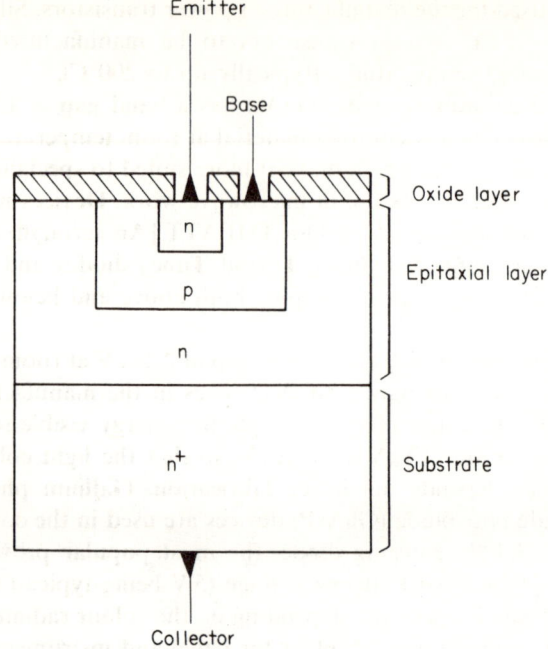

Fig. 1.2 A section through a planar epitaxial BJT.

n-type regions diffused into it. The diffusions are introduced by selectively masking the surface of the semiconductor slice by an oxide layer and subjecting the surface to suitable dopants conveyed by a carrier gas, the slice being raised to a temperature of about 1200°C during this process. Modern technology has reached a stage where it is possible to manufacture transistors with a lateral base thickness (between the collector and emitter) of 0.5 μm. As a comparison, the wavelength of light is in the range 0.4 to 0.7 μm. Finally, contacts are made to the exposed regions by evaporating metal on to the whole of the surface of the chip, and then the unwanted metal is removed by a chemical process. In all, up to about thirty processes are involved in manufacturing a planar BJT.

The discovery that silicon oxide (glass) on the surface of a silicon slice prevented certain dopants from diffusing into the slice gave impetus to the development of silicon transistors. Unfortunately, germanium oxides do not have this property and, in addition, are not chemically stable. The relative simplicity of the silicon planar production process more than outweighs the advantage of germanium transistors in efficient operation. As a consequence, silicon planar epitaxial transistors are the first choice for performance and cost, other materials being selected on their merits in specialized applications.

1.4   Inversion channels in planar BJTs

An unavoidable consequence of silicon planar technology is the introduction of positive charges (ions) into the bulk material of the oxide, shown in
Fig. 1.3. This is due largely to the presence of sodium ions in the oxide. In
planar BJTs, the collector region under the oxide is very lightly doped so
that the number of mobile (majority) charge carriers is not too large. Also
in the collector region are a number of thermally generated minority charge
carriers (electrons in p-type material). The positive charges on and in the
oxide layer attract the electrons to the oxide-semiconductor interface, and
the consequent concentration of electrons can result in an *induced inversion
channel* of n-type material. This channel can cause an effective spreading of
the base region, which may result in unpredictable behaviour of the BJT,
thereby limiting the maximum collector voltage to a low value. The
inversion channel is put to good use in *insulated-gate field-effect transistors*
(FET).

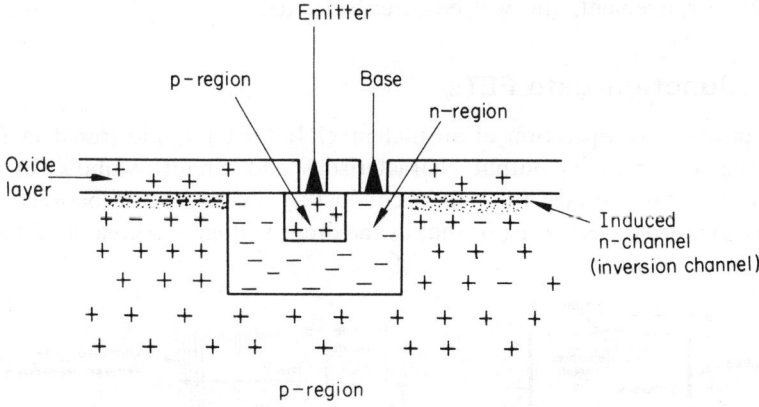

Fig. 1.3   The formation of an induced inversion channel in a BJT.

During manufacture, one method of overcoming the inversion channel is
to diffuse a ring of heavily doped (p^+-type) semiconductor around the
transistor, shown in Fig. 1.4. The heavily doped guard ring prevents the
base region from 'spreading' beyond the annular ring, thereby limiting
inversion channel effects.

1.5   Field-effect transistors (FET)

There are two identifiable types of FET which are

  (a) The junction-gate FET (JUGFET)
  (b) The insulated-gate FET (IGFET or MOSFET).

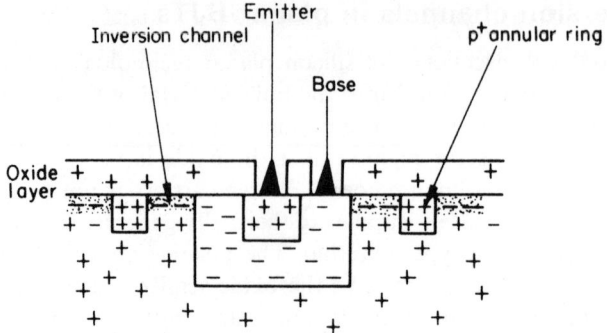

Fig. 1.4   A method of overcoming the effects of the inversion channel.

Insulated-gate FETs are sometimes described as MOSFETs due to their structure, since the geometry between the control electrode (the *gate*) and the conducting *channel* is in the form of a *metal-oxide-semiconductor* (MOS) arrangement, and will be described later.

1.6   Junction-gate FETs

The principle of operation of an n-channel JUGFET is illustrated in Fig. 1.5, together with its output characteristics and circuit symbols. In the absence of gate voltage ($v_{GS}=0$), a conducting channel exists between the *source* and *drain* electrodes, so that as the drain voltage is increased initially

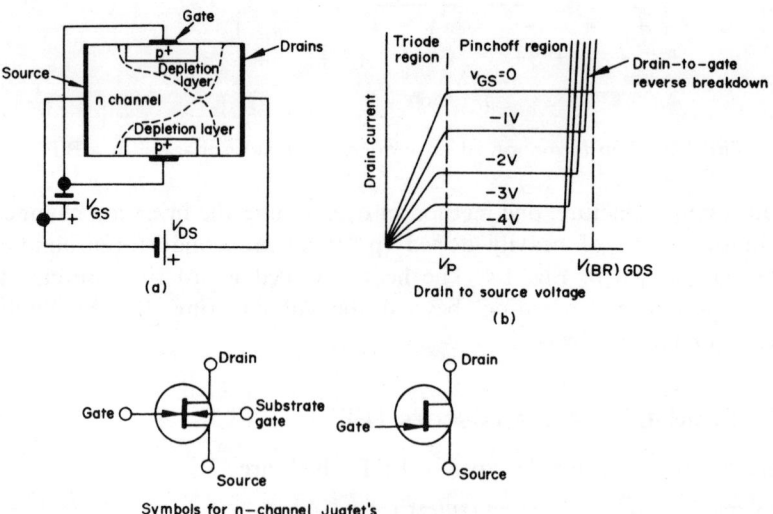

Fig. 1.5   (a) An n-channel JUGFET and (b) the output characteristics.

then the drain current increases in a more or less linear fashion up to the *pinch-off voltage* V_P. Up to this point, the increase in longitudinal p.d. due to the increasing drain voltage causes a depletion layer of increasing depth to be formed between the gate region and the channel, the depth of the depletion region being the greatest at the drain end of the channel. When pinch-off occurs, the drain current is 'pinched' into a thin sheet and beyond that value of v_D the drain current remains approximately constant.

If, now, we apply a negative voltage to the gate we increase the reverse bias applied to the gate junction. We see from Fig. 1.5 that for an increasing negative bias voltage (with an n-channel device) the drain current reduces until, finally, the drain current is cut off entirely. This occurs when v_{GS} is of the same order as V_P. Up to the point of cut-off, the mode of operation is known as *depletion-mode*, since increased (negative) bias increases the depth of the depletion region and depletes the drain current. In this operating state, the gate junction is reverse biased so that the input resistance is very high, typically several hundred megohms.

It is possible to apply a small forward bias (a few hundred millivolts) to the gate before it becomes forward biased. This has the effect of reducing the depth of the depletion layer and increasing the drain current above the $v_{GS}=0$ characteristic. This enhances the drain current and is known as *enhancement-mode operation.*

Generally speaking, devices such as the JUGFET which pass a steady drain current for zero gate voltage are described as *depletion-mode devices.*

The operation of the p-channel JUGFET is generally similar to that described above with the exception that the gate regions are n^+ diffusions and that the drain voltage is negative and the bias voltage is positive. (Note: In both p-channel and n-channel JUGFETs the drain voltage polarity is opposite to the channel type and the gate polarity is the same as the channel type.)

For either channel device in normal operation with the channel pinched off (i.e., constant current operation) the drain current i_D is given by

$$i_D = I_{DSS}\left(1 - \frac{v_{GS}}{V_P}\right)^2 \tag{1.1}$$

where I_{DSS} is the drain-to-source current in the pinch-off region with $v_{GS}=0$ (gate shorted to source). The *transconductance* (or *mutual*) *characteristic* of an n-channel device is shown as curve A in Fig. 1.6.

Due to the manufacturing spread of parameters in any batch of FETs, there is a range of characteristics for that type of device. The limiting curves may be, for instance, curves A and B in Fig. 1.6, so that a FET of a given type has a characteristic which either lies on one of the curves shown, or between the two.

The *transconductance* g_{fs} (or *mutual conductance* g_m) is given by the

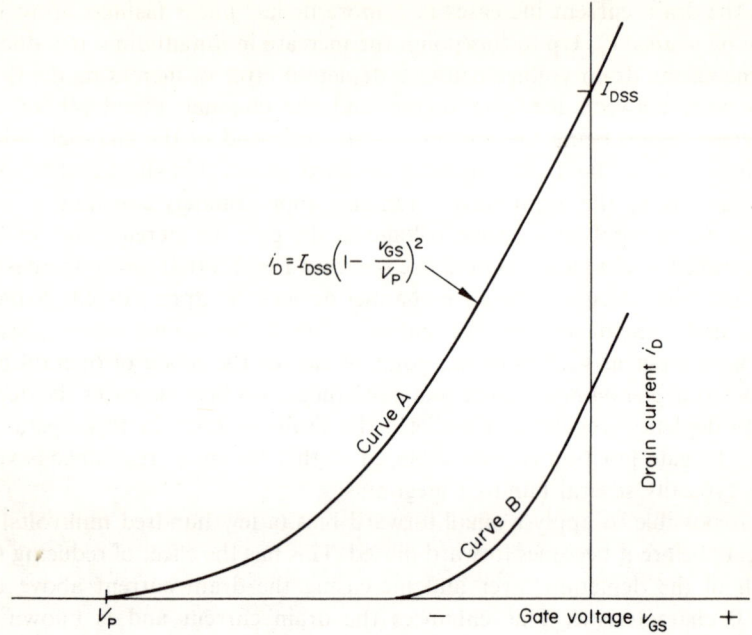

$$i_D = I_{DSS}\left(1 - \frac{v_{GS}}{V_P}\right)^2$$

Fig. 1.6   The transconductance characteristic of an n-channel JUGFET. Curves A and B indicate the limits of the characteristic 'spread' in a typical device.

relationship

$$g_{fs} = \frac{di_D}{dv_{GS}} = -\frac{2I_{DSS}}{V_P}\left(1 - \frac{v_{GS}}{V_P}\right) \tag{1.2}$$

a result obtained by differentiating eq. (1.1). The value of transconductance given in data sheets is normally measured at $v_{GS}=0$, and if we define this value as g_{fso} ($= -2I_{DSS}/V_P$) then the transconductance at any bias voltage V_G is given by

$$g_{fs} = g_{fso}(1 - V_G/V_P) \tag{1.3}$$

If, for example, $g_{fso}=2$ mS and the quiescent bias voltage is equal to $V_P/2$, then the working transconductance is 1 mS. Typical values of g_{fso} lie between about 0.05 mS and 10 mS.

Since the JUGFET is a semiconductor device, any change in ambient temperature affects its operation. Temperature change principally affects the gate leakage current and the drain saturation current I_{DSS}. An increase in temperature causes the gate leakage current to increase in value in much the same manner as in normal p-n junction diodes. In terms of actual values, the gate leakage current lies between a fraction of a nanoampere and a few hundred nanoamperes, but when this flows through a bias resistor of more than a megohm it can cause a significant change in bias voltage. The effect of an increase in temperature on the main conducting channel is to increase its resistance and to modify the mobility of the charge carriers. As a consequence, I_{DSS} reduces with increasing temperature by about 0.3 per cent per centigrade degree.

1.7 Choice of JUGFET operating point

Suppose the characteristic of the FET to be used in an amplifier is as shown in curve A in Fig. 1.7. Clearly, for least distortion we would select the most linear part of the characteristic, which can be seen to occur at a low quiescent value of gate voltage. The value of bias voltage chosen must not, however, be so low as to allow the applied signal to cause the gate junction to be forward biased, otherwise signal clipping occurs. In this region of the characteristic, the value of g_{fs} is at its highest (see eq. (1.2)), and where low distortion is important then this region also provides high gain. Since the drain current consists of charge carriers arriving in a rather random fashion,

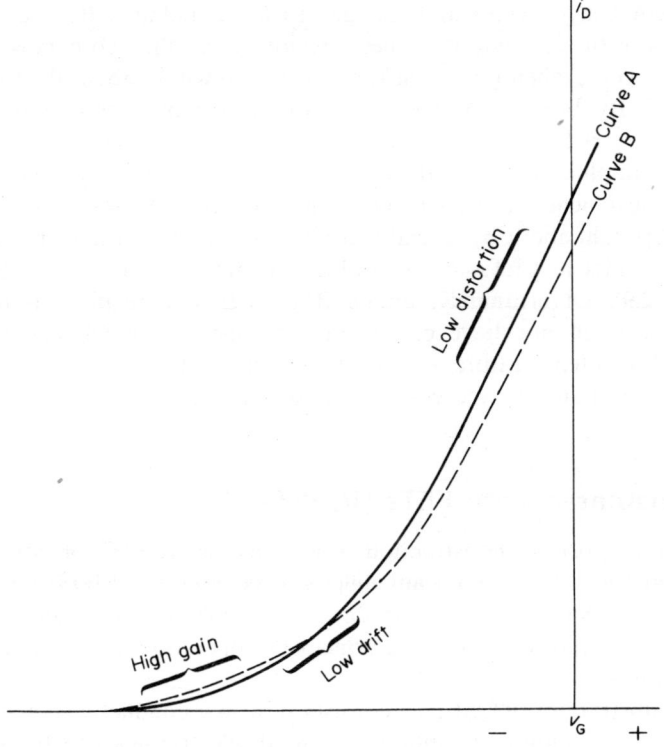

Fig. 1.7 The effect of temperature change on the transconductance curve.

an electrical *noise* signal is generated inside the FET. The larger the drain current, the greater the noise. In most applications JUGFETs are less 'noisy' than BJTs, and for low noise circuits it is necessary to employ a low quiescent drain current.

Where distortion is not of primary interest, the highest values of gain are obtained by biasing the FET nearly to cut-off. That high gains are achieved under these conditions is not apparent at first sight, the reasons being as follows. The operating point of an amplifier working under class A conditions is chosen so that the *quiescent* p.d. across the drain load is more-or-less constant irrespective of the load resistance chosen. That is, if the quiescent drain current is reduced by a factor of, say, ten as a result of increased bias, then a load resistor of ten times the previous value can be used. Now, even if g_{fs} is reduced to one-half or even one-third by the change of operating point, the overall gain is increased since the voltage gain is approximately equal to $-g_{fs}R_L$ (see chapter 2, section 2.3.2). Due to the curvature of this region of the characteristic the output signal will be subjected to some distortion.

The effect of an increase in temperature is to cause the transconductance curve to shift from curve A to curve B in Fig. 1.7, due primarily to variation in depletion layer width and change in carrier mobility. Between the two extremes of the characteristics lies a region where the two curves coalesce, and this occurs when the transistor is biased to within about 0.6 to 0.7 V of pinch-off. By biasing to this region the effects of drift due to thermal changes are reduced to a minimum. This arrangement is convenient in a 'one-off' amplifier in which the components can be carefully selected, but it is not convenient for mass production techniques. Reduced thermal drift can be brought about by special bias circuits (see chapter 2) or by the use of matched pairs of FETs in long-tailed pair amplifiers (see also chapter 2, section 2.9). Unfortunately, unlike BJTs, FETs fabricated on the same silicon slice do not have characteristics which *track* (change with one another) with temperature. As a result, matched FETs are obtained only by testing large numbers and comparing parameters.

1.8   Insulated-gate FETs (IGFET)

A simplified form of construction of a p-channel IGFET or MOSFET is shown in Fig. 1.8. A significant difference between the MOSFET and the JUGFET is that the gate electrode in the MOSFET is isolated from the substrate by an oxide layer. The principle of operation of the device in Fig. 1.8 is as follows.

The substrate of MOSFETs consists of a low conductivity material, so that the application of a suitable potential (electric field) to the semiconductor surface can result in an inversion layer being formed at the surface,

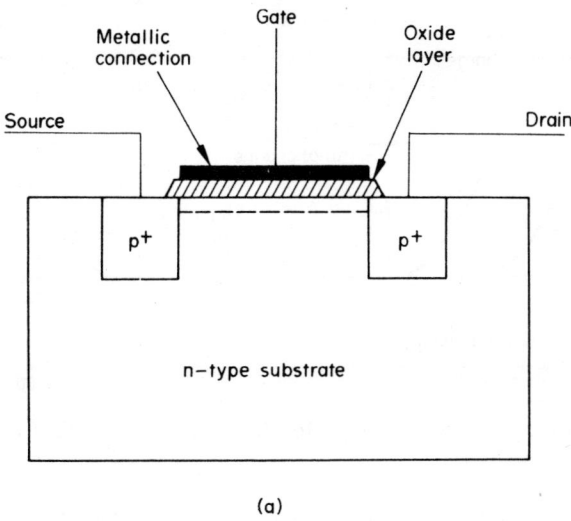

(a)

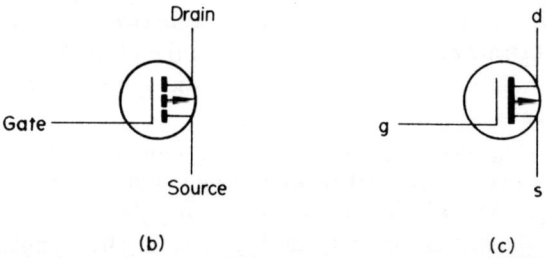

(b)   (c)

Fig. 1.8 (a) A p-channel MOSFET with the circuit symbol for an enhancement-mode device (b) and a depletion-mode device (c).

as described in section 1.4. In the case of an n-type substrate, the application of a negative voltage to the gate causes the minority carriers (holes) to be attracted to the area under the gate electrode, resulting in a conducting p-type channel between the p$^+$-type source and drain regions. The greater the negative gate voltage, the greater the current flow through the FET.

An ideal transconductance characteristic for a p-channel MOSFET is shown as curve A in Fig. 1.9(a). Here we define the *threshold voltage*, designated V_T, as the gate voltage at which drain current flow just commences (the measurement is taken at about 1 μA in practice). As stated earlier, a feature of silicon planar construction is the presence of a positive

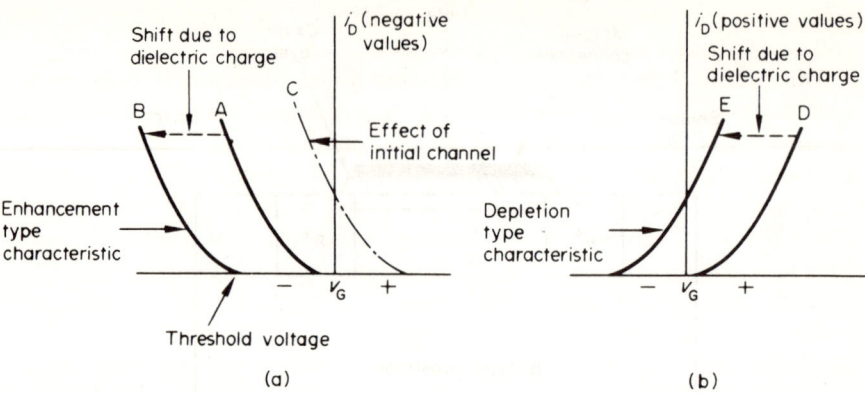

Fig. 1.9 Transconductance curves for (a) p-channel MOSFETs and (b) n-channel MOSFETs.

charge in the oxide layer. This dielectric charge has the effect of shifting the characteristic to the left, giving curve **B** in Fig. 1.9(a), the explanation being that the dielectric charge is equivalent to a positive voltage applied to the gate. Since i_D is zero at $v_G=0$ in both cases and the drain current can only be increased (enhanced) by increasing v_G, then the *natural form* of a p-channel MOSFET is as an *enhancement-mode device*. The net result of increasing the substrate conductivity is to reduce the proportion of minority charge carriers available, so that the characteristic would be shifted further to the left by such a modification.

P-channel *depletion mode* MOSFETs can be manufactured by introducing an *initial p-channel* into the structure during manufacture, shown by the dotted line in Fig. 1.8(a) which links the source and the drain. The resulting characteristic is shown as curve C in Fig. 1.9(a). The application of a positive potential to the gate of the depletion MOSFET causes the drain current to be reduced below the value for $v_{GS}=0$, and a negative gate potential causes an increase in i_D above that value.

The characteristic of an n-channel MOSFET (using a p-type substrate) is shown in curve D in Fig. 1.9(b). In this type of device a positive gate potential is required to attract the minority carriers (electrons) to the oxide-semiconductor interface to form the inversion channel. Basically, the ideal device is an enchancement-mode element since i_D is zero when v_G is zero. However, the positive charge in the dielectric has the effect of shifting the characteristic to the left, to curve E, so that it becomes a depletion-mode device. As a consequence, the *natural form* of an n-channel MOSFET is a depletion-mode device.

Increasing the conductivity of the substrate of the n-channel device has the effect of shifting the characteristic to the right, that is towards curve D, to allow the manufacture of n-channel enhancement-mode devices. The introduction of an initial n-channel would have the effect of shifting the transconductance curve E further to the left.

Due to the higher mobility of electrons when compared with holes, n-channel devices have better high frequency characteristics than p-channel devices. When discrete devices are used in amplifiers they tend therefore to be n-channel depletion devices rather than p-channel MOSFETs. However, in monolithic integrated circuits (see section 1.12) p-channel enhancement mode MOSFETs are used for a number of reasons. These include the fact that the threshold voltage is fairly stable, simple circuit configurations are possible, and the isolation of elements is simplified. Also, the negative gate voltage does not drive the positive sodium ions in the oxide into the oxide-semiconductor interface (as is the case in n-channel enhancement and p-channel depletion devices).

The characteristics of MOSFETs are generally similar in shape to those of JUGFETs, typical sets being shown in Fig. 1.10. The characteristics turn sharply upwards at a high voltage (typically 50 V to 80 V) when breakdown of the drain-to-substrate junction occurs.

Summary: In principle it is possible to manufacture both p- and n-channel depletion and enhancement MOSFETs, but due to the dielectric charge the most readily manufactured types are *p-channel enhancement* and *n-channel depletion* types. The former is more widely used currently because of its application to monolithic integrated circuits.

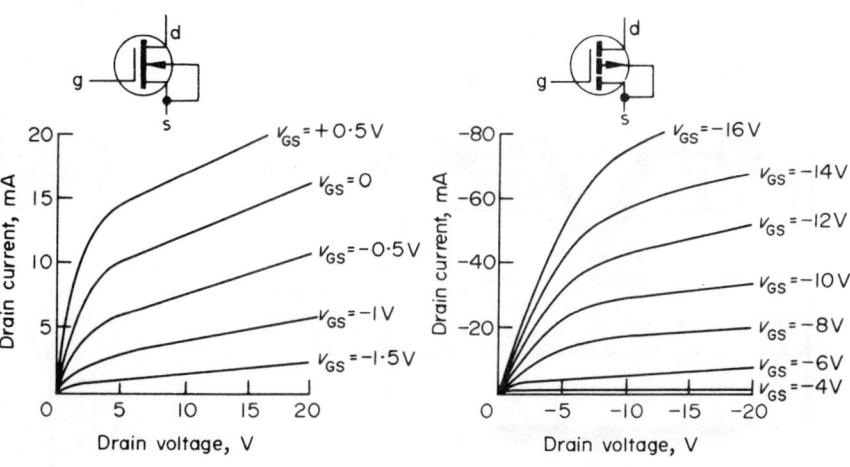

n−channel depletion−type MOSFET output characteristics

p−channel enhancement−type MOSFET output characteristics

(a)

(b)

Fig. 1.10 Static drain characteristic for (a) an n-channel depletion MOSFET and (b) a p-channel enhancement MOSFET.

1.9 Vertical-MOS (VMOS) FETs — power FETs

A major limitation of the FETs described so far is that they are incapable of handling large values of current. With 'horizontal' geometry FETs, i.e., those having current flow parallel to the upper face of the chip, the current density is much less than in devices which utilize 'vertical' current flow, i.e., current flow perpendicular to the upper face of the chip (bipolar transistors fall into the latter category). For a given current rating and using conventional 'horizontal' FET design, the chip area of a high current FET is greater than that of an equivalent bipolar transistor; for a given current, this leads to a higher cost and (due to production problems) a lower production yield.

Vertical-MOS or VMOS technology (developed by Siliconix) has overcome the above problem, enabling MOS transistors to handle voltage and current levels comparable with many bipolar devices (the reader should note that they cannot handle the highest current and voltage levels [typically 400 A and 2 kV] of which bipolar devices are capable).

A section through an n-channel enhancement-mode VMOS FET is shown in Fig. 1.11(a) together with its circuit symbol in diagram (b). The gate region is in the form of a 'V' which is produced between and isolated from two n^+-type source regions (these regions act as the source of negative

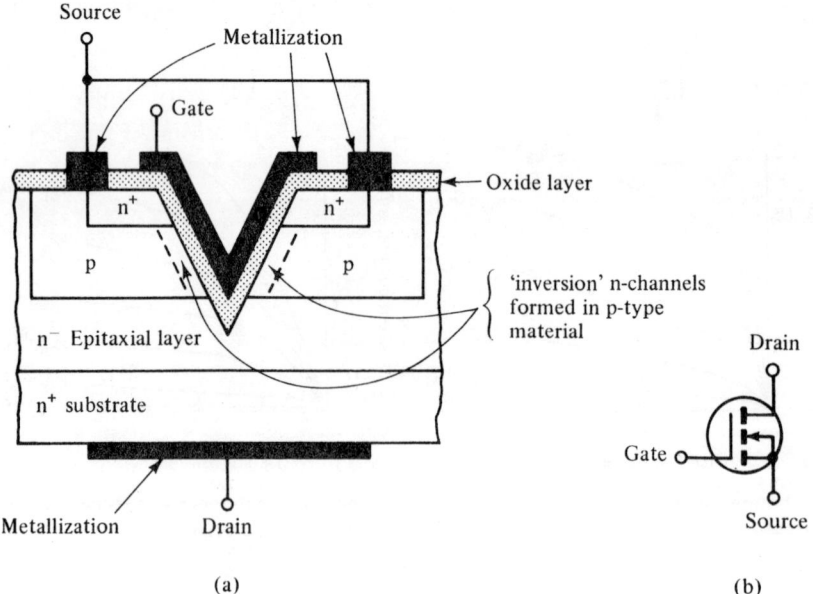

Fig. 1.11(a)   Construction of an n-channel-made VMOS FET and (b) its circuit symbol.

charge carriers). The n-type drain region comprises an n^+ substrate and an n^- epitaxial region; the $n^+ - n^-$ structure not only increases the breakdown voltage of the device but also reduces the feedback capacitance of the FET.

One result of the use of the V-type gate region is to provide two parallel paths for current flow to the drain, one path from each source electrode. This is one reason why the current-carrying capacity of the device is increased when compared with a 'horizontal' FET.

Since the VMOS FET in Fig. 1.11 is an enhancement-mode device, only leakage current flows (typically 10–500 μA) between the source and the drain when the gate voltage is zero. A finite threshold voltage (typically 1.0 V) is necessary to produce an inversion channel in the p-region (see Fig. 1.11(a)); when this occurs, conduction between the source and drain commences. A fairly large value of gate voltage (typically 10 V) is needed to provide full conduction, but the mutual conductance or transconductance, g_m or g_{fs}, of VMOS FETs is fairly constant from about one-half full current up to full current. This means that the output characteristics are uniformly separated over a large part of the drain current – drain voltage graph. The value of g_{fs} depends on the device, and can typically have a value in the range 250 mS to 2.0 S.

The principal advantages of VMOS FETs over power bipolar devices include the following.

(a) They have a very high input impedance and draw a gate current of less than about 100 nA.
(b) Since MOS devices are majority carrier devices, they do not exhibit minority carrier storage time. The time taken to switch a large current is therefore very small; the 2N6657 VMOS FET can, for example, switch 1.0 A either on or off in 4 ns (which is much faster than a comparable bipolar transistor).
(c) Secondary breakdown does not occur (see below).
(d) Current 'hogging' between parallel-connected FETs does not occur (see below).

As mentioned earlier, FETs are voltage-driven devices and do not draw a very large gate current. Consequently, the current gain of a VMOS FET has a value typically in the range 10^9 to 10^{12}.

Secondary breakdown is a phenomenon which can occur in a bipolar transistor and is explained in the following. The resistance–temperature coefficient of a bipolar transistor has a negative value, and when the current density in one part of the transistor increases, the net effect is a further increase in the current density in that part of the transistor (resulting in local current 'hogging'). If this effect continues, the transistor may break down due to local overheating. The resistance–temperature coefficient in VMOS devices is positive, so that local heating results in a redistribution of current away from the hot spot. This provides automatic equalization of current distribution.

For much the same reason, a large value of current can be shared between many parallel-connected VMOS FETs; there is no need to employ either resistors or reactors to ensure current-sharing between parallel-connected VMOS FETs. In fact, many VMOS power FETs comprise many smaller VMOS FETs connected in parallel on the same chip; the device density may be as large as 100 000 per cm^2.

The VMOS FET has a number of disadvantages when compared with a junction device, which include

(a) High cost.
(b) The saturation voltage is high (typically 2 V to 6 V at a current of 2 A to 4 A).
(c) It cannot operate at very high frequency.

1.10   Compatible FET and bipolar transistors for high current

Advances in semiconductor technology have allowed manufacturers to produce FETs and bipolar transistors on the same chip; thus devices can be produced which have the advantages of both types. The principal features of this type of device (described by various trade names including BIFET, SUPERFET, and BIPMOS) include a high input impedance (and therefore high current gain) and low saturation voltage.

A typical circuit arrangement is shown in Fig. 1.12. The input signal is applied between the gate of the FET and the emitter of the bipolar transistor. Resistor R has a low value (typically 10 Ω), and its function is to prevent unwanted turn-on of the bipolar transistor by transients.

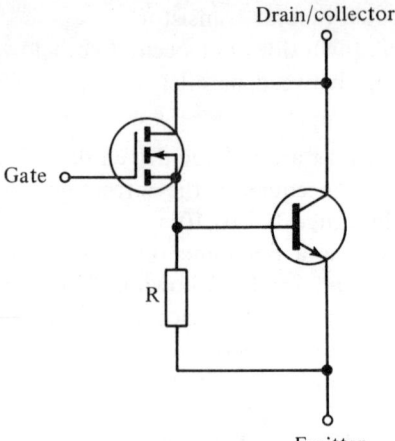

Fig. 1.12   A hybrid FET and bipolar transistor.

The circuit in Fig. 1.12 is produced in a single chip, and the speed-to-power performance of the circuit cannot be reproduced by connecting together the equivalent discrete components, the reason being that the parasitic inductance, resistance, and capacitance of the interconnection of the equivalent discrete component circuit limits the switching speed.

1.11 Developments in MOS technology

Early types of MOS devices had relatively large threshold voltages and low switching speeds. Developments in manufacturing techniques have reduced these limitations, the more important techniques being outlined below.

The threshold voltage of MOSFETs is related to the thickness of gate insulation. By employing insulation which has a higher dielectric strength than silicon oxide, the thickness of the gate dielectric can be reduced. One material which may be used is silicon nitride but, unfortunately, it introduces unwanted changes in the transistor characteristics. A compromise solution used to reduce the effects of the nitride is to deposit a layer of SiO_2 (about $0.06\,\mu m$ thick), followed by a layer of nitride (about $0.04\,\mu m$ thick). The metallic gate electrode is then deposited on the nitride. The resulting structure is known as a *metal-nitride-oxide-semiconductor* (MNOS) transistor, and has a threshold voltage in the region of 2 V.

Another construction is the *silicon-gate* MOS structure. The gate electrode in this structure is a heavily doped silicon region (p-type material in a p-channel enhancement device to reduce the threshold voltage). The gate oxide and silicon gate electrode are deposited before the source and the drain are diffused; this allows the gate region to be used as a 'mask' for the gate and drain diffusions. The net result is a smaller transistor than can be produced by conventional methods with a low value of gate capacitance, which results in a higher switching speed.

Yet another method of increasing the switching speed is by 'implanting' the source and drain regions *after* the aluminium gate electrode has been deposited. Ions can be implanted into silicon by accelerating them to energies of 50 to 100 keV in a mass spectrometer and aiming them at the required area. By this means the ions are implanted to a depth of about $1\,\mu m$. The aluminium gate electrode is used as a mask, and this eliminates overlap of the gate onto the source and drain. The *ion implantation* technique reduces the gate capacitance to allow a higher switching speed to be obtained. Ion implantation is also used to implant gold ions into BJT devices to improve their switching performance (see also section 5.12.1).

1.12 Integrated circuits (IC)

Integrated circuits can be divided into three broadly identifiable groups, namely:

(a) Film circuits
(b) Monolithic circuits
(c) Hybrid circuits.

Film circuits consist of layers or films of conducting and non-conducting materials on a passive (insulating) substrate. In this way passive elements (resistors, inductors, and capacitors) can be fabricated. The technique permits the manufacture of small values of capacitance and inductance (in the form of spiral windings). The values of resistance so manufactured can be controlled very accurately during manufacture by etching techniques.

Film circuits are further divided into thick-film and thin-film circuits on the basis of the thickness of the film deposited on the substrate. Transistors in the form of FETs can also be manufactured in film integrated circuit form, and are known as *thin-film transistors*.

Monolithic integrated circuits are manufactured in a single chip of silicon by a process of selective diffusions of p-type and n-type impurities into an epitaxial layer which is laid down on a substrate of semiconducting material.

A monolithic form of integrated circuit resistor is shown in Fig. 1.13. Here the (p-type) resistor is isolated from the p-type substrate by a diffused n-type isolation region. Unfortunately this process results in the introduction of a parasitic p-n-p transistor in the structure. The area of the chip taken up by the resistor depends on its value, and as a rough guide the area covered by a 2 kΩ resistor is approximately equal to that required by a BJT. Clearly, high values of bias resistor are uneconomic in monolithic ICs in terms of the chip area used. In the manufacture of ICs using BJTs the number of processes involved in preparing a single slice is more or less

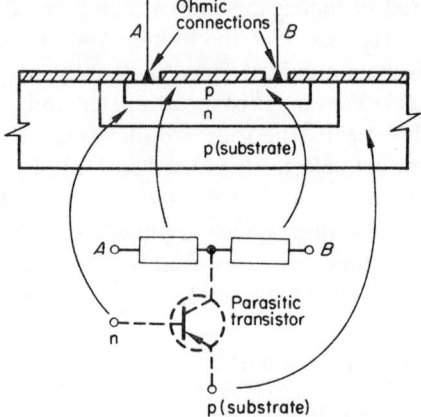

Fig. 1.13 A monolithic integrated circuit resistor.

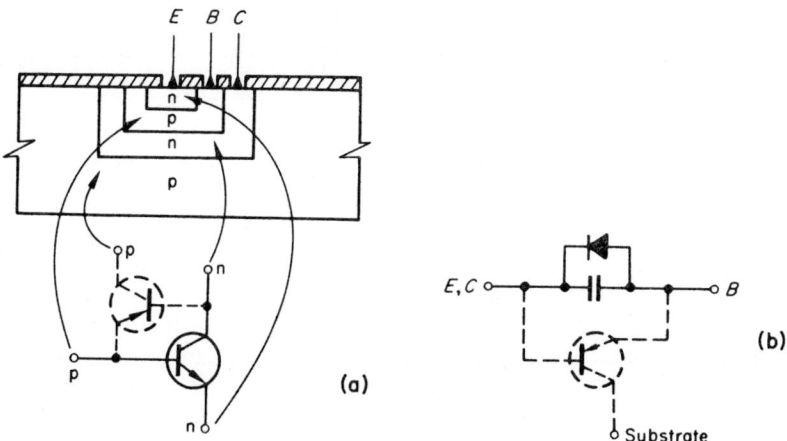

Fig. 1.14 (a) An n-p-n transistor is manufactured by a multiple diffusion process. (b) A capacitor can be formed by reconnecting the transistor as shown.

constant whatever the circuit, so that the cost per slice is constant. It is evident that economies can be brought about by using techniques which use the smallest possible chip area.

An n-p-n transistor is produced by a process of multiple diffusions, as shown in Fig. 1.14(a), while a capacitor can be constructed from the same configuration by reconnecting the transistor as a diode, shown in Fig. 1.14(b), and then to use the capacitance of the diode in reverse bias. In both cases parasitic transistors are introduced into the structure, shown by the broken lines. Small values of capacitance can also be formed by utilizing the oxide layer as a dielectric, the upper plate being a metallic film deposited on the surface of the dielectric and the lower plate being a suitable doped region of the chip. This type of capacitor is sometimes known as a MOS (metal-oxide-semiconductor) capacitor.

Inductors are not normally produced by monolithic techniques and must either be added to the circuit as discrete components or, alternatively, the circuit must be redesigned to operate without inductors.

One form of monolithic IC is shown in Fig. 1.15 in which circuit (a) is manufactured in the form shown in (b). An interesting feature of this circuit is the necessity to provide isolation regions between the circuit elements by means of perpendicular p-regions between the elements. The n-region surrounding the resistor isolates the p-type resistor from the substrate. Further isolation is provided by connecting the substrate to the point of lowest potential in the circuit, thereby reverse biasing the substrate to the rest of the circuit.

Bipolar technology finds application in both digital (ON/OFF) and

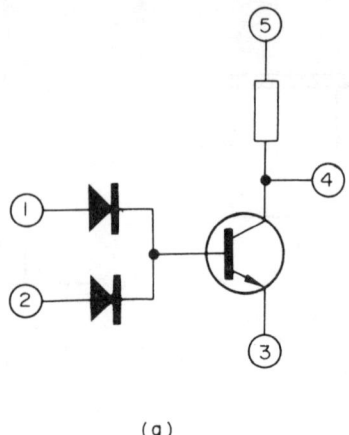

(a)

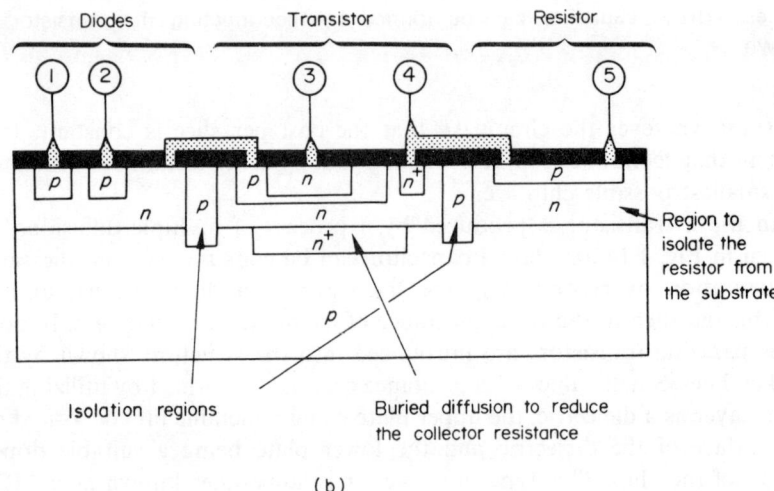

(b)

Fig. 1.15   The circuit in (a) can be manufactured in the monolithic form (b).

linear (proportional) systems and, when compared with FET circuits, can operate at very high frequencies (several thousand megahertz as compared with a few hundred megahertz for the FET). However, in digital electronics MOS elements have several advantages including:

(a) MOS devices do not require an isolation diffusion between them because the junctions are reverse biased.
(b) The chip area required by a MOS device is only a fraction of that of a BJT, the ratio being typically 1:30.
(c) The input resistance of MOS elements is several thousand megohms.

(d) MOSTs can be used as active load resistors, thereby taking up much less area than a diffused resistor.

(e) The production of a MOS IC requires as little as one-fifth the number of production steps as does the equivalent BJT circuit.

As a result of these factors, MOS circuits frequently require fewer components than similar BJT circuits, thereby simplifying production problems and increasing production yield. The saving in chip area by using MOS technology can be as much as 50:1 when compared with bipolar technology. Moreover, in p-channel enhancement MOS circuits (used in digital ICs) the cut-off current can be as low as 1 nA, so that the power loss in this state is only a few nanowatts.

The circuit in Fig. 1.16(a) is that of a 2-input CMOS NOR gate, and that in Fig. 1.16(b) is of a 2-input CMOS NAND gate. These devices use a positive supply voltage and operate in the positive logic notation. Due to the use of n-channel elements in the structure, the propagation time is reduced by a factor of about one-half when compared with p-MOS logic. Moreover when the n-MOS devices conduct, the p-MOS devices in the circuit are cut off, and vice versa; since the resistance of a non-conducting MOS transistor is of the order of 5000 MΩ, the current drain per gate in either logic state is very small. The resistance of the conducting transistor is about 750 Ω, hence the output resistance for either output logic state is relatively low.

In operation, a 'high' value of input voltage causes the p-MOS device to

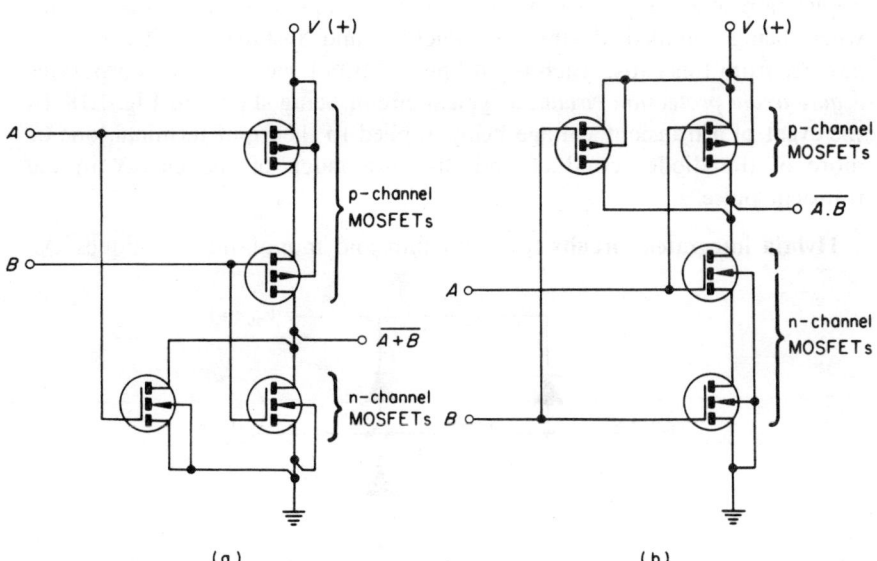

Fig. 1.16   (a) CMOS NOR gate, and (b) CMOS NAND gate

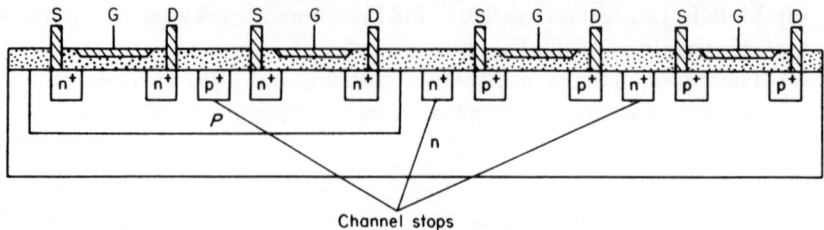

Fig. 1.17   Basic CMOS geometry

be cut off, and the n-MOS device conducts. Only when the output logic level is changing state do both the p- and n-MOS devices conduct simultaneously for a very short period of time; however, the peak magnitude of the current pulse drawn from the supply does not usually exceed about a few hundred microamperes per gate.

A section through a CMOS structure is illustrated in Fig. 1.17. The p-MOS devices are diffused into high-resistivity n-type material, and the n-MOS devices are formed in a 'tub' of p-type semiconductor. Additionally, heavily doped p^+ and n^+ diffusions are introduced between the FETs; these act to prevent undesirable MOSFET action in the regions between the two n-MOS devices and the two p-MOS elements, and are known as *channel stops*. If the channel stops were not introduced, unwanted inversion channels would form between the devices.

Because the gate insulation is very thin, it may easily be ruptured by the application of a comparatively low voltage such as those it may experience when being handled during manufacture and installation. To prevent damage from this cause, each input line of MOS logic families incorporates a *gate-oxide protection circuit*, a typical circuit being shown in Fig. 1.18. In the event of a transient voltage being applied to the input terminal, one or more of the diodes conduct and dissipate much of the energy in the transient pulse.

Hybrid integrated circuits use both film and monolithic techniques. An

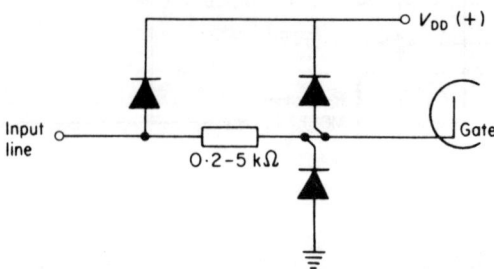

Fig. 1.18   One form of CMOS gate-oxide protection circuit

example would be a circuit in which active elements in the form of BJTs (or even a complete monolithic IC) are mounted on a film circuit. This form of construction permits the manufacture of complete ICs both with film circuit close-tolerance passive components and complex monolithic circuit elements.

The rapid growth of integrated circuit electronics has given rise to its own jargon, and the number of complete logic gates in a single IC package is described as follows.

SSI (small scale integration) — up to 10 gates per IC package
MSI (medium scale integration) — 10 to 100 gates per IC
LSI (large scale integration) — 100 to 1000 gates per IC
VLSI (very large scale integration) — greater than 1000 gates per IC.

1.13 Device configurations

BJTs may be used in any one of three configurations or connections, namely *common-emitter, common-base,* or *common-collector.* The name of the configuration used implies that the electrode mentioned, i.e., the emitter in common-emitter, is the common point in the circuit between the input and output *so far as signal frequencies are concerned.* Similarly, FETs can be operated in either *common-source, common-gate,* or *common-drain* configurations.

1.14 Low frequency equivalent circuits

In this section a number of linear circuits are developed which are electrically equivalent to transistors and other devices when operating as small-signal amplifiers. Active devices can be regarded as having two sections, namely an input circuit and an output circuit.

Before proceeding to discuss circuit details, it is as well to define our general terms of reference. We can identify two distinct types of generator or equivalent circuit, which are the *voltage generator circuit* and the *current generator circuit.* The two basic forms are shown in Fig. 1.19. The voltage generator circuit, Fig. 1.19(a), is found to be particularly convenient when the generator resistance is relatively low (i.e., a few hundred or a few thousand ohms), as in the case of the input circuit of the BJT. The current generator circuit, Fig. 1.19(b), is more convenient when the generator resistance is high, as in the case of the input circuits of FETs and in the output circuits of BJTs, and FETs. The parameters of the two circuits in Fig. 1.19 are related as follows

$$I_S = E_S / r \qquad (1.4)$$
$$R = r \qquad (1.5)$$

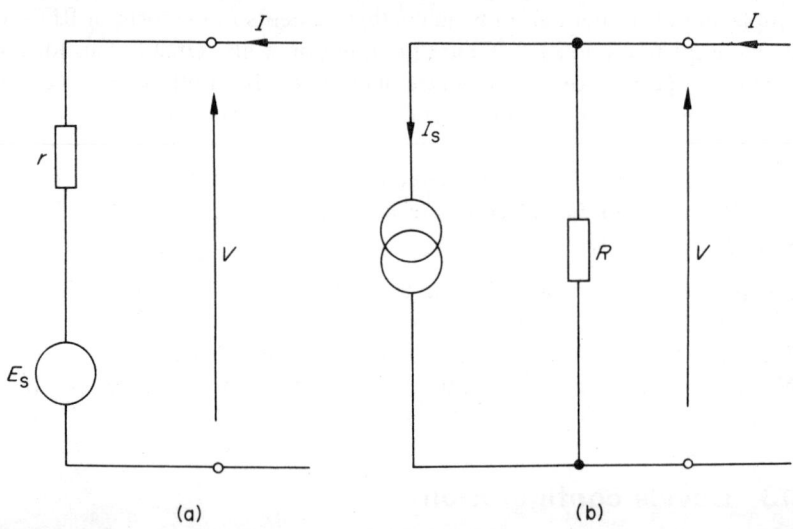

Fig. 1.19   (a) A basic voltage generator equivalent circuit and (b) that for a current source.

1.15   BJT hybrid equivalent circuit

The common-emitter input characteristics of a BJT are shown in Fig. 1.20(a) and it is found on test that the slope resistance lies between a few ohms in power transistors and a few thousand ohms in small signal transistors. It is therefore convenient to use the voltage generator equivalent circuit to represent the input circuit. The output characteristics, shown in Fig. 1.20(b), have a slope resistance between about $50\,\mathrm{k}\Omega$ and $200\,\mathrm{k}\Omega$ (depending on the transistor type), so that the current generator equivalent circuit is used to represent the output circuit.

Although the characteristics described above refer only to the common-emitter mode, the curves for other modes are generally similar. The *general* small-signal equivalent circuit *for any configuration* is shown in Fig. 1.21(a) for which the circuit equations are

$$V_1 = h_i I_1 + h_r V_2 \qquad (1.6)$$

$$I_2 = h_f I_1 + h_o V_2 \qquad (1.7)$$

in which V_1 and I_1 are the input voltage and current, and V_2 and I_2 are the output voltage and current. The subscripts represent the following relationships

i = Input parameter (a resistance) measured with the output on a.c. short-circuit ($V_2 = 0$)

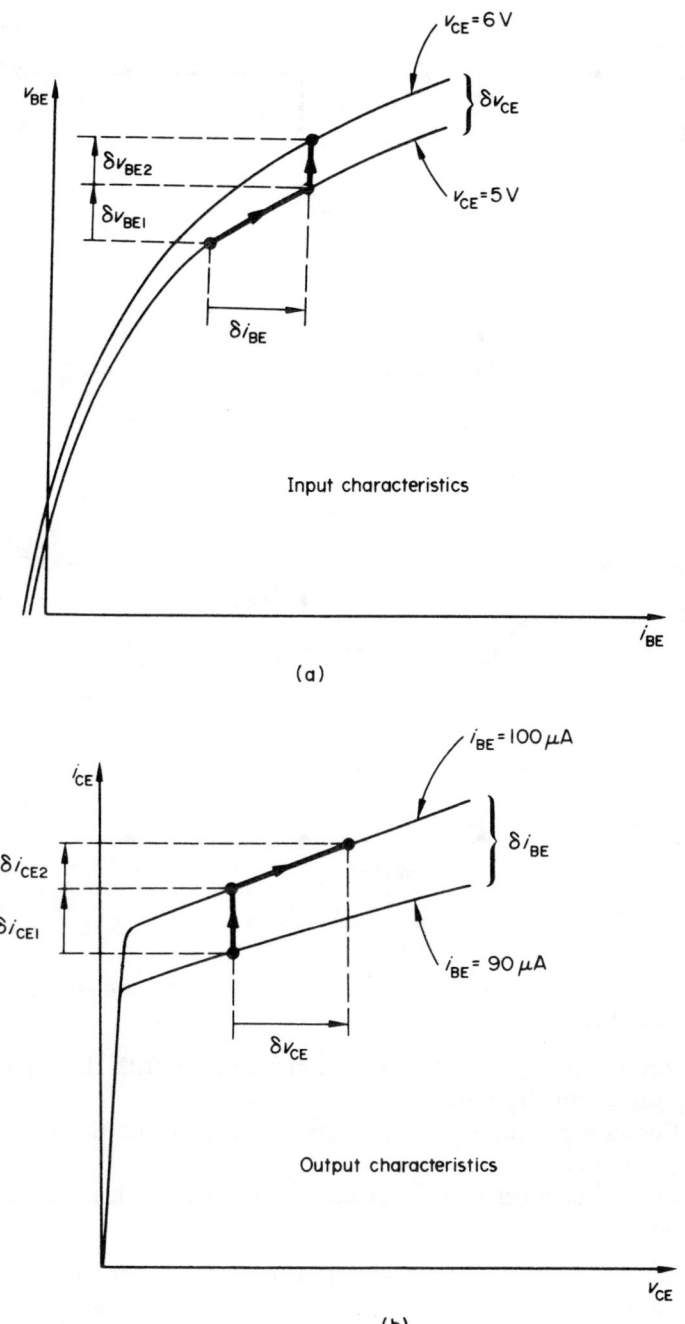

Fig. 1.20 Common-emitter BJT characteristics (a) at the input (base) and (b) at the output (collector).

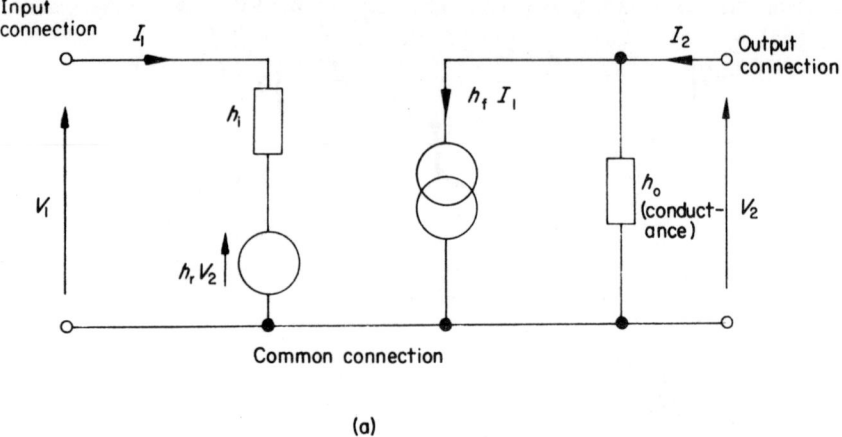

(a)

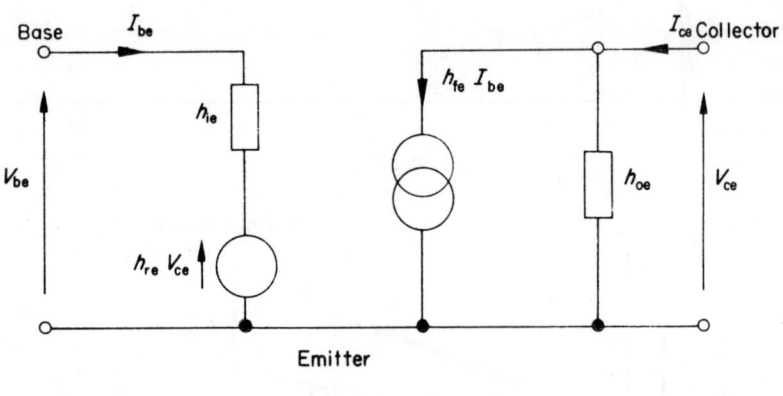

(b)

Fig. 1.21 (a) The general hybrid equivalent circuit for the BJT and (b) the common-emitter version of the circuit.

    r = Reverse parameter (dimensionless) measured with the input on a.c. open-circuit $(I_1 = 0)$

    f = Forward parameter (dimensionless) measured with the output on a.c. short-circuit.

    o = Output parameter (a conductance) measured with the input on a.c. open-circuit.

Since the dimensions of the parameters are mixed, they are described as *hybrid parameters* (*h*-parameters). Depending on the circuit configuration, other subscripts are given to the parameters as follows:

    e = Common-emitter   b = Common-base   c = Common-collector.

Using the above notation, the resulting common-emitter h-parameter small-signal equivalent circuit of the BJT is shown in Fig. 1.21(b), for which the circuit equations are

$$V_b = h_{ie} I_b + h_{re} V_c \tag{1.8}$$

$$I_c = h_{fe} I_b + h_{oe} V_c \tag{1.9}$$

where V_b and I_b are the base (input) voltage and current, and V_c and I_c are the collector (output) voltage and current. A method of evaluating the common-emitter parameters from the static characteristics is shown in Fig. 1.20. From the input characteristics

$$h_{ie} = \delta v_{BE1} / \delta i_{BE} \qquad \text{at constant } v_{CE}$$
$$h_{re} = \delta v_{BE2} / \delta v_{CE} \qquad \text{at constant } i_{BE}$$

and from the output characteristics

$$h_{fe} = \delta i_{CE1} / \delta i_{BE} \qquad \text{at constant } v_{CE}$$
$$h_{oe} = \delta i_{CE2} / \delta v_{CE} \qquad \text{at constant } i_{BE}$$

An interesting point to note in Fig. 1.20(b) is that the output characteristics tend to separate from one another as v_{CE} increases. This is known as the *Early effect* and is an increase in current gain with collector voltage. The reason is that as the collector voltage increases, the width of the collector junction depletion layer increases, thereby reducing the effective base width which in turn leads to increased current gain. In basic circuit analysis its effect can be ignored although, in practice, in common-emitter circuits it may introduce distortion.

1.16   FET small-signal equivalent circuit

The common-source equivalent circuit of the FET is shown in Fig. 1.22(a). In practice, the values of the input parameters y_{is} and y_{rs} are small, and can be omitted from the circuit without significant loss of accuracy. These two terms represent the input conductance of the device which, as stated elsewhere, is practically zero (i.e., the input resistance is practically infinity). The output conductance parameter y_{os} is often small (corresponding to a high output resistance) and is omitted from the simplified equivalent circuit in Fig. 1.22(b). This parameter is also described as g_{os} and r_d (where $r_d = 1/y_{os}$). Parameter y_{fs} is the *transconductance* of the FET and is also described as g_{fs} or g_m (mutual conductance). The simplified equivalent circuit in Fig. 1.22(b) is adequate for most basic applications.*

---

*The *y-parameters* of the FET are described as *admittance* parameters and the *g-parameters* as *conductance parameters*. The admittance of any element is a complex mixture of conductance and susceptance, but since the susceptance of the FET parameters is very small, the terms 'y' and 'g' are freely interchanged in FET analysis. To all intents and purposes, and without any great loss of accuracy, $y_{is} = g_{is}$, $y_{rs} = g_{rs}$, $y_{os} = g_{os}$, etc.

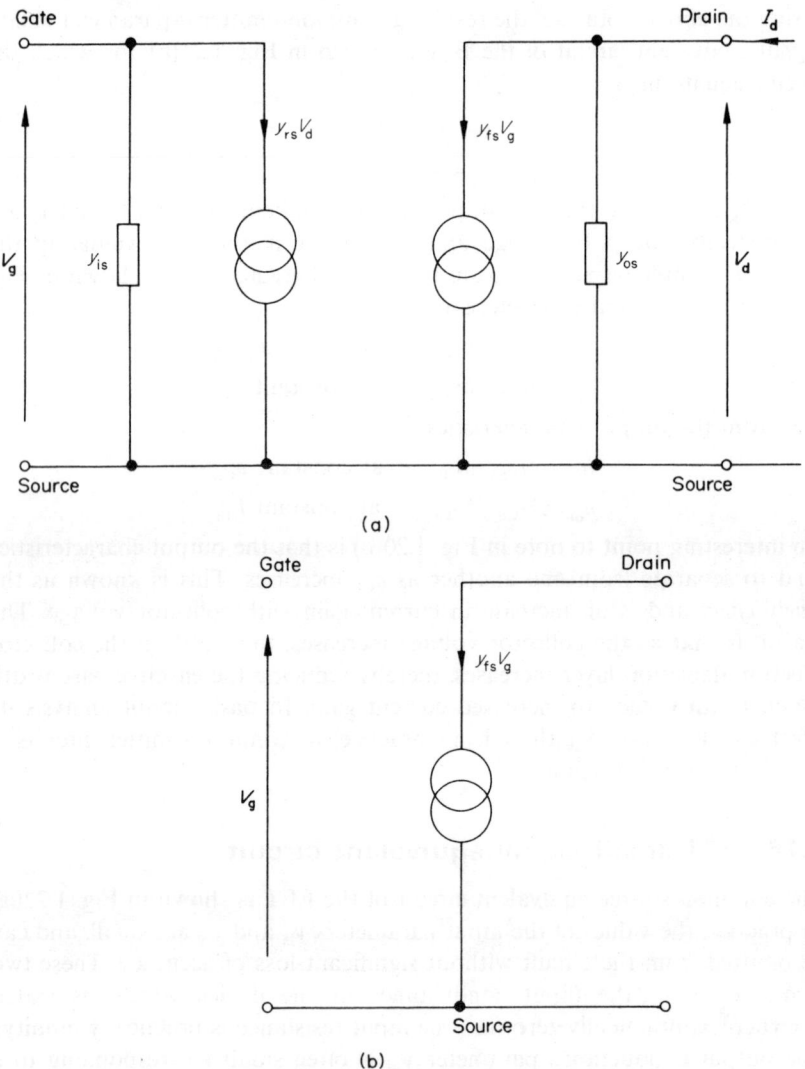

Fig. 1.22   (a) One form of low-frequency equivalent circuit of the FET and (b) a simplified equivalent circuit.

1.17   The BJT at high frequencies

The h-parameters of BJTs are frequency conscious, and at high frequencies the internal capacitive effects can no longer be ignored. Also, the internal capacitance is distributed so that the simple h-parameter circuit does not satisfactorily describe the operation of the BJT at high frequencies, and the *hybrid-π* circuit (Fig. 1.23) is more appropriate.

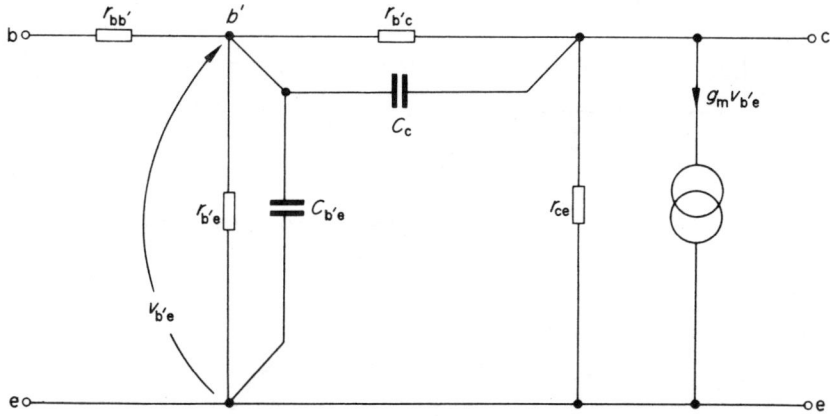

Fig. 1.23   The hybrid-π equivalent circuit of the BJT.

In this circuit, point b is the point at which connection is made to the base electrode, but point b′ is the *effective* base node. The link between b and b′ is a resistive connection $r_{bb'}$ whose value lies between about 100 and 400 Ω. The effective resistance $r_{b'e}$ of the base region has a value between about 1 and 4 kΩ. The feedback resistor $r_{b'c}$ has a value of several megohms, and the value of output resistance r_{ce} lies between about 50 kΩ and 200 kΩ.

Capacitance $C_{b'e}$ is the sum of two capacitances known as the *junction capacitance* C_e and the *diffusion capacitance* C_D. The junction capacitance is dependent on the area of the junction and the voltage across it, whereas the diffusion capacitance is related to the number of charge carriers flowing into and out of the base region. The effective value of $C_{b'e}$ may be up to several thousand picofarad. The collector-base capacitance C_C is the sum of the collector junction depletion capacitance ($\simeq 10\,\text{pF}$) and another small value of capacitance due to internal feedback between the regions of the transistor. Additionally, a capacitance exists between the collector and emitter (not shown in the equivalent circuit) which accounts for other factors such as header lead capacitance and the capacitance of the outer edges of the collector region which are not related to C_C. A further limitation on high frequency performance is the *transit time*, or the time taken for charge carriers to travel from the emitter to the collector.

The effect of the above on the common-emitter current gain of BJTs is illustrated in Fig. 1.24. The general expression for the current gain h_{fe} at any frequency f is given by

$$h_{fe} = \frac{h_{FE}}{1 + jf/f_{hfe}} \tag{1.10}$$

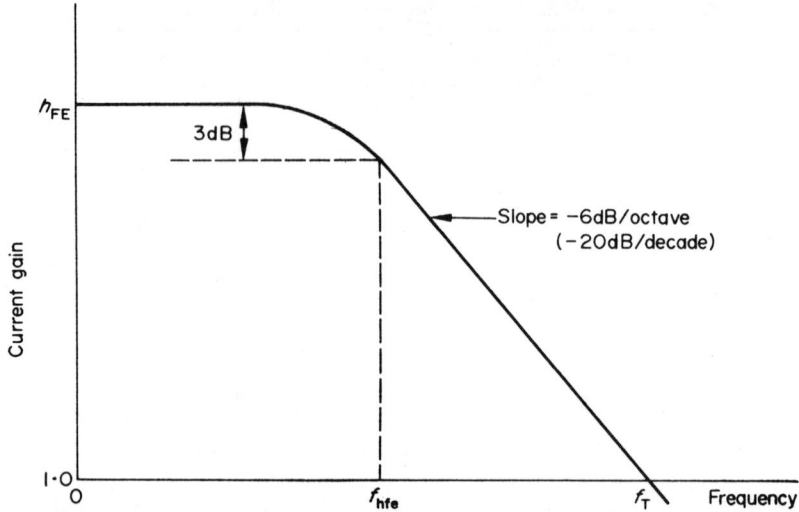

Fig. 1.24 Variation of the common-emitter current gain with frequency.

where h_{FE} is the zero frequency (d.c.) current gain, and f_{hfe} is the frequency at which the (output) a.c. short-circuit current gain has fallen by 3 dB below the low frequency value, i.e., when $h_{fe} = h_{FE}/\sqrt{2}$. This frequency is also known as f_β or the *beta cut-off frequency*.

When $f/f_{hfe} \gg 1$ (i.e., at a frequency above 5 to 10 times f_{hfe}), then the magnitude of eq. (1.10) reduces to

$$h_{fe} = h_{FE}/(f/f_{hfe}) = h_{FE}f_{hfe}/f$$

or

$$fh_{fe} = h_{FE}f_{hfe} \qquad (1.11)$$

Equation (1.11) simply tells us that at frequencies beyond the beta cut-off the numerical value of the *gain-bandwidth product* is constant. That is, if $f_{hfe} = 10$ MHz and $h_{FE} = 100$, then at a frequency of 100 MHz the current gain of the transistor is $100 \times 10/100 = 10$.

The frequency at which h_{fe} falls to unity is known as the *transition frequency* f_T, and from eq. (1.11)

$$f_T \times 1 = h_{FE}f_{hfe} = \text{Gain-bandwidth product}$$

that is, the numerical value of f_T is equal to the gain-bandwidth product of the transistor. Frequency f_T is the maximum frequency at which voltage amplification is possible, and is also approximately equal to the *common-base cut-off frequency* known as f_{hfb}, f_α or the *alpha cut-off frequency*. From the above it can be seen that common-base circuits can operate at much higher frequencies than can common-emitter circuits.

1.18 The FET at high frequencies

In normal operation, the gate junction of the JUGFET is reverse biased so that a depletion capacitance appears between the gate and all regions of the channel. In MOSFETs the oxide insulation between the gate and the channel acts as the dielectric of a capacitor, and capacitive effects appear in a similar manner. It is this capacitance which restricts the high frequency performance of FETs to values below those of BJTs. The values of capacitance are not in themselves particularly large when compared with BJTs, but the input *resistance* of FETs is so large that the effective input time constant is very large indeed, resulting in relatively low cut-off frequencies. However, high frequency FETs working in the common-gate mode can provide useful voltage gains at 400 to 800 MHz.

The capacitive effects on the equivalent circuit of the FET are dealt with in chapter 2.

1.19 Power dissipation and thermal resistance

In power amplifiers we are concerned with the power handling capacity of transistors, so that we must investigate the steady-state relationships which exist in thermal circuits.

At low values of ambient temperature the power dissipation is independent of temperature as shown in Fig. 1.25(a), but above about 25°C transistors have to be derated in order to limit the maximum junction temperature to a safe value. The limiting junction temperature in the case of

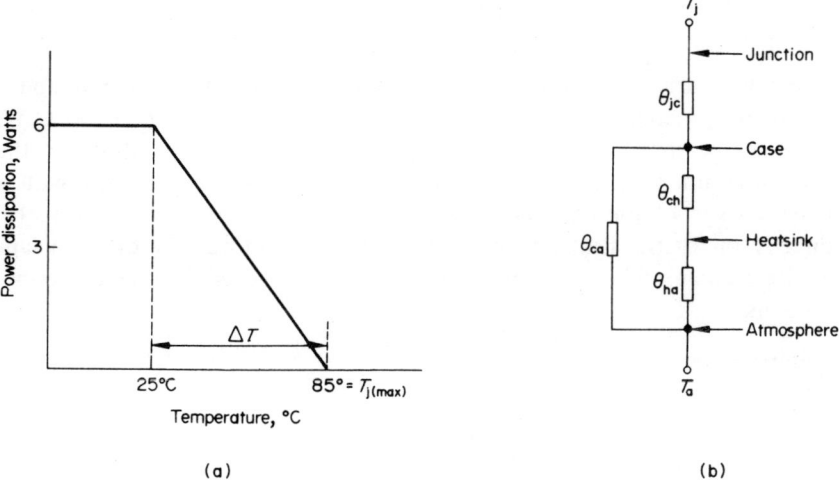

Fig. 1.25 (a) Thermal derating curve for a transistor and (b) an equivalent circuit for heat flow from the transistor.

germanium devices lies between $75°$ and $100°C$ and for silicon devices it is between $150°$ and $200°C$. The *derating factor* is derived from the slope of the derating curve as follows

$$\text{Derating factor} = \frac{\text{Maximum allowable collector dissipation}}{\Delta T}$$

$$= 6 \text{ W}/60°C = 0.1 \text{ W}/°C$$

Typical values range from 1 to 0.001 W/°C

Thermal circuits are analogous to linear electric circuits, the *thermal resistance* θ of an element being given by

$$\theta = \frac{\text{Temperature drop}}{\text{Heat dissipated in watts}} = \frac{\Delta T}{P} \tag{1.12}$$

That is, the thermal resistance of the transistor is the reciprocal of the derating factor. Thus, if the junction temperature is T_j and the ambient temperature is T_a, then the steady-state equation of the thermal circuit is

$$T_j - T_a = \theta P \tag{1.13}$$

where θ is the *total* thermal resistance between the junction and the surrounding air and P is the power dissipated in the transistor. Individual thermal resistors are dealt with in much the same way as linear resistors, and a typical thermal circuit for heat flow from a transistor is shown in Fig. 1.25(b). Here θ_{jc} is the thermal resistance from the junction to the case of the transistor, θ_{ch} is the resistance from the case to the heat sink, and θ_{ha} is the resistance from the heat sink to atmosphere. A path θ_{ca} from the case to air also exists. The principles are now illustrated by means of a worked example.

Example: A transistor has a maximum allowable dissipation of 5 W and a maximum junction temperature of $100°C$. If θ_{jc} is $2°C/W$, θ_{ca} is $30°C/W$, estimate the maximum allowable ambient operating temperature for the transistor mounted (a) on a heatsink of thermal resistance $6°C/W$ with a mica washer of thermal resistance $2°C/W$ interposed between them, (b) directly on to the heatsink without the washer, (c) directly on to a fan-cooled heatsink which can be approximated to a heatsink of zero thermal resistance.

Solution:

Case (a)
$$\theta_{jc} = 2°C/W$$
$$\theta_{ca} = 30°C/W$$
$$\theta_{ch} = 2°C/W$$
$$\theta_{ha} = 6°C/W$$

The total thermal resistance is

$$\theta = 2 + \{30 \times (2+6)/(30+2+6)\} = 8.32°C/W$$

From eq. (1.13)

$$T_a = T_{j(max)} - \theta P = 100 - (8.32 \times 5) = \mathbf{58.4°C}$$

Case (b)

$$\theta = 2 + (30 \times 6/36) = 7°C/W$$
$$T_a = 100 - (7 \times 5) = \mathbf{65°C}$$

Case (c)

$$\theta = 2°C/W$$
$$T_a = 100 - (2 \times 5) = \mathbf{90°C}$$

1.20   Gunn effect devices

Pulses of energy at microwave frequencies are generated in n-type gallium arsenide, indium phosphide (InP), and cadmium telluride (CdTe) semiconductors when the electric field strength across the semiconductor exceeds a critical value (which is in the region of 300 to 350 V/mm). The operation of these devices cannot be explained by any mechanism so far mentioned, and a simplified theory to describe their operation is given here.

The energy band structure is, in fact, more complex than has hitherto been suggested since conduction electrons can either lie in the conduction band or in a satellite band, the minimum energy levels between the two bands differing only by a fraction of an electron volt. Electrons in the conduction band have a higher mobility than those in the satellite band, and under normal equilibrium conditions nearly all the free electrons are situated in the conduction band.

When the electric field strength exceeds the critical value, some electrons transfer to the satellite band, and in so doing give rise to a negative resistance region on the characteristic. Also, at this time, the electric field in the crystal no longer remains homogeneous, and small domains of high field strength are formed. These travel through the material with a velocity of 10^8 mm/s, giving rise to the *travelling domain effect*. The frequency of the microwave oscillations generated in this manner is dependent upon the transit time of the domains, and the length of a sample for a 10 GHz oscillator is 10 μm.

1.21   The unijunction transistor (UJT)

The UJT or *double-based diode* has two ohmic connections to a bar of extrinsic semiconductor material, which is n-type in Fig. 1.26(a). The two connections are known as *base-one* (B1) and *base-two* (B2); a p-n junction is

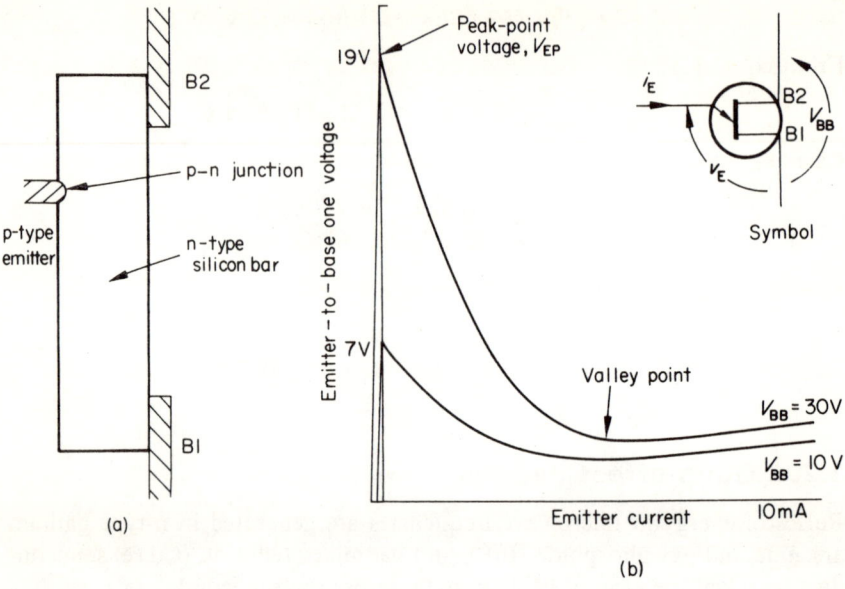

Fig. 1.26   Construction and characteristics of a UJT.

formed between the p-type *emitter* and the bar. When connected to the supply (B2 positive with respect to B1), the bar acts as a potential divider and a p.d. between about $0.4V_{BB}$ and $0.8V_{BB}$ ($V_{BB}=$inter-base voltage) appears between the emitter and base-one. The coefficient above is known as the *intrinsic standoff ratio η*.

When the emitter voltage V_E is less than ηV_{BB}, the emitter junction is reverse biased and no current flows in the emitter circuit. Conduction takes place when the emitter voltage exceeds the *peak-point voltage* V_{EP} (see Fig. 1.26(b)), where

$$V_{EP}=V_j+\eta V_{BB}$$

V_j is the forward bias voltage required for a silicon diode and is typically 0.7 V. Once the UJT commences conducting the resistance between the emitter and base-one falls to a low value. The main field of application of UJTs is to capacitor discharge circuits such as are encountered in pulse generator and timebase circuits.

Other forms of UJT including those with p-type bars are also manufactured.

1.22   Bidirectional breakdown diodes (Diac)

This device is a multi-layer semiconductor device which has a very high resistance between terminals 1 and 2 (see Fig. 1.27) when the voltage applied initially is less than the breakdown voltage V_{BR}. When the applied voltage is equal to V_{BR} the diode exhibits a low resistance and, as in the case of the UJT, can be used in capacitor discharge circuits.

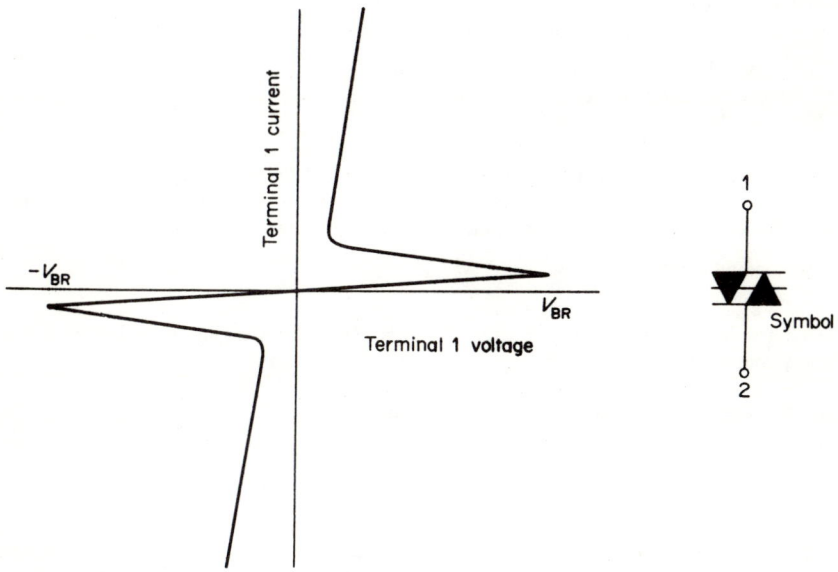

Fig. 1.27   Bidirectional breakdown diode or *diac*.

1.23   Schottky barrier diodes

In these diodes, rectification occurs at a metal-to-semiconductor junction. Several types of metal may be used including gold, molybdenum, titanium, chromium, nickel, nichrome, and aluminium in conjunction with either p-type or n-type semiconductors. The latter type of semiconductor is more frequently used due to the higher mobility of electrons leading to higher operating frequencies.

The principle of operation of Schottky barrier diodes differs from that of p-n junction diodes in that it utilizes the property that certain solids emit electrons under the influence of a strong electric field. Schottky barrier diodes are widely used in switching applications as they do not exhibit carrier charge storage effects (see section 5.11), as do junction diodes. This enables switching speeds of less than 0.1 ns to be achieved.

Problems

1.1 (a) Describe with a sketch the construction of either (i) a planar transistor, or (ii) an alloy junction transistor.

(b) State four differences in the electrical properties between (i) a germanium alloy junction transistor, and (ii) a silicon epitaxial transistor of comparable size.

Why is silicon used more than germanium in the construction of modern transistors?

1.2 Describe the steps taken in constructing a silicon n-p-n epitaxial planar transistor, using sketches as necessary.

Briefly explain why this transistor will have the following electrical properties

(i) high collector-to-emitter breakdown voltage
(ii) low collector-to-emitter leakage current
(iii) low collector-to-emitter saturation voltage.

1.3 Sketch a test circuit for obtaining the collector current-collector voltage characteristic of a transistor. Give a brief description, with typical resulting curves, of an experiment determining these characteristics for a transistor connected in the following configurations

(a) common-emitter
(b) common-base.

Sketch a simple transistor amplifier circuit for *either* (a) *or* (b), showing the polarities of the battery connections.

1.4 The input characteristics of a low power transistor are

$i_{BE}(\mu A)$	v_{BE} (mV) atV_{CE} of	
	5 V	10 V
−5	68	75
0	103	108
5	125	128
10	140	142
15	152	155
20	162	165
25	171	174
30	180	183

The output characteristics are linear between the following points

v_{CE} (V)	2	10	
i_C (mA)	1.9	2.9	when $I_B=20\,\mu A$
i_C (mA)	2.7	3.75	when $I_B=30\,\mu A$

Estimate the values of the four *h*-parameters of this transistor if the quiescent base current and voltage are 25 μA and 171 mV, respectively, and the quiescent collector current is 2.7 mA.

1.5 A 500 μF capacitor is charged to 20 V and then connected as shown in Fig. 1.28. If the emitter current is adjusted to 0.505 mA before the capacitor is connected, draw a graph of

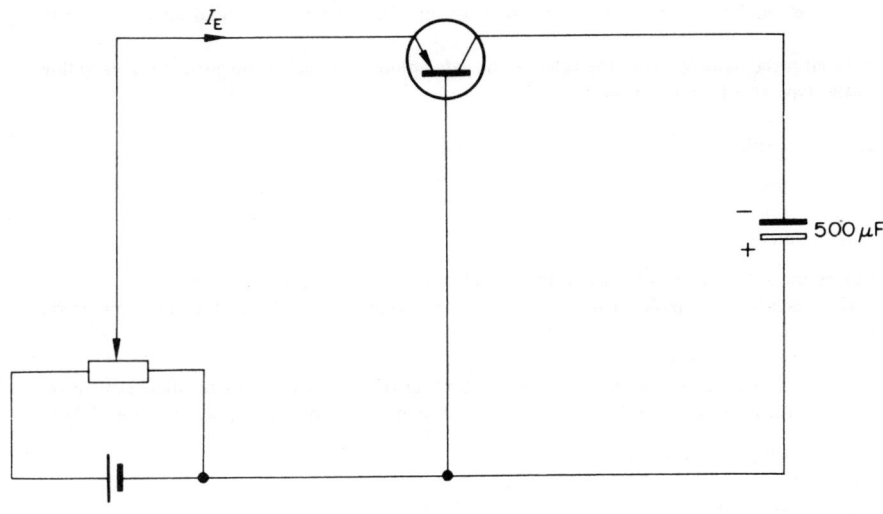

Fig. 1.28

capacitor voltage with respect to time from the instant that the connection is made. The 'h' parameters of the transistor are $h_{FB} = -0.99$, $h_{OB} = h_{RB} = 0$. Show all calculations. What would be the effect of the output voltage if h_{OB} had a finite value?

1.6 Sketch the construction and describe the principle of operation of a junction-gate field-effect transistor. Explain what is meant by *pinch-off* and show its effect on the drain characteristic curves.

Draw a circuit diagram of an automatically biased junction-gate FET used as a class A amplifier in the common-source mode.

Compare its input resistance and voltage gain with

(a) a junction transistor as a common-emitter voltage amplifier.
(b) a junction transistor as a common-base voltage amplifier.

1.7 Describe a test circuit suitable for the determination of the output characteristics of a junction-gate field-effect transistor. State which types of instrument you would use in the circuit, and give reasons for their selection. Indicate on the circuit diagram the polarities of the gate and source voltages.

1.8 In a test on a junction-gate FET the following results were obtained. At a constant gate voltage of -1 V, an increase in drain voltage from 8 to 10 V caused the drain current to increase from 2.8 to 3 mA. The drain voltage was then held constant and when the gate voltage was changed from -1 to -0.5 V, the drain current changed from 3 to 4.8 mA. Estimate the values of the parameters y_{fs} (mutual conductance [g_m or g_{fs}]) and y_{os} (g_{os}) (reciprocal of the output slope resistance).

1.9 Describe the principle of operation and sketch the standard symbol of *either* of the following insulated-gate field-effect transistors (MOST'S)

(i) depletion type
(ii) enhancement type.

Illustrate your answer with appropriate sketches and draw typical gate and drain characteristic curves.

What is the name given to the value of the gate voltage at which drain current ceases to flow in the type you have described?

1.10 (a) Explain, with appropriate sketches, how

(i) a capacitor
(ii) a resistor
(iii) a diode

can be formed on a silicon slice in the manufacture of an integrated circuit.

(b) Describe any method which can be used to isolate electrically each component on the wafer.

1.11 With the aid of suitable illustrations showing the electrical characteristics, explain the important features of the following diodes, mentioning *one* common application of each type:

(a) silicon junction diode
(b) Zener diode
(c) tunnel diode
(d) thyristor.

1.12 Explain why the a.c. input resistance of a transistor, when used in an amplifier, differs in value from h_{ie}.

1.13 A power transistor which has θ_{jc} of $1.0°C/W$ is mounted on a heatsink of thermal resistance $3.0°C/W$. The maximum allowable junction temperature is $90°C$ and the thermal resistance between the case and heatsink is $1.0°C/W$. How much power can be dissipated when the ambient temperature is $25°C$? What is the rating of the transistor?

2. Amplifiers

2.1 Equivalent circuits

The high output impedance of many types of electronic devices leads to the wide use of current generator equivalent circuits, and it is these which are largely used in circuit analysis in this book.

2.2 *h*-Parameter analysis

The hybrid parameters are derived from a set of equations which define the operation of transistors in terms of the voltages and currents at the input and output of the device. Consequently, it is possible to derive a general set of equations for important features such as current gain, voltage gain, input impedance, output impedance, etc., which are applicable to any configuration, e.g., common-emitter, common-base.

2.2.1 Complete *h*-parameter analysis

The general hybrid equations of the transistor are

$$V_1 = h_i I_1 + h_r V_2 \tag{2.1}$$

$$I_2 = h_f I_1 + h_o V_2 \tag{2.2}$$

where the terms have the meanings defined in chapter 1. A block diagram of the amplifier under analysis is shown in Fig. 2.1(a) in which R_S is the resistance of the signal source and R_L is the resistance of the load. In this as with other amplifier circuits, it is assumed that all currents flow *into* the amplifier terminals. This is a convenient fiction, and in fact we may find in the final analysis that one or more values are negative. This means that the

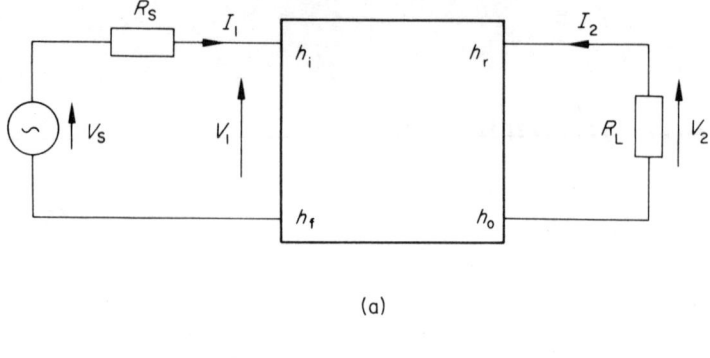

(a)

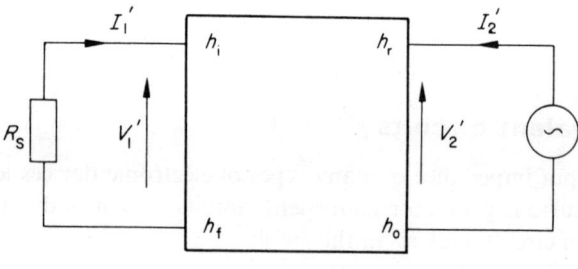

(b)

Fig. 2.1   Amplifier analysis using the h-parameters.

current or voltage concerned actually has the opposite direction or polarity to the one we have chosen.

In Fig. 2.1(a), since the output current flows upwards in R_L, the output voltage is

$$V_2 = -I_2 R_L \tag{2.3}$$

Substituting eq. (2.3) into eq. (2.2) yields

$$I_2 = h_f I_1 + h_o(-I_2 R_L)$$

or

$$I_2(1 + h_o R_L) = h_f I_1$$

giving a current gain of

$$A_i = I_2/I_1 = h_f/(1 + h_o R_L) \tag{2.4}$$

The first step in estimating the input resistance of the amplifier ($R_{in} = V_1/I_1$) is to substitute eq. (2.3) into eq. (2.1).

$$V_1 = h_i I_1 + h_r(-I_2 R_L)$$

But, since

$$I_2 = A_i I_1$$

then
$$V_1 = h_i I_1 - h_r R_L A_i I_1 = I_1 (h_i - h_r R_L A_i)$$

hence
$$R_{in} = V_1 / I_1 = h_i - h_r R_L A_i \qquad (2.5)$$

The overall voltage gain of the amplifier is V_2 / V_1. Now $V_1 = I_1 R_{in}$ and, from eq. (2.3), $V_2 = -I_2 R_L$, then the voltage gain A_v is

$$A_v = \frac{V_2}{V_1} = -\frac{I_2}{I_1} \cdot \frac{R_L}{R_{in}} = -A_i \frac{R_L}{R_{in}} \qquad (2.6)$$

The power gain A_p of the stage is the ratio of the signal power developed in the load $(I_2^2 R_L)$ to the signal power applied to the amplifier $(I_1^2 R_{in})$, hence

$$A_p = \frac{I_2^2 R_L}{I_1^2 R_{in}} = \left(\frac{I_2}{I_1}\right)^2 \frac{R_L}{R_{in}} = A_i^2 \frac{R_L}{R_{in}} \qquad (2.7)$$

Now, since the magnitude of the voltage gain $|A_v|$ without regard to sign is $A_i R_L / R_{in}$, it is possible to rewrite eq. (2.7) in the form

$$A_p = A_i (A_i R_L / R_{in}) = |A_i A_v| \qquad (2.8)$$

The output resistance of the amplifier is derived from Fig. 2.1(b). To determine the output resistance, the input voltage is reduced to zero and the amplifier is energized at its output terminals. The output resistance is given by $R_{out} = V'_2 / I'_2$. The voltage between the input terminals is $V'_1 = -I'_1 R_S$ and substituting this into eq. (2.1) gives

$$-I'_1 R_S = h_i I'_1 + h_r V'_2$$

or
$$I'_1 = -h_r V'_2 / (h_i + R_S) \qquad (2.9)$$

From eq. (2.2) and eq. (2.9)

$$I'_2 = h_f [-h_r V'_2 / (h_i + R_S)] + h_o V'_2 = V'_2 [h_o - h_f h_r / (h_i + R_S)]$$

therefore

$$R_{out} = V'_2 / I'_2 = 1/[h_o - h_f h_r / (h_i + R_S)] \qquad (2.10)$$

The above equations apply to any transistor configuration, i.e., common-emitter, common-base, or common-collector provided that the appropriate parameters are inserted into the equations. Since the common-collector (or emitter follower) amplifier is regarded as a common-emitter circuit with negative feedback applied to it, we need not concern ourselves here with the common-collector equations. The equations for the common-emitter and the common-base configurations are given in Table 2.1.

To illustrate the way in which the various values change with load and source resistance in the common-emitter mode, their variations are plotted

Table 2.1

Basic equations of common-emitter and common-base amplifiers

	Common-emitter	Common-base
A_i	$h_{fe}/(1+h_{oe}R_L)$	$h_{fb}/(1+h_{ob}R_L)$
R_{in}	$h_{ie}-h_{re}R_LA_i$	$h_{ib}-h_{rb}R_LA_i$
A_v	$-A_iR_L/R_{in}$	$-A_iR_L/R_{in}$
A_p	$A_i^2R_L/R_{in}=|A_iA_v|$	$A_i^2R_L/R_{in}=|A_iA_v|$
R_{out}	$1/[h_{oe}-h_{fe}h_{re}/(h_{ie}+R_S)]$	

in Fig. 2.2 for a transistor with $h_{ie}=1.5\,k\Omega$, $h_{re}=0.0002$, $h_{fe}=50$, and $h_{oe}=10^{-5}\,S$. The limiting values of the constants of the common-emitter amplifier depend on the maximum and minimum allowable values of R_L and R_S; a list of absolute limits for the common-emitter configuration is given in Table 2.2.

Example 2.1: Given that the common-emitter parameters of a transistor are $h_{fe}=100$, $h_{ie}=1.75\,k\Omega$, $h_{oe}=50\,\mu S$, and $h_{re}=6.25\times10^{-4}$ at its operating point, calculate the current gain, the input resistance, the voltage gain, the power gain, and the output resistance if $R_L=10\,k\Omega$ and the source resistance is $250\,\Omega$. The bias circuit has no significant effect on the input resistance.

Solution: Using the equations in Table 2.1

$$A_i=h_{fe}/(1+h_{oe}R_L)=100/[1+(50\times10^{-6}\times10^4)]$$
$$=66.7$$
$$R_{in}=h_{ie}-h_{re}R_LA_i=1750-(6.25\times10^{-4}\times10^4\times66.7)$$
$$=1333\,\Omega$$
$$A_v=-A_iR_L/R_{in}=-66.7\times10^4/1333=-500$$
$$A_p=|A_iA_v|=66.7\times500=33\,350$$
$$R_{out}=1/[h_{oe}-h_{fe}h_{re}/(h_{ie}+R_S)]$$
$$=1/[(50\times10^{-6})-100\times6.25\times10^{-4}/(1750+250)]\,\Omega$$
$$=53.4\,k\Omega$$

The relationship between the common-emitter and common-base parameters is of academic interest and it will suffice here to show their relationship in Table 2.3.

Example 2.2: A transistor similar to the one used in example 2.1 is employed in a common-base amplifier with $R_L=10\,k\Omega$ and $R_S=250\,\Omega$. Calculate the values of the amplifier constants.

Solution: From Table 2.3

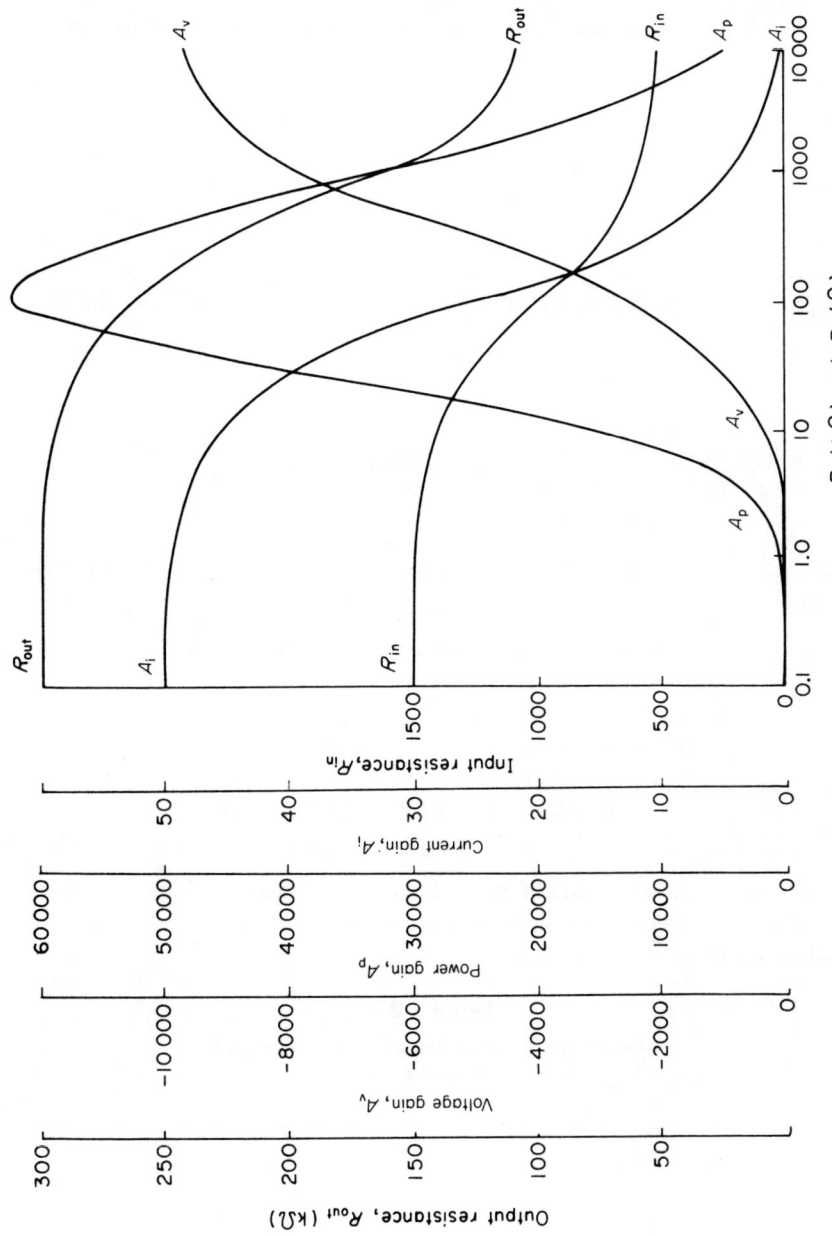

Fig. 2.2 Variation in amplifier 'constants' with R_L and R_s.

Table 2.2

Limiting values of the common-emitter amplifier constants

	Maximum value	Minimum value	Condition
A_i	h_{fe}	0	$R_L \to 0$ $R_L \to \infty$
R_{in}	h_{ie}	$h_{ie} - h_{fe}h_{re}/h_{oe}$	$R_L \to 0$ $R_L \to \infty$
A_v	$-h_{fe}/(h_{oe}h_{ie} - h_{fe}h_{re})$	0	$R_L \to \infty$ $R_L \to 0$
R_{out}	$1/(h_{oe} - h_{fe}h_{re}/h_{ie})$	$1/h_{oe}$	$R_S \to 0$ $R_S \to \infty$

$$h_{ib} = 1750/(1+100) = 17.3 \, \Omega$$
$$h_{rb} = [1750 \times 50 \times 10^{-6}/(1+100)] - 6.25 \times 10^{-4}$$
$$= 2.4 \times 10^{-4}$$
$$h_{fb} = -100/(1+100) = -0.99$$
$$h_{ob} = 50 \times 10^{-6}/(1+100) \simeq 0.5 \times 10^{-6}$$

and from Table 2.1

$$A_i = h_{fb}/(1 + h_{ob}R_L) = -0.99/[1 + (0.5 \times 10^{-6} \times 10^4)] \simeq -0.99$$
$$R_{in} = h_{ib} - h_{rb}R_L A_i = 17.3 - [2.4 \times 10^{-4} \times 10^4 + (-0.99)] = 19.7 \, \Omega$$
$$A_v = -A_i R_L/R_{in} = -(-0.99) \times 10^4/19.7 = 503$$
$$A_p = |A_i A_v| = 0.99 \times 503 = 498$$
$$R_{out} = 1/[h_{ob} - h_{fb}h_{rb}/(h_{ib} + R_S)]$$
$$= 1/[0.5 \times 10^{-6} - (-0.99) \times 2.4 \times 10^{-4}/(17.3 + 250)] \, \Omega$$
$$= 720 \, k\Omega$$

The above examples highlight some features of interest where common-emitter and common-base amplifiers are concerned. The main points are listed opposite.

Table 2.3

Relationships between the common-emitter and common-base h-parameters

$h_{ib} = h_{ie}/(1 + h_{fe})$	$h_{ie} = h_{ib}/(1 + h_{fb})$
$h_{rb} = \dfrac{h_{ie}h_{oe}}{1 + h_{fe}} - h_{re}$	$h_{re} = \dfrac{h_{ib}h_{ob}}{1 + h_{fb}} - h_{rb}$
$h_{fb} = -h_{fe}/(1 + h_{fe})$	$h_{fe} = -h_{fb}/(1 + h_{fb})$
$h_{ob} = h_{oe}/(1 + h_{fe})$	$h_{oe} = h_{ob}/(1 + h_{fb})$

(a) The current gain of the common-base amplifier is approximately unity and has a negative sign associated with it. The latter implies that *either* the emitter current *or* the collector current flows into the transistor, but not both.

(b) The input resistance of the common-base amplifier is low.

(c) The voltage gains in both common-base and common-emitter amplifiers are of the same order.

(d) The power gain of the common-base amplifier is less than that of the common-emitter amplifier.

(e) The output resistance of the common-base amplifier is much greater than that of the common-emitter amplifier.

The above statements are based on the results of examples 2.1 and 2.2. With the normal range of load and source resistances associated with the amplifiers, the statements are generally valid.

2.2.2 Simplified h-parameter analysis

In both common-emitter and common-base configurations, the parameters h_r and h_o are small, so that $h_r V_2 \ll h_i I_1$ and $h_o V_2 \ll h_f I_1$ and the basic equations are reduced to

$$V_1 = h_i I_1 \tag{2.11}$$

$$I_2 = h_f I_1 \tag{2.12}$$

The amplifier constants are obtained as follows. From eq. (2.12)

$$\text{Current gain} = A_i = I_2/I_1 = h_f$$

and from eq. (2.11)

$$\text{Input resistance} = R_{in} = V_1/I_1 = h_i$$

Also from Fig. 2.1(a), $V_2 = -I_2 R_L$ and $V_1 = I_1 R_{in}$, hence

$$\text{Voltage gain} = A_v = V_2/V_1 = -I_2 R_L/I_1 R_{in}$$

$$= -h_f R_L/R_{in} \tag{2.13}$$

$$\simeq -h_f R_L/h_i \tag{2.14}$$

In many cases eq. (2.13) is more accurate than eq. (2.14) since the input resistance may be less than h_i due to the bias circuit. This also applies in the case of eq. (2.16) below.

$$\text{Power gain} = A_p = I_2^2 R_L/I_1^2 R_{in} = A_i^2 R_L/R_{in} \tag{2.15}$$

$$\simeq h_f^2 R_L/h_i \tag{2.16}$$

and

$$\text{Output resistance} = R_{out} = \infty \tag{2.17}$$

Table 2.4

Results of the simplified h-parameter analysis

	Common-emitter	Common-base
A_i	h_{fe}	h_{fb}
R_{in}	h_{ie}	h_{ib}
A_v	$-h_{fe}R_L/R_{in} \simeq -h_{fe}R_L/h_{ie}$	$-h_{fb}R_L/R_{in} \simeq -h_{fb}R_L/h_{ib}$
A_p	$|A_iA_v| \simeq h_{fe}^2 R_L/h_{ie}$	$|A_iA_v| \simeq h_{fb}^2 R_L/h_{ib}$
R_{out}	∞	

since $h_0 = 0$ and $h_r = 0$ (see eq. (2.10)). The results for both common-base and common-emitter configurations are listed in Table 2.4.

Example 2.3: Compute the approximate values of the common-emitter current gain, input resistance, voltage gain, power gain, and output resistance for an amplifier using a transistor with $h_{ie} = 1.75$ kΩ, $h_{fe} = 100$. The load resistance is 10 kΩ.

Solution:
Current gain $= A_i = h_{fe} = 100$
Input resistance $= R_{in} = h_{ie} = 1.75$ kΩ
Voltage gain $= A_v = -h_{fe}R_L/h_{ie} = -100 \times 10^4/1750 = -572$
Power gain $= A_p = |A_i A_v| = 57\,200$
Output resistance $= R_{out} = \infty$

These values should be compared with the values of 66.7, 1.33 kΩ, -500, 33 350, and 53.4 kΩ, respectively, from example 2.1. The principal reason for the differences in value between the two examples is the relatively high value of h_{oe} in example 2.1, which cannot be completely ignored.

Example 2.4: A transistor similar to the one in example 2.3 is used in a common-base amplifier with a load resistance of 10 kΩ. Calculate the values of the amplifier constants.

Solution: From the relationships in Table 2.3 we deduce that $h_{ib} = 17.3\,\Omega$ and $h_{fb} = -0.99$.

Current gain $= A_i = h_{ib} = -0.99$
Input resistance $= R_{in} = h_{ib} = 17.3\,\Omega$
Voltage gain $= A_v = -h_{fb}R_L/h_{ib} = -(-0.99) \times 10^4/17.3 = 572$
Power gain $= A_p = |A_i A_v| = 0.99 \times 572 = 567$
Output resistance $= R_{out} = \infty$

These results should be compared with those of example 2.2, which refer to a transistor in which h_{rb} and h_{ob} were finite. It is seen that, as absolute numerical quantities, the approximate results are not too far in error.

2.2.3 A note on amplifier output resistance and current gain

The output resistance evaluated in the preceding work is, in fact, the output resistance between the collector and emitter (for the common-emitter mode) of the transistor. In practice, it is necessary to include resistor R_L in the collector circuit in order that the amplifier shall work at all! It is more frequently the case that the actual load applied to the amplifier is connected in the a.c. equivalent circuit in parallel with R_L. So far as the external load is concerned, the *effective output resistance of the amplifier* is the parallel combination of R_L and R_{out}, the latter having the value calculated above.

Also, the current gain evaluated above is the ratio of the current in R_L to that flowing into the amplifier. In practice, we may be more concerned with the current flowing in the external load connected in parallel with R_L. Suppose this load is R_ℓ and the current flowing in it is I_ℓ. The effective current gain of the circuit I_ℓ/I_1 has the value $A_i R_L/(R_L+R_\ell)$, where A_i is the current gain in the absence of R_ℓ.

2.3 y-Parameter analysis*

As stated in chapter 1, the y-parameters are convenient for use in FET and multi-electrode valve amplifiers. The general y-parameters y_i and y_r in the input circuit are usually very small and for most practical purposes can be ignored. For completeness, the general relationships are listed below.

$$\text{Current gain} = A_i = y_f/[y_i(1+y_oR_L)-y_ry_fR_L] \tag{2.18}$$

$$\text{Input resistance} = R_{in} = (1+y_rR_LA_i)/y_i \tag{2.19}$$

$$\text{Voltage gain} = A_v = -A_iR_L/R_{in} \tag{2.20}$$

$$\text{Power gain} = A_p = A_i^2R_L/R_{in} = |A_iA_v| \tag{2.21}$$

$$\text{Output resistance} = R_{out} = 1/[y_o-y_fy_r/(y_i+1/R_S)] \tag{2.22}$$

The parameters appropriate to the mode of operation, i.e., common-source or common-gate, can then be inserted into the equations.

2.3.1 Common-source y-parameter analysis (neglecting y_{is} and y_{rs})

The equivalent circuit of the amplifier with y_{is} and y_{rs} omitted is shown in Fig. 2.3. As a result of neglecting the input parameters the input current is

---

*y-parameters are sometimes known as g-parameters. The parameter y_{os} is sometimes described as $1/r_d$, where r_d is the 'drain-source' resistance parameter (Ω). The parameter y_{fs} is sometimes described as g_m, where g_m is the 'mutual conductance' of the FET.

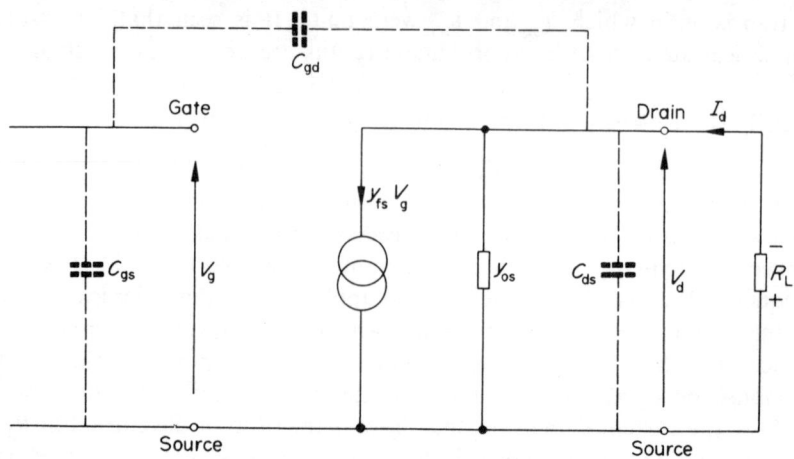

Fig. 2.3 Common-source y-parameter equivalent circuit (neglecting y_{is} and y_{rs}).

zero, hence   Current gain $= \infty$

and   Input resistance $= \infty$

The signal voltage V_d appearing between the drain and source regions is developed across y_{os} and R_L in parallel with one another, that is across a conductance $(y_{os} + 1/R_L)$. Since the current flowing through the parallel circuit is $y_{fs} V_g$, then

$$V_d = -V_g y_{fs}/(y_{os} + 1/R_L) = -V_g y_{fs} R_L/(1 + y_{os} R_L)$$

hence the voltage gain is

$$A_v = V_d/V_g = -y_{fs} R_L/(1 + y_{os} R_L) \tag{2.23}$$

and   Power gain $= |A_i A_v| = \infty$

Since we have neglected the feedback effect between the output and the input, the output resistance of the amplifier is

$$R_{out} = 1/y_{os} \tag{2.24}$$

In FET circuits we must account for the amplification of the capacitance C_{gd} between the gate and the drain, since this is increased in magnitude by the amplification of the stage. This is known as the *Miller effect*. The voltage across C_{gd} while the circuit acts as an amplifier is

$$V_g - V_d = V_g - A_v V_g = V_g(1 - A_v)$$

Thus, the current flowing in C_{gd} is

$$V_g(1 - A_v)/(1/\omega C_{gd}) = V_g \omega [C_{gd}(1 - A_v)]$$

So far as the input circuit is concerned, the gate-to-drain capacitance appears to have a value of $C_{gd}(1-A_v)$. This gives an effective input capacitance to the amplifier of

$$C_{in}=C_{gs}+C_{gd}(1-A_v)$$

In a typical JUGFET, $C_{gs}=5\,\text{pF}$ and $C_{gd}=1.5\,\text{pF}$. With a voltage gain of -19 the input capacitance is

$$C_{in}=5+1.5[1-(-19)]=35\,\text{pF}$$

If the resistance of the signal source is R_S, the input capacitance of the amplifier causes the high-frequency gain to be reduced by 3 dB at a frequency of

$$f=1/2\pi R_S C_{in}\,\text{Hz}$$

If $R_S=1\,\text{k}\Omega$ and $C_{in}=35\,\text{pF}$, this frequency is 4.55 MHz. The effective capacitance shunting the output of the FET is $(C_{ds}+C_{gd})$; for most practical purposes we may assume C_{ds} to be zero so that the effective capacitance shunting the load is C_{gd}. The output capacitance is of course supplemented by any stray capacitance in the circuit.

2.3.2 Simplified y-parameter analysis (neglecting y_{is}, y_{rs}, and y_{os})

In many instances, the output parameter y_{os} is much smaller than the conductance of the load, so that y_{os} can be neglected without significant loss of accuracy. The resulting equivalent circuit is shown in Fig. 2.4. In this case, the following relationships hold good:

Current gain $=\infty$
Input resistance $=\infty$
Power gain $=\infty$
Output resistance $=\infty$

Now, the drain current is $y_{fs}V_g$, hence

$$V_d=-I_d R_L=-y_{fs}R_L V_g$$

giving a voltage gain of

$$A_v=V_d/V_g=-y_{fs}R_L \qquad (2.25)$$

As before, the input capacitance of the amplifier is

$$C_{in}=C_{gs}+C_{gd}(1-A_v)=C_{gs}+C_{gd}(1+y_{fs}R_L)$$

Example 2.5: A field-effect transistor having the parameters $y_{fs}=1\,\text{mS}$, $y_{os}=10\,\mu\text{S}$ is used in a common-source amplifier with a load resistance of

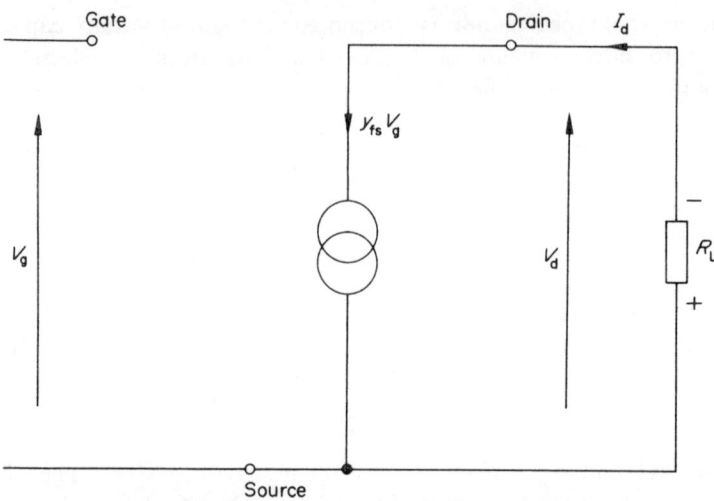

Fig. 2.4 Simplified y-parameter equivalent circuit (neglecting y_{is}, y_{rs}, and y_{os}).

$10\,\mathrm{k}\Omega$. Estimate the voltage gain and the input capacitance if $C_{gs} = 4\,\mathrm{pF}$ and $C_{gd} = 1.3\,\mathrm{pF}$.

Solution: Inserting the values in $\mathrm{k}\Omega$ and mS in eq. (2.23)

$$A_v = -y_{fs}R_L/(1 + y_{os}R_L)$$
$$= -1 \times 10/(1 + 0.01 \times 10) = -9.1$$

and $C_{in} = 4 + 1.3(1 + 9.1) = 17.1\,\mathrm{pF}$

Note: Using the results developed in section 2.3.2, the approximate gain is $A_v = -y_{fs}R_L = -1 \times 10 = -10$, and $C_{in} = 18.3\,\mathrm{pF}$.

2.4 Transistor bias circuits

The value of the bias current injected into the base of a bipolar transistor depends on the point on the characteristics at which we wish to operate the transistor. In class A operation the quiescent collector voltage should be about one-half of the available supply voltage. This places the *quiescent point* in the region of the characteristics in which the collector current is linearly related to base current. This goes some way to ensuring that the output waveform is relatively undistorted by the amplifier.

Basic bias circuits

A basic bipolar transistor bias circuit is shown in Fig. 2.5 in which the quiescent bias current I_B is derived from the supply voltage via resistor R_B.

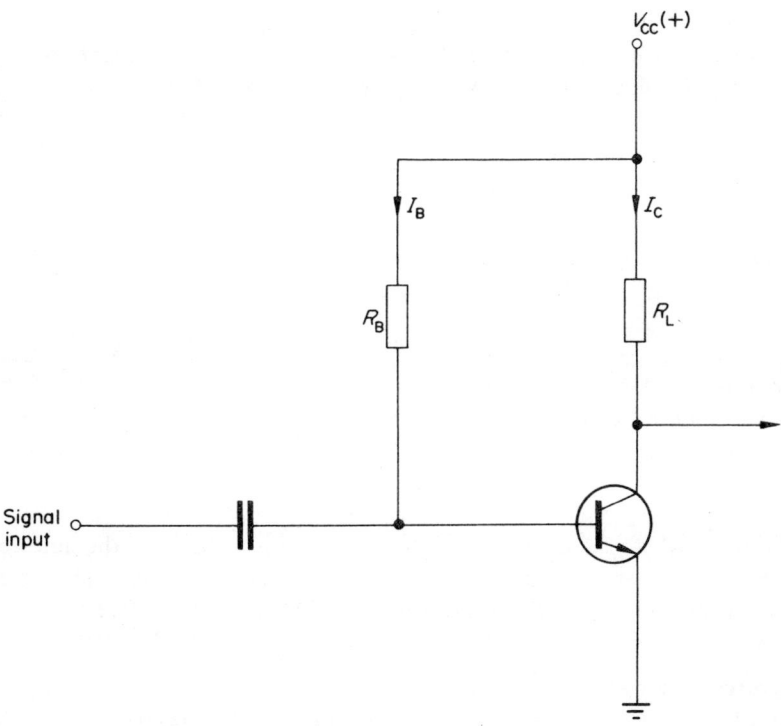

Fig. 2.5 Basic bipolar transistor bias circuit.

The equation for current I_B is

$$I_B = (V_{CC} - V_{BE})/R_B$$

where voltage V_{BE} is the forward conduction p.d. across the emitter junction and is of the order 0.4 to 0.8 V, depending on the transistor. When compared with V_{CC} this voltage can often be neglected, so that $I_B \simeq V_{CC}/R_B$. If the large-signal current gain of the transistor is h_{FE}, then

$$I_C = h_{FE} I_B \simeq h_{FE} V_{CC}/R_B \qquad (2.26)$$

Under conditions of zero input signal, the quiescent collector voltage for class A operation should be about $V_{CC}/2$, that is

$$I_C R_L = V_{CC}/2 \qquad (2.27)$$

Solving for I_C between eq. (2.26) and eq. (2.27) yields

$$\frac{V_{CC}}{2R_L} = \frac{h_{FE} V_{CC}}{R_B}$$

or
$$R_B = 2h_{FE} R_L \qquad (2.28)$$

Example 2.6: Estimate the value of the bias resistor required in a circuit similar to that in Fig. 2.5, given that $V_{CC}=20$ V and the quiescent collector current and voltage are 10 mA and 10 V respectively. Parameter $h_{FE}=80$.

What value of base bias resistor would be required if the quiescent collector voltage was to be 12 V? Assume that the same value of collector resistance is used in both cases.

Solution: The p.d. across R_L when it carries 10 mA is 10 V, hence $R_L=10$ V/10 mA $=1$ kΩ. Hence, from eq. (2.28),

$$R_B=2h_{FE}R_L=2\times 80\times 1=160\,\text{k}\Omega$$

Equation (2.28) was developed for a quiescent collector voltage of $V_{CC}/2$, but the value of R_B can be computed from first principles as follows. Since $h_{FE}=80$, then $I_B=10/80=0.125$ mA. Now $I_B\simeq V_{CC}/R_B$, therefore

$$R_B\simeq V_{CC}/I_B=20\text{ V}/0.125\text{ mA}=160\,\text{k}\Omega$$

In the case where the quiescent collector voltage is to be 12 V, the quiescent collector current is (20–12) V/1 kΩ $=8$ mA. This requires a base bias current of 8 mA/80 $=0.1$ mA, and a bias resistor of 20 V/0.1 mA $=200$ kΩ.

Improved bias circuit

While the bias circuit in Fig. 2.5 is very simple, its main disadvantage is that it does not compensate for the effects of temperature change on the transistor parameters. We are particularly concerned with the effects on the current gain (mainly in silicon devices), on the leakage current (mainly in germanium devices), and on the base-emitter voltage. The net effect of a temperature change on the collector voltage of Fig. 2.5 is a slow variation of voltage (known as *drift*). In voltage amplifiers this has the effect of shifting the quiescent point, which leads to a change in the amplifier constants, e.g., gain, input impedance, etc. In power amplifiers the increased collector current can lead to a rise in junction temperature, followed by a further increase in current. If the heat is generated at a faster rate than it is dissipated into the atmosphere, the effect becomes cumulative and is known as *thermal runaway* and may damage the transistor.

A circuit which offers some improvement in thermal stability is shown in Fig. 2.6. The improvement is brought about by including resistor R_E in the emitter lead. Since we are considering direct current changes only, we shunt alternating currents from R_E through capacitor C_E which is an effective short-circuit to alternating currents at the lowest frequency of interest. To achieve this end we make $X_C\leqslant R_E/10$, so that if ω_ℓ is the lowest frequency to be amplified then $C\geqslant 10/\omega_\ell R_E=10/2\pi f_\ell R_E$. If $R_E=1$ kΩ and the lowest frequency to be amplified is 318 Hz, then $C\geqslant 5\,\mu$F.

In Fig. 2.6, as the temperature increases the net effect of the parameter

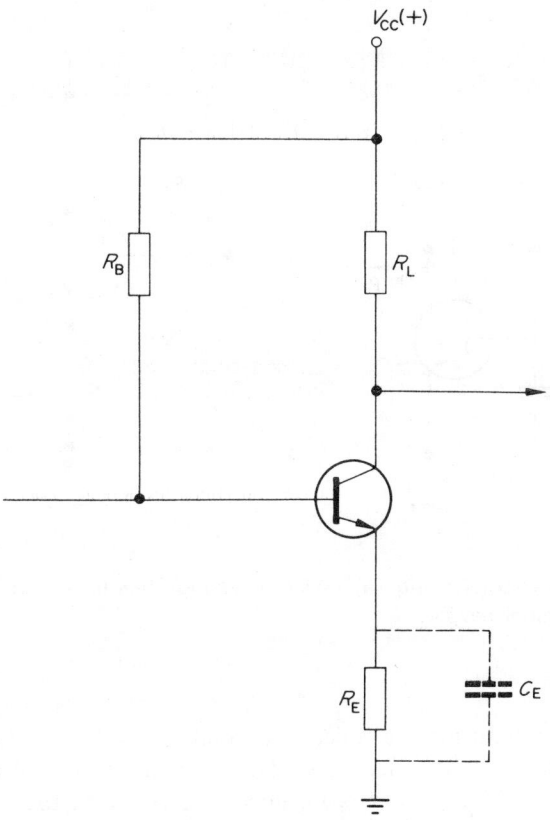

Fig. 2.6   A circuit with improved thermal stability.

changes is for the collector current to increase, thereby increasing the p.d. across R_E. This reduces the base current, which in turn reduces the collector current to a value which is less than it would otherwise be without R_E. The circuit therefore shows better *thermal stability* than the circuit in Fig. 2.5.

Collector feedback bias circuit
The bias circuit in Fig. 2.7(a) employs resistive feedback between the collector and base. For the moment we will ignore the effects of capacitor C_1. In this circuit the base bias resistor is effectively $(R_1+R_2)=R'_B$, say. Making the approximation that V_{BE} is small compared with the quiescent collector voltage V_C, the bias current is approximately

$$I_B=V_C/R'_B \tag{2.29}$$

Let us now consider the effect of temperature change on the circuit. As the ambient temperature rises, the collector current tends to increase and the collector voltage to decrease. From eq. (2.29), this leads to a reduced base

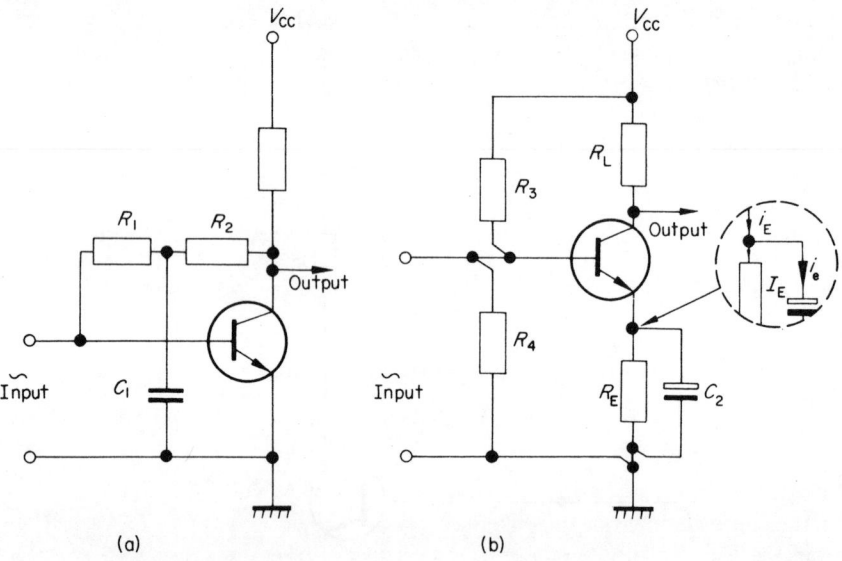

Fig. 2.7 Bias circuits using (a) collector voltage feedback and (b) potential divider and emitter resistor.

current thereby reducing the collector current to a value less than it would be without this method of feedback. The thermal stability of Fig. 2.7(a) is further improved if a resistor is included in series with the emitter, as in Figs. 2.6 and 2.7(b).

Unfortunately, R_1 and R_2 provide a.c. feedback as well as d.c. feedback, and this has the effect of reducing the overall gain (see chapter 3). To limit the effects of feedback at signal frequencies, the junction of R_1 and R_2 is decoupled to earth by C_1. This effectively causes R_1 to shunt the amplifier input and R_2 to shunt the output.

The basic equations of the circuit are

$$I_C = h_{FE}I_B = h_{FE}V_C/R'_B$$

$$V_C = V_{CC} - I_C R_L = V_{CC} - h_{FE}V_C R_L/R'_B$$

hence

$$V_C(1 + h_{FE}R_L/R'_B) = V_{CC}$$

Solving for R'_B yields

$$R'_B = h_{FE}R_L \left/ \left\{ \frac{V_{CC}}{V_C} - 1 \right\} \right. \tag{2.30}$$

So, if the quiescent collector voltage is $0.5V_{CC}$, then $R'_B = h_{FE}R_L$, which is one-half of the value computed for the basic bias circuit.

Example 2.7: Compute the values of the quiescent collector current and voltage of a transistor amplifier of the type in Fig. 2.7(a) given that $R_L = 4.7 \text{ k}\Omega$, $R_1 + R_2 = 560 \text{ k}\Omega$, $h_{FE} = 100$, and $V_{CC} = 10 \text{ V}$.

Solution: From eq. (2.30)

$$\frac{V_{CC}}{V_C} - 1 = \frac{h_{FE}R_L}{R'_B} = 100 \times 4.7/560 = 0.84$$

or $$V_{CC} = 1.84 V_C$$

Therefore, $$V_C = 10/1.84 = 5.44 \text{ V}$$

$$I_C = (10 - V_C)/4.7 = (10 - 5.44)/4.7 = 0.97 \text{ mA}$$

Potential divider and emitter resistor bias circuit

Further improvement in thermal stability is brought about by the use of the circuit in Fig. 2.7(b). Capacitor C_2 is an a.c. bypass capacitor and is only effective at signal frequencies; at very low frequencies (drift frequencies) or zero frequency (d.c.) it is effectively an open-circuit and can be ignored. The base bias current is supplied by the resistor chain $R_3 R_4$, while the emitter resistor provides an emitter voltage which increases with temperature rise in the manner of the circuit in Fig. 2.6. The current drawn by the bias circuit should be of the order of ten times the quiescent base current; this ensures that the potential at the junction of R_3 and R_4 remains fairly constant even when the quiescent base current changes slightly.

When the ambient temperature rises, the emitter current tends to rise initially due to the increase in collector current, but the base voltage remains constant for the reason given above. As a result, the base-emitter voltage is reduced with increased temperature, so compensating for the rise in temperature by reducing the base current. This forces the quiescent collector conditions to return to a near-normal state.

The design of this type of bias circuit is often a compromise between several factors. For good thermal stability the value of R_E should be large, but this leads to increased power loss and a limited output voltage swing. If R_E has a low value then temperature stability suffers.

With a supply voltage of about 10 V a p.d. of about 1 V across R_E is generally adequate, otherwise a value between $0.1 V_{CC}$ and $0.2 V_{CC}$ should be used. As a general rule of thumb, R_4 should not be much more than ten to twenty times greater than R_E.

The thermal stability of Fig. 2.7(b) can be further increased by including a forward biased diode in series with R_4. One effect of an increase in temperature on a diode is to reduce its forward p.d. by about 2.5 mV/°C, so that the voltage at the junction of R_3 and R_4 reduces with increasing temperature. The diode used should be constructed of the same material as the transistor used in the amplifier.

Example 2.8: Estimate the values of the components in an amplifier circuit of the type in Fig. 2.7(a) given that $V_{CC}=10$ V, $h_{FE}=100$, and that the quiescent collector current is 1 mA. The lowest frequency to be amplified is 31.8 Hz.

Solution: Let us assume that a p.d. of 1 V across R_E is adequate. Since $h_{FE}=100$, we may reasonably assume that $I_E \simeq I_C$, so that

$$R_E = 1 \text{ V}/I_E \simeq 1 \text{ V}/I_C = 1 \text{ V}/1 \text{ mA} = 1 \text{ k}\Omega$$

If a silicon transistor is used $V_{BE} \simeq 0.6$ V, giving a base voltage V_B of approximately 1.6 V. The base current is equal to $I_C/h_{FE}=1/100=0.01$ mA, so that R_4 should pass about 0.1 mA. Therefore

$$R_4 = 1.6 \text{ V}/0.1 \text{ mA} = 16 \text{ k}\Omega$$

This value can be reduced to give improved thermal stability at the expense of increased current drain from the supply. We will continue the problem taking R_4 to be 16 kΩ. The current in R_3 is $(0.1 \text{ mA} + I_B)=0.11$ mA, hence

$$R_3 = 8.4 \text{ V}/0.11 \text{ mA} = 76.4 \text{ k}\Omega$$

The value of C_2 is estimated from the relationship $X_{C_2} \leqslant R_E/10$ at 31.8 Hz. This gives

$$C \geqslant 1/2\pi \times 31.8 \times (1000/10) = 50 \ \mu\text{F}$$

FET bias circuits
N-channel JUGFETs can use an automatic bias circuit, as shown in Fig. 2.8. The actual value of the bias voltage used needs to be carefully considered, as a bias of a few tenths of a volt is all that is necessary to provide the maximum gain. However, the effects of temperature change on the JUGFET indicate that bias to about 0.6 V of pinch-off is best for thermal stability. The latter means that the FET must operate in a non-linear region of its characteristics, and one in which the gain may be low. The input resistance of the circuit in Fig. 2.8 is approximately equal to R_G, and if this is to be large then the effects of gate leakage current may be significant. Figure 2.8(b) illustrates the effect of the spread of the transistor curves for a 'family' of transistors. The quiescent drain current may have any value between I_{D1} and I_{D2}. The spread of quiescent drain current values is minimized by the use of the circuit in Fig. 2.9, in which an additional voltage divider network is used. The spread of drain current values in the latter case is from I'_{D1} to I'_{D2}, and is smaller than in the case of Fig. 2.8. The values of R_1 and R_2 are small compared with that of R_3, so that the input resistance is approximately equal to the value of R_3.

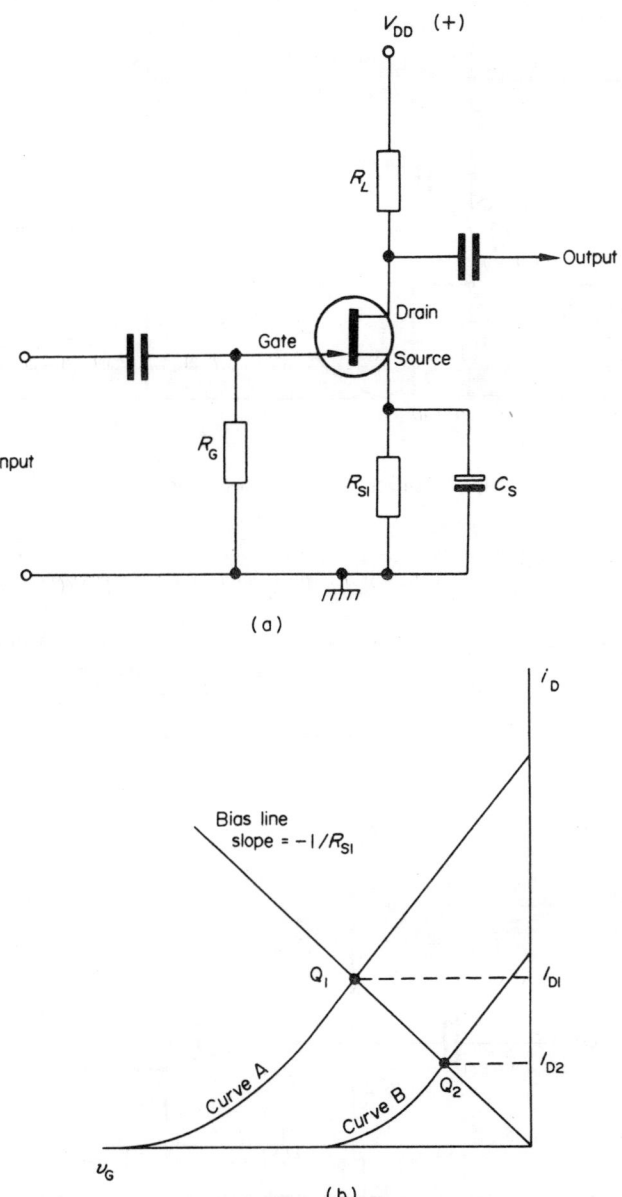

Fig. 2.8   A common-source JUGFET bias circuit.

Enhancement mode IGFETs must be biased on to their characteristics, and one method of doing this is shown in Fig. 2.10. Here the bias voltage at the junction of R_1 and R_2 is applied to the gate of the FET via R_3, which has a large value (say 10 MΩ). Capacitor C acts to decouple the junction of R_1 and R_2 to earth at signal frequencies.

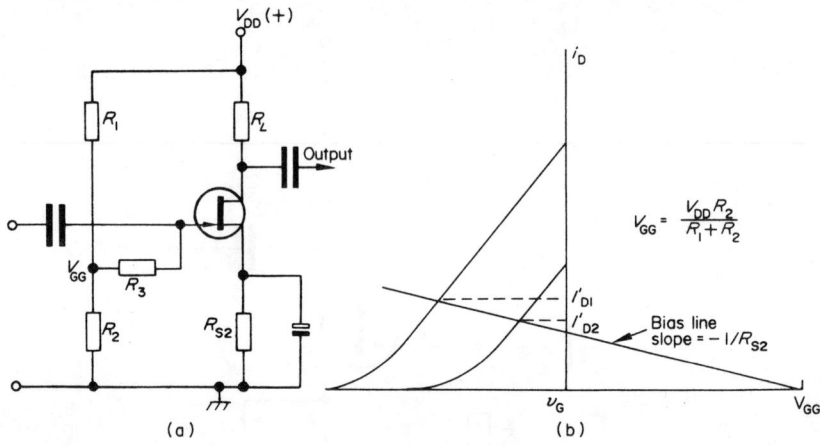

Fig. 2.9   A common-source JUGFET circuit with improved stability.

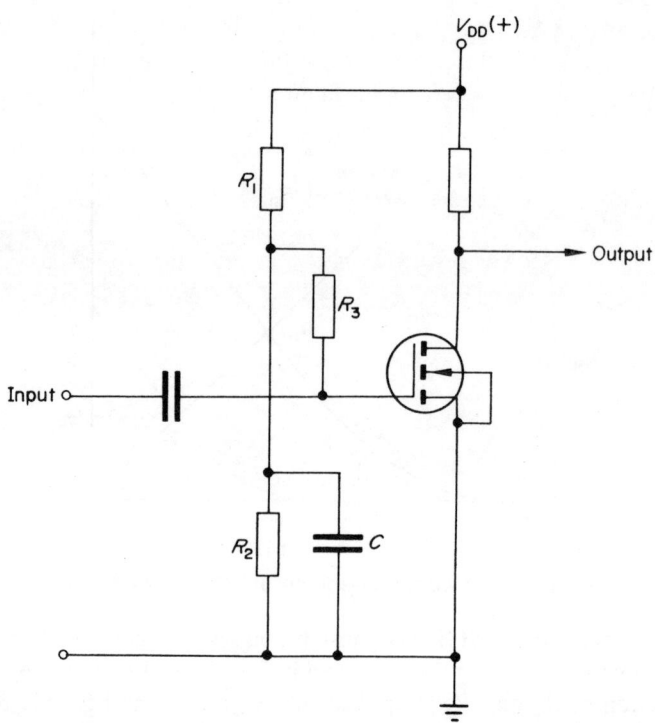

Fig. 2.10   A bias circuit for an enhancement-mode insulated-gate FET.

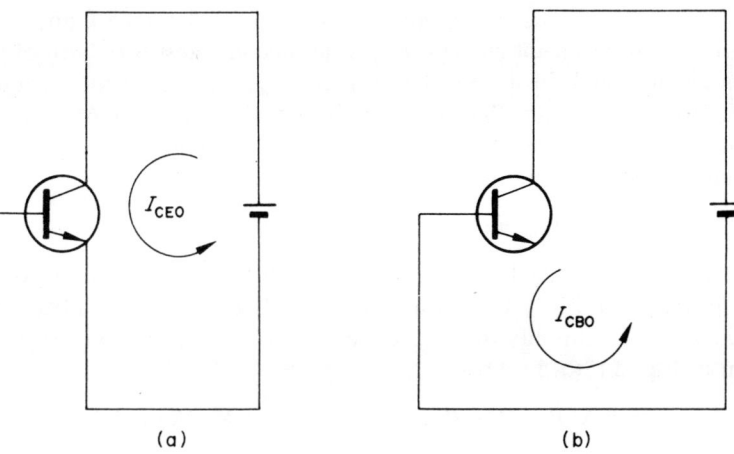

Fig. 2.11   Transistor leakage currents.

2.5   Thermal stability factor

In many circuits, notably those using germanium transistors, the leakage current increases with temperature. In silicon transistors, the parameter h_{FE} changes with temperature and has much the same effect as a change in leakage current in germanium transistors.

The two values of leakage current normally specified are I_{CEO} and I_{CBO} (sometimes stated as I_{CO}), shown in Figs. 2.11(a) and (b), respectively. Current I_{CEO} is the collector-to-emitter leakage current with the base

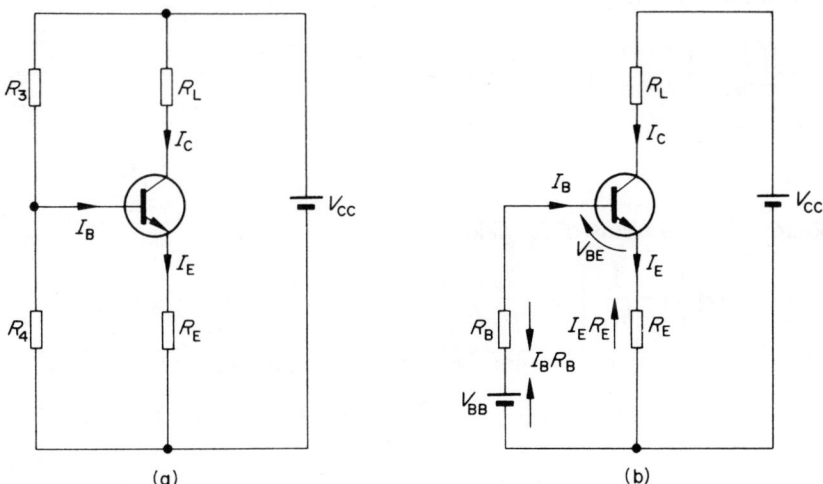

Fig. 2.12   The basic d.c. circuit in (a) is modified in (b) to show the Thévenin equivalent circuit of the bias network.

effectively open-circuited (the collector-emitter cut-off current), and I_{CBO} is the collector-base cut-off current. In planar silicon transistors both of these values are low and, in a 2N929 transistor I_{CBO} has a maximum value of 10 nA. For comparison, the maximum value of I_{CBO} in an ACY 17 germanium p-n-p junction transistor is $10 \mu A$. For a given transistor, the relationship between the two leakage currents is

$$I_{CBO} = (1 + h_{FB})I_{CEO} = I_{CEO}/(1 + h_{FE})$$

The thermal stability factor of the circuit in Fig. 2.7(b) can be deduced from the circuit in Fig. 2.12(a), which shows only the flow of steady currents in the circuit. To simplify the base circuit, its Thévenin equivalent circuit is drawn in Fig. 2.12(b) in which

$$R_B = R_3 R_4/(R_3 + R_4) \quad \text{and} \quad V_{BB} = V_{CC}R_4/(R_3 + R_4)$$

The loop equation of the base circuit in Fig. 2.12(b) is

$$V_{BB} - I_B R_B - V_{BE} - I_E R_E = 0 \tag{2.31}$$

also
$$I_E = I_C + I_B \tag{2.32}$$

Now, the total collector current is

$$I_C = h_{FE}I_B + I_{CEO} = h_{FE}I_B + (1 + h_{FE})I_{CBO}$$

Therefore

$$I_B = \frac{I_C}{h_{FE}} - \left[\frac{1 + h_{FE}}{h_{FE}}\right]I_{CBO} \tag{2.33}$$

Substituting eq. (2.32) and (2.33) into eq. (2.31) yields

$$V_{BB} - \left(\frac{I_C}{h_{FE}} - \left[\frac{1 + h_{FE}}{h_{FE}}\right]I_{CBO}\right)R_B - V_{BE}$$

$$- \left(I_C + \left\{\frac{I_C}{h_{FE}} - \left[\frac{1 + h_{FE}}{h_{FE}}\right]I_{CBO}\right\}\right)R_E = 0$$

Rearranging in terms of I_C yields

$$I_C\left(\left[1 + \frac{1}{h_{FE}}\right]R_E + \frac{R_B}{h_{FE}}\right) = \frac{I_{CBO}(R_E + R_B)(1 + h_{FE})}{h_{FE}} + V_{BB} - V_{BE}$$

The equation is further simplified by multiplying throughout by $h_{FE}/(1 + h_{FE})$.

$$I_C[R_E + R_B/(1 + h_{FE})] = I_{CBO}(R_E + R_B) + (V_{BB} - V_{BE})h_{FE}/(1 + h_{FE}) \tag{2.34}$$

If V_{BE}, V_{BB}, and h_{FE} remain constant, then the relationship between a change

∂I_{CBO} in leakage current and the corresponding change ∂I_C in collector current is

$$\partial I_C[R_E + R_B/(1 + h_{FE})] = \partial I_{CBO}(R_E + R_B)$$

Therefore, the *stability factor S* is

$$S = \frac{\partial I_C}{\partial I_{CBO}} = \frac{R_E + R_B}{R_E + R_B/(1 + h_{FE})} \qquad (2.35)$$

where $R_B = R_3 R_4/(R_3 + R_4)$. Equation (2.35) can also be obtained by partial differentiation of eq. (2.34) with respect to I_{CBO}, when S is once more expressed as $\partial I_C/\partial I_{CBO}$. Also, since $I_{CEO} = (1 + h_{FE})I_{CBO}$, then the *thermal stability factor k* which relates the change in I_C with the change in I_{CEO} is

$$k = \partial I_C/\partial I_{CEO} = S/(1 + h_{FE}) \qquad (2.36)$$

The worst case of thermal stability (or more correctly thermal instability) occurs when no emitter resistor is used ($R_E = 0$) when $S = 1 + h_{FE}$ (or $k = 1$). Best thermal stability is obtained when $R_E \gg R_B$ when $S = 1$ [or $k = 1/(1 + h_{FE})$]; in this case the input resistance may be low and the output voltage is small. The actual value of S is a compromise between the two values, and typical values lie between about $0.1 h_{FE}$ and $0.2 h_{FE}$, i.e., k has a value between about 0.1 and 0.2.

Equation (2.35) can be simplified if $R_E \gg R_B/(1 + h_{FE})$ when

$$S \simeq (R_E + R_B)/R_B = 1 + R_B/R_E \qquad (2.37)$$

or

$$R_E \simeq R_B/(S - 1) \qquad (2.38)$$

Equations (2.37) and (2.38) are adequate for basic design purposes. An analysis of the feedback circuit in Fig. 2.7(a) will not be undertaken here, but it can be shown that for this circuit

$$S = \frac{\{(R_1 + R_2) + R_L\}(1 + h_{FE})}{(R_1 + R_2) + R_L(1 + h_{FE})} \qquad (2.39)$$

In the discussion of the circuit in Fig. 2.7(a), it was shown that a practical value of $(R_1 + R_2)$ was about $h_{FE}R_L$. Substituting this value into eq. (2.39) gives a stability factor S of $0.5 h_{FE}$; as a result we see that the feedback circuit in Fig. 2.7(a) is inferior in the respect of thermal stability to the circuit shown in Fig. 2.7(b).

Example 2.9: Determine the stability factor of a circuit of the type shown in Fig. 2.7(b) in which $R_E = 1 \text{ k}\Omega$, $R_3 = 76.5 \text{ k}\Omega$, $R_4 = 16 \text{ k}\Omega$, and $h_{FE} = 100$.

Solution: $R_B = 76.5 \times 16/(76.5 + 16) = 13.25 \text{ k}\Omega$. From eq. (2.35)

$$S = \frac{1 + 13.25}{1 + (13.25/101)} = 12.6$$

Note: The approximate solution from eq. (2.37) is

$$S = 1 + 13.25/1 = 14.25$$

2.6 Direct current amplifiers

The principal difference between d.c. amplifiers and a.c. amplifiers is that in the former the bandwidth of the amplifier extends down to zero frequency. This feature is advantageous in certain instrument amplifiers, e.g., thermo-couple amplifiers and in analogue computing amplifiers. Direct coupling is also used in many forms of class B audio-frequency power amplifiers. Direct current amplifiers fall into two main categories:

(a) Direct coupled amplifiers
(b) Chopper amplifiers.

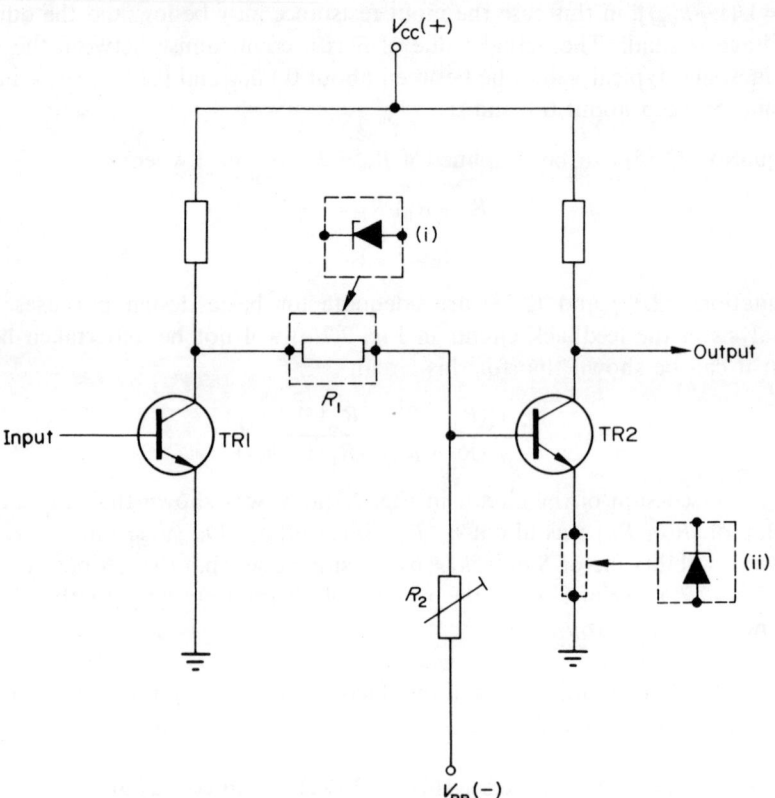

Fig. 2.13 The basis of direct coupling between stages.

In *direct coupled amplifiers*, the output signal from one stage is directly coupled to the input of the following stage. Since every form of amplifier requires a power supply, the output signal is superimposed upon a quiescent voltage; in direct coupled amplifiers the quiescent voltage must be compensated for in some way in order that it does not saturate the following stage. One method of overcoming this difficulty is shown in Fig. 2.13, in which resistor chain $R_1 R_2$ is taken to a negative bias voltage; under zero input conditions the base bias voltage of TR2 is adjusted to give the correct quiescent conditions for that stage. A disadvantage of this arrangement is that R_1 reduces the magnitude of the signal voltage at the base of TR2 in addition to reducing the quiescent voltage. The former effect can be minimized by replacing R_1 by the Zener diode in inset (i). Providing that the Zener diode breakdown voltage is equal to the p.d. formerly developed across R_1, it does not alter the circuit quiescent conditions. Alternatively, a Zener diode can be connected in series with the emitter of TR2, shown in inset (ii), to raise its quiescent emitter voltage.

One form of d.c. amplifier is shown in Fig. 2.14(a). In this circuit the collector voltage of TR1 must be relatively low in order that the voltage swing at the collector of TR2 shall be reasonably large. The circuit in Fig. 2.14(b) uses a pair of complementary transistors in the common-emitter mode. In this circuit the voltage developed across R_{L_1} is applied directly to

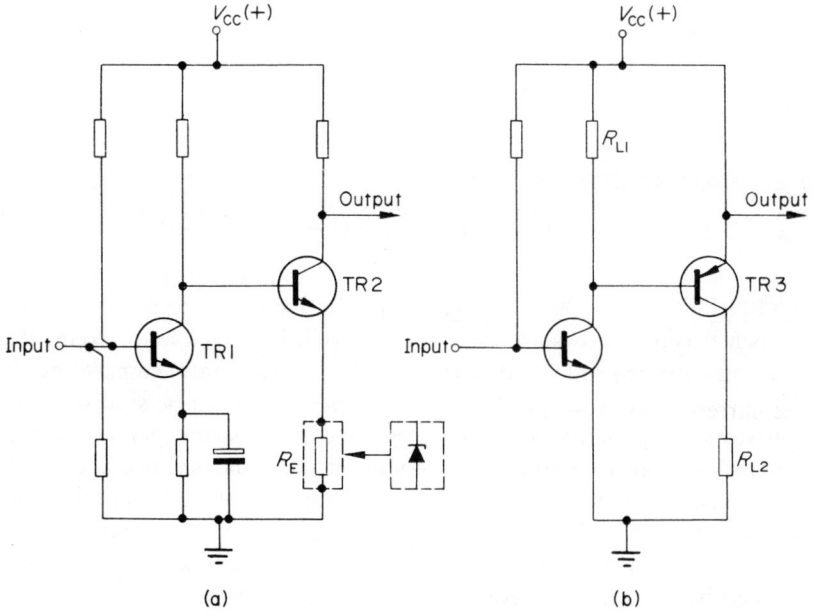

(a)   (b)

Fig. 2.14   (a) A d.c. amplifier using bias resistor R_E to offset the quiescent input voltage while (b) uses complimentary common-emitter amplifiers.

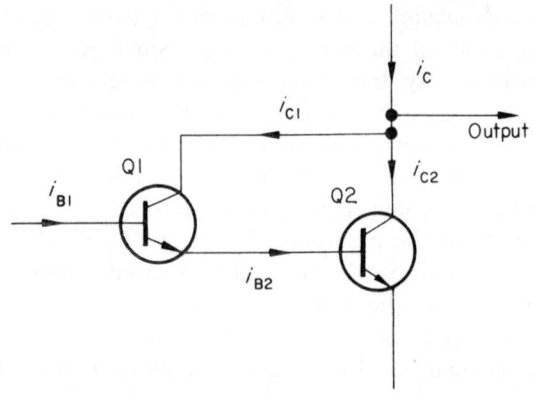

Fig. 2.15   The basic Darlington (super-alpha) circuit.

the base-emitter junction of TR3, and the circuit output is developed at the collector of TR3.

Another frequently used configuration is the *Darlington connection* or *super-alpha pair* in Fig. 2.15. If the current gain of Q1 is h_{FE1},

then
$$i_{C1} = h_{FE1} i_{B1}$$

and
$$i_{B2} = i_{E1} = i_{B1} + i_{C1} = i_{B1}(1 + h_{FE1})$$

The current gain of Q2 is h_{FE2}, hence

$$i_C = i_{C1} + i_{C2} = h_{FE1} i_{B1} + h_{FE2} i_{B2}$$
$$= i_{B1}[h_{FE1} + h_{FE2}(1 + h_{FE1})]$$

giving a combined current gain of

$$i_C/i_{B1} = h_{FE1} + h_{FE2}(1 + h_{FE1})$$

If $h_{FE1} = 99$ and $h_{FE2} = 100$, then the current gain of the pair of transistors is $99 + 100(1 + 99) = 10\,099$, which gives an effective α for the pair of 0.999 99.

When two transistors are used in this way it is advisable to mount them on a common heatsink so that they operate at the same temperature; this causes variations in temperature-sensitive parameters to be similar in both transistors. A preferred alternative is to use transistor pairs which are manufactured in a common semiconductor chip and encapsulated in the same canister. A typical high power Darlington pair of the latter type would have an effective h_{FE} of about 2500 at a collector current of 3 A.

In the Darlington configuration, a degree of thermal stability can be obtained by including resistors R_1 and R_2, shown in Fig. 2.16. In the basic circuit the thermal leakage current of TR1 enters the base of TR2, and is effectively multiplied by the current gain of TR2; in germanium transistors

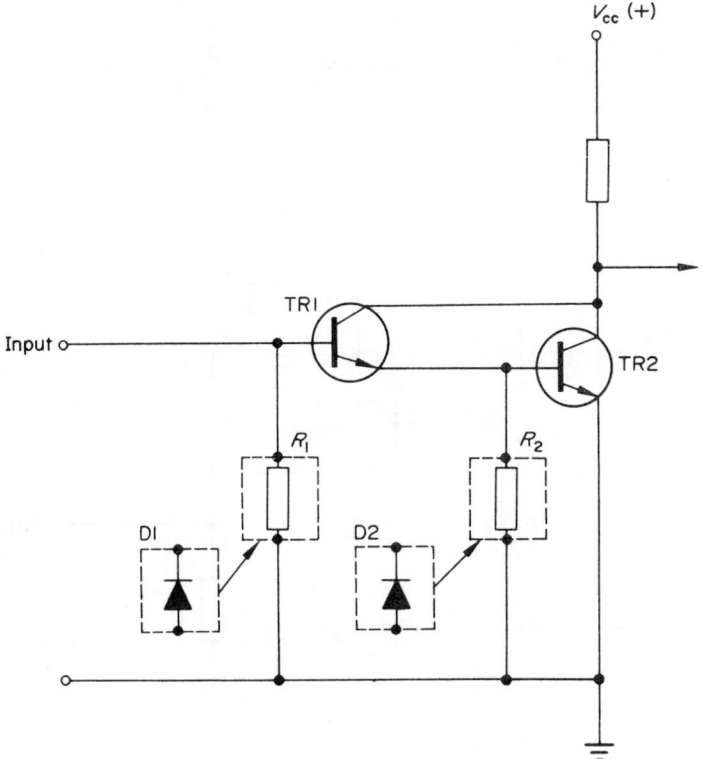

Fig. 2.16 Methods of improving the thermal stability of Darlington-connected transistors.

the output current from this source can be very large. By including resistors R_1 and R_2, a path is provided for part of the leakage current so that its effect is minimized. The resistors will, of course, shunt part of the signal current and will effectively reduce the current gain. Fortunately, the small-signal input resistance of the emitter junction is lower than the static input resistance over the operating range, so that the leakage currents are reduced to a greater extent than are the signal currents.

Alternatively, reverse biased diodes can be used to replace the resistors, as shown in the insets in the figure. The reverse saturation current of D1 should be equal to the I_{CBO} of TR1, and the reverse saturation current of D2 should be equal to the I_{CBO} of TR2. Since the diodes are reverse biased they do not have any great effect on the current gain of the stage.

A circuit sometimes described as a *complementary compound Darlington circuit* is shown in an emitter follower circuit (see also chapter 3) in Fig. 2.17. The operation of the transistors is similar to that of the basic transistor pair, with the exception that the output is taken from the

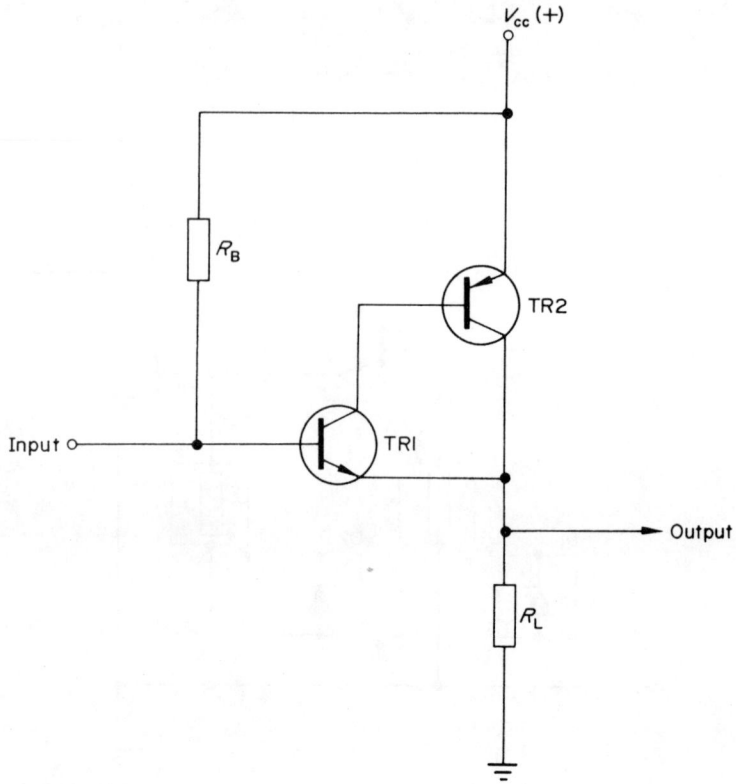

Fig. 2.17   A complementary compound Darlington connection.

collector of TR2. An advantage of the compound circuit over the basic circuit is that in this application there is only one base-emitter voltage drop between the input and output terminals. Resistor R_B is included for bias purposes.

A typical transformerless class B audio-frequency output stage is shown in Fig. 2.18. In this circuit the driver transistor TR1 is directly coupled to the output transistors which comprise a Darlington pair TR3, TR5 and a complementary Darlington pair TR2, TR4. Diodes D1 and D2 are included to provide thermal stability.

2.7   Offset voltage in direct coupled amplifiers

In some of the amplifiers described in section 2.5 it is found that, for zero input voltage, the output voltage does not have the value one would normally expect. This is often due to changes in the parameters of passive and active elements in the amplifier. This voltage is said to be due to an imaginary generator (the *offset voltage* generator) at the input of the

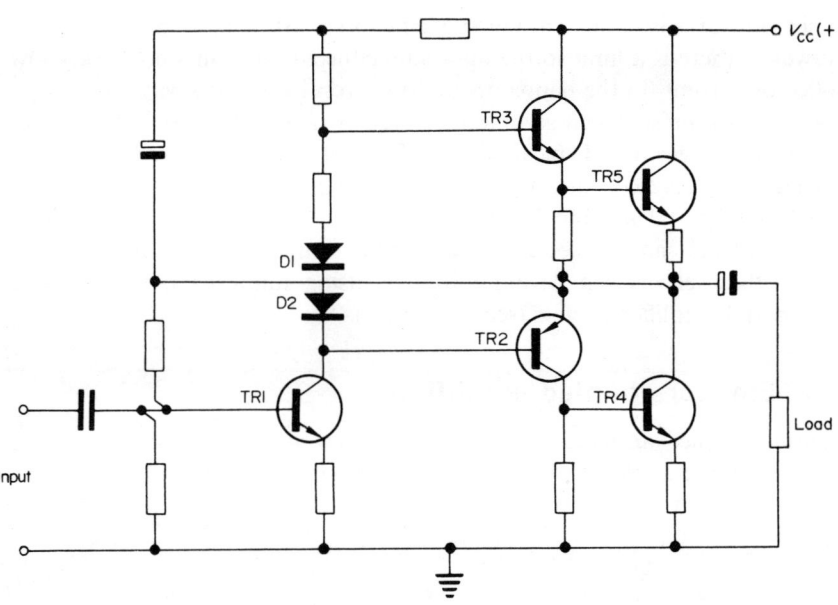

Fig. 2.18   A class B transformerless amplifier stage.

amplifier. In bipolar transistor amplifiers the offset is expressed in terms of an offset current; a typical offset current in a general purpose high-gain ($\simeq 50\,000$) direct coupled amplifier is about ± 10 nA.

2.8   Drift in direct coupled amplifiers

Offset voltage is not stable over long periods of time due to unpredictable changes in component parameters, and permanent compensation for the changes is not possible. The low change in offset voltage gives rise to a slow variation in output voltage, and is known as *drift*. In bipolar transistors the main reasons for drift with time and temperature are:

(a) Change in base-emitter voltage V_{BE}
(b) Change in leakage current
(c) Change in current gain h_{FE}.

In terms of the transistor itself, very little can be done about minimizing the effect of variations in V_{BE}. Equations of the transistor also show that the drift voltage is dependent upon the relationship $R_S I_C / h_{FE}^2$, where R_S is the source resistance. If the value of this function is to be small, then

(a) R_S must be small
(b) I_C must be small
(c) h_{FE} must be large.

Where low drift is important the above factors must be built into the circuit. However, there is a limit to the minimum collector current since a very low collector current (in the nanoampere range) results in a low value of h_{FE}, so that a compromise which gives a low value of I_C with a reasonable value of h_{FE} must be selected. Typical values of h_{FE} at various values of collector current for a 2N2639 silicon planar transistor for d.c. amplifiers are 150 at 10 μA, 170 at 100 μA, and 200 at 1 mA.

Drift effects can also be introduced as a result of supply voltage variations; these effects can be minimized by using stabilized power supplies and special amplifier circuits (see section 2.9).

2.9 Emitter coupled amplifiers

Emitter coupled amplifiers or *long-tailed pairs* are very important in the field of industrial electronics since the circuit reduces inherent drift effects

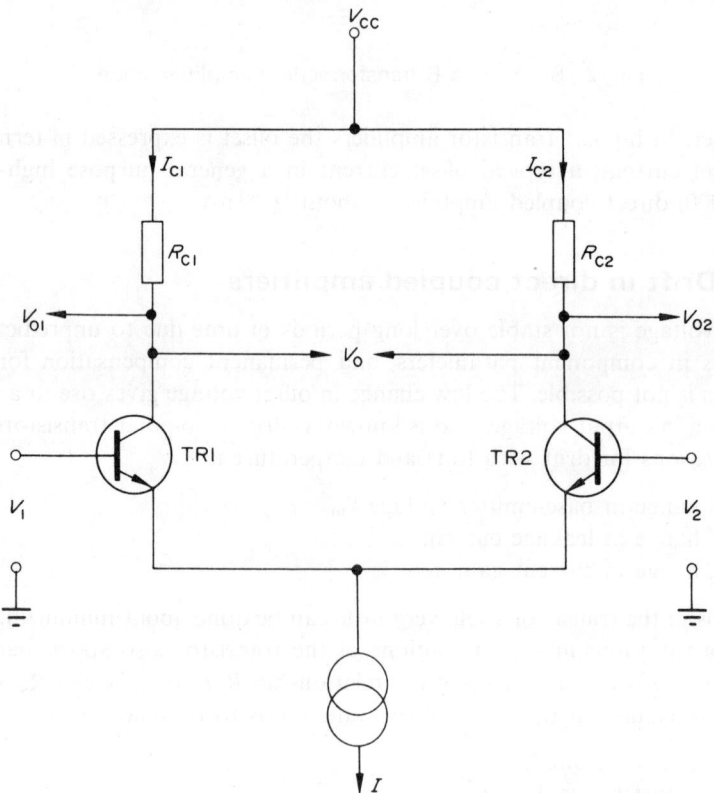

Fig. 2.19 A basic emitter coupled amplifier.

due to transistor parameter changes and supply voltage variations. The basis of the circuit is shown in Fig. 2.19, in which two transistors have their emitters coupled together and the current through them is supplied from a constant current generator. This forces the sum of the emitter currents (and therefore collector currents) to be constant so that

$$I = I_{C_1} + I_{C_2} = \text{a constant}$$

Thus, an increase in one of the collector currents, say I_{C_1}, causes I_{C_2} to be reduced by an equal amount. As a result the changes in voltages at the two collectors are antiphase to one another. To minimize drift effects the transistors used in emitter coupled amplifiers should be manufactured in the same chip of semiconductor material.

The output from the amplifier in Fig. 2.19 is generally taken as the voltage difference V_0 between the collectors, and for this reason the circuit is also known as a *difference amplifier* or *differential amplifier*. Zero input voltage and equal values of load resistance result in $I_{C_1} = I_{C_2}$ and V_{C_1} and V_{C_2}, giving a differential output voltage V_0 of zero. The constant current source in the emitter circuit is, in practice, not perfect so that changes in supply voltage or circuit parameters do cause small changes in I. Should the supply voltage change, the collector currents change by similar amounts so that the differential output voltage remains unchanged. Also, since the transistors are either on the same chip of semiconductor or, if separate transistors are used, are mounted on a common heatsink, then temperature effects produce equal changes in both collector currents so that V_0 is unchanged. By this means the circuit minimizes drift effects.

If input signals V_1 and V_2 are in phase with one another, the collector currents try to change in phase with one another but are prevented from doing so by the constant current generator in the emitter lead. As a result, input signals which are in phase with one another give a differential output voltage of practically zero. When the input signals are in phase with one another they are described as *common-mode input signals*, since the input signal is common to both transistors. In difference amplifiers, common-mode input signals are generally electrical noise signals, and a high *common-mode rejection* figure is very desirable.

An application of a difference amplifier as a strain gauge instrument is shown in Fig. 2.20. The active strain gauges in the bridge circuit are mounted on the test piece, and the bridge is energized by means of battery E. The voltage developed between points A and B on the bridge circuit is applied to the terminals of the amplifier. Signals E_1 and E_2 are electrical noise signals resulting from battery voltage variations and induced voltages in the wiring, respectively. Even though the voltage between points A and B on the bridge as a result of strain in the test piece may only be a few millivolts, and E_1 and E_2 may be much larger than this, the amplifier rejects

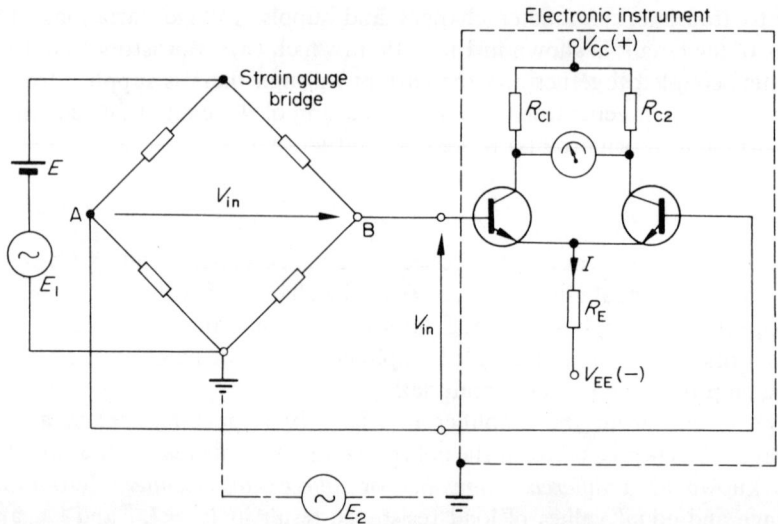

Fig. 2.20   Using a long-tailed pair to overcome common-mode noise.

the common-mode noise signals E_1 and E_2 and amplifies only the *differential mode input signal* V_{in}.

Due to the mode of operation the potential of the emitters remains approximately constant, so that an increase in the base current of one transistor causes a reduction in the base current of the other transistor. It therefore appears that, so far as the input signal is concerned, the base current flows into one base region and out of the other. As a result, the effective input resistance to differential mode signals is approximately

$$R_{\text{in}} = 2h_{\text{ie}}$$

that is, it is twice the value of the input resistance of an otherwise similar common-emitter amplifier.

Also, the output current has to flow from one transistor into the other, so that the differential output resistance is twice the value of the output resistance of a conventional common-emitter amplifier. Due to the higher value of input resistance, the net input current for a given signal voltage is halved when compared with a common-emitter amplifier. But, since the effects of the two halves of the amplifier are additive to differential inputs, the differential current gain is therefore of the same order as that of a similar common-emitter amplifier.

Due to the mode of connection, the input signal voltage is shared between the two transistors, and the voltage gain for each half of the amplifier is one-half that of a similar common-emitter amplifier. However, the differential voltage gain is twice that for one stage, since the collector voltages are antiphase to one another, giving a differential voltage gain of approximately $h_{fe} R_C/h_{ie}$. Amplifiers of this type can be cascaded to give very high values of gain since drift effects are not amplified.

In Fig. 2.20 the constant current source is simply a resistor R_E which is taken to supply voltage V_{EE}. If we assume that the potential of the emitter point is approximately at earth potential, then

$$I \simeq V_{EE}/R_E$$

Provided that R_E (and V_{EE}) are large enough, then each transistor is forced to pass a quiescent current of $I/2$. In some circuits R_E is replaced by a semiconductor constant current source, thereby ensuring excellent thermal stability and common-mode rejection.

Common-mode voltage rejection is usually expressed in the form of the *common-mode rejection ratio* (CMRR) which is

$$\text{CMRR} = \frac{\text{Voltage gain for differential input signals}}{\text{Voltage gain for common-mode input signals}} \tag{2.40}$$

The equivalent voltage gain for common-mode input signals is the same as for a common-emitter amplifier with load resistance R_C and unbypassed emitter resistor R_E (see chapter 3 for details), and is approximately R_C/R_E. Hence

$$\text{CMRR} = (h_{fe} R_C/h_{ie})/(R_C/R_E) = h_{fe} R_E/h_{ie} \tag{2.41}$$

A typical figure for the CMRR of a well designed amplifier would be about 2000, with a maximum figure of about 10^6.

Example 2.10: A 2N2639 dual n-p-n silicon transistor with $h_{FE} = 100$, $h_{fe} = 100$, and $h_{ie} = 1.5\,\text{k}\Omega$ is used in a differential amplifier. The power supplies are $V_{CC} = +12\,\text{V}$ and $V_{EE} = -10\,\text{V}$, and the collector current is to be 50 μA. Estimate the values of components used in the circuit. Compute also the values of the differential voltage gain and the common-mode rejection ratio.

Solution: The low value of collector current in this problem ensures that thermal effects are minimized, and that the electrical noise introduced by recombination effects in the transistor is small.

Since $I_C = 0.05\,\text{mA}$, then $I = 0.1\,\text{mA}$. If both signal inputs are zero then the emitter voltage is approximately zero if V_{BE} is negligible in comparison with V_{EE}. Hence

$$R_E \simeq V_{EE}/I = 10/0.1 = 100\,\text{k}\Omega$$

A quiescent collector voltage of about 3 V is suitable in this application so that

$$I_C R_C = 12 - 3 = 9 \text{ V}$$

Therefore $$R_C = 9 \text{ V}/0.05 \text{ mA} = 180 \text{ k}\Omega$$

The base bias required for a collector current of $50 \, \mu\text{A}$ is $50 \, \mu\text{A}/100$ $= 0.5 \, \mu\text{A}$. If the bias current is derived from V_{CC} it would require a bias resistor of $24 \, \text{M}\Omega$, which is rather high. The base bias can, alternatively, be fed from a lower voltage through a much lower value of resistance (say $1 \, \text{M}\Omega$).

The differential voltage gain is

$$A_v = h_{fe} R_C / h_{ie} = 100 \times 180\,000 / 1500 = 12\,000$$

and $$\text{CMRR} = h_{fe} R_E / h_{ie} = 100 \times 100\,000 / 1500 = 6666$$

Since the calculated value of the collector resistance is very high, a high value of load impedance must be used if the full voltage gain is to be developed. One method of satisfying this problem with a low impedance load is illustrated in Fig. 2.21. Here TR3 and TR4 are emitter followers each

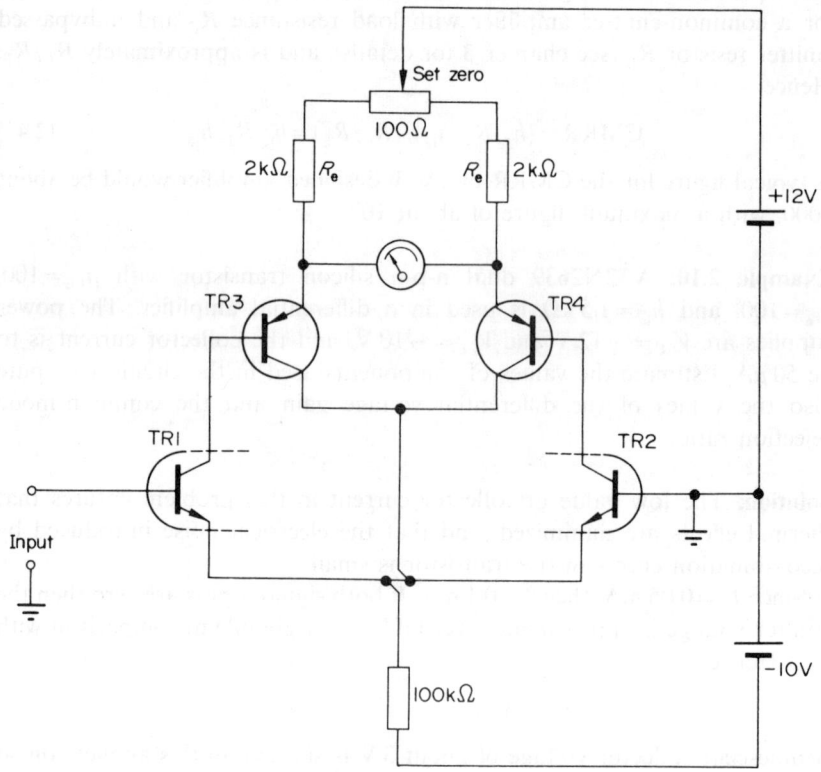

Fig. 2.21 A low-drift electronic instrument circuit.

having an input resistance of approximately $(h_{ie} + h_{fe} R_e)$ [see chapter 3], which has the value of approximately $180\,\text{k}\Omega$ if $h_{fe} = 90$. The low output resistance of the emitter follower allows a low resistance meter to be connected to the amplifier. By earthing the base of TR2, the input voltage of any point relative to earth is applied differentially between the bases of TR1 and TR2. The set zero control is a means of balancing the output meter when the input voltage is zero.

2.10 Chopper amplifiers

In chopper amplifiers, the d.c. (or very low frequency) input signal is 'chopped' into a series of pulses by a switching device which may be either a semiconductor element or a relay. This causes the input signal to be converted into a series of pulses. The general circuit arrangement is shown in Fig. 2.22(a). Here switch S1 is known as a *shunt chopper*, and serves to apply a short-circuit to the input terminals. A shunt chopper is best used where the source resistance is high so that the short-circuit current is small. A *series chopper* using switch S3 shown in the inset is sometimes used, and this is best employed with low resistance sources. In some cases, *series-shunt choppers* using both S1 and S3 are employed; the performance of series-shunt choppers is at least as good as or better than either the series or shunt chopper for almost any combination of source and load resistance. Chopping is also known as *modulating*.

The resulting signal is then amplified by an a.c. amplifier and reconverted into a unidirectional signal by switch S2, which is known as a *demodulator*. The operation of S1 (also S3) and S2 are synchronized so that when S1 is closed, S2 (and S3) is open and vice versa.

Waveforms at various points in the circuit are shown in Fig. 2.22(b). Voltage V_X is the voltage across S1; during the time interval t_1 to t_2 switch S1 is open and V_1 appears across it, and during time interval t_2 to t_3 the switch is closed so that V_X is zero. Capacitor C_1 blocks the d.c. level of V_X from the amplifier so that V_T is the alternating component of V_X. This voltage is then amplified by an a.c. amplifier which is drift-free as a result of the a.c. inter-stage coupling, and is applied to the output via capacitor C_2. Since switch $S2$ operates in synchronism with S1, it reduces the output voltage during one-half of the output cycle to zero and allows the voltage in alternate half-cycles to be transmitted to the output terminals.

The output signal is therefore a series of unidirectional pulses which are either averaged by a moving-coil instrument to give a direct current output reading or are smoothed by an R–C network if a voltage output is required. In either case, the output time constant is usually large, and the resulting amplifier bandwidth is small.

Choppers include photoconductive devices in which a photoconductive cell and a flashing light source (e.g., a cold-cathode glow tube driven from

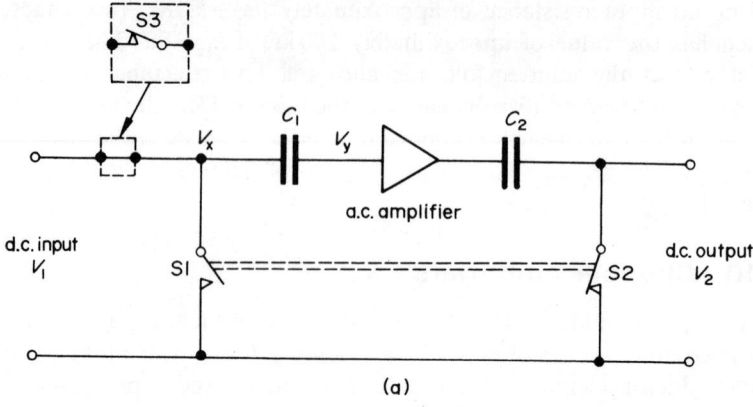

(a)

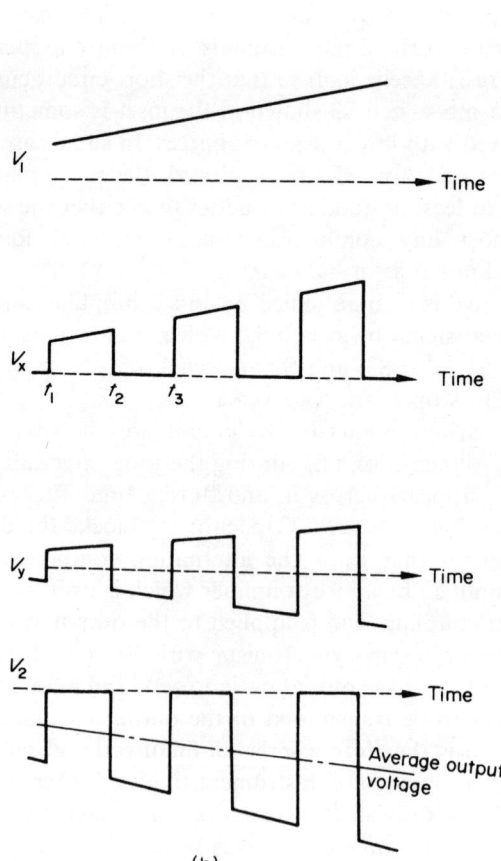

(b)

Fig. 2.22 (a) The basis of the chopper amplifier and (b) waveforms at various points in the circuits.

an a.c. source) are enclosed in a light-tight enclosure. When the cell is illuminated its resistance is low, and when not illuminated its resistance is high. Semiconductor choppers include both bipolar transistors and FETs, and their operation is described in the following section.

2.11 Semiconductor choppers

A transistor chopper operates for most of its time either in the ON (saturated) state or in the OFF (cut-off) state. A simplified circuit of the transistor in these modes is shown in Fig. 2.23.

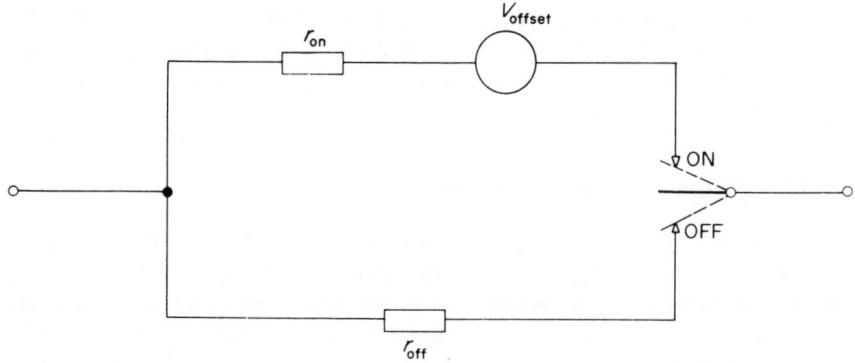

Fig. 2.23 Simplified equivalent circuit of a semiconductor chopper.

When used as a chopper, parameters r_{on} (the ON resistance) and V_{offset} (the offset voltage) should be small and r_{off} (the cut-off resistance) should be large. Typical values for a number of transistors are given in Table 2.5. To achieve good results, bipolar transistors are used in an inverted mode, that is the normal emitter region is used as a collector and the normal collector as an emitter. The figures in Table 2.5 correspond to this mode of operation. In the table, the offset voltage for the two FETs is quoted as zero, but a few micro-volts may appear across the transistor in practice

Table 2.5

Parameters of transistor choppers

Type	Construction	r_{on} (Ω) (max.)	V_{offset} (μV) (max.)	r_{off} $(M\Omega)$ (typical)
2N2432	n-p-n silicon planar	20	1000	100
3N74	n-p-n double-emitter silicon planar	40	50	100
2N3970	n-channel JUGFET	25	Zero	1000
BSV22	n-channel depletion IGFET	500	Zero	100

2.12 Generation of transient spikes in chopper circuits

Transistor choppers are switched on and off by applying a square wave to the base region or the gate electrode. Both bipolar and field-effect transistors have inter-region capacitances (C_{cb}, C_{ce}, C_{gd}, etc.), and a rapid change of voltage at the control region causes a transient spike to be transmitted to the output (or input in some cases) of the chopper element. This voltage can be of the order of 20 to 150 mV. If the input voltage to the amplifier is very small, the voltage spike may exceed the value of the input signal and cause an offset voltage to appear at the output. In some cases the spike may cause the amplifier to overload.

One method of minimizing the amplitude of the spike is to drive the chopper with a voltage wave form with sloping edges; also by using a voltage which does not appreciably exceed that needed to switch the transistor should be used.

2.13 Basic chopper circuits

A simple bipolar shunt chopper with the transistor operating in the inverted mode is shown in Fig. 2.24(a). In Fig. 2.24(b) a double-emitter transistor is used as a series chopper. When used in this mode the offset voltages of the two emitters are approximately equal in magnitude, and the emitters are so connected that the offset voltages are in opposition to one another. As a result, the offset voltage of a double-emitter transistor is much lower than that of an otherwise equivalent single-emitter type.

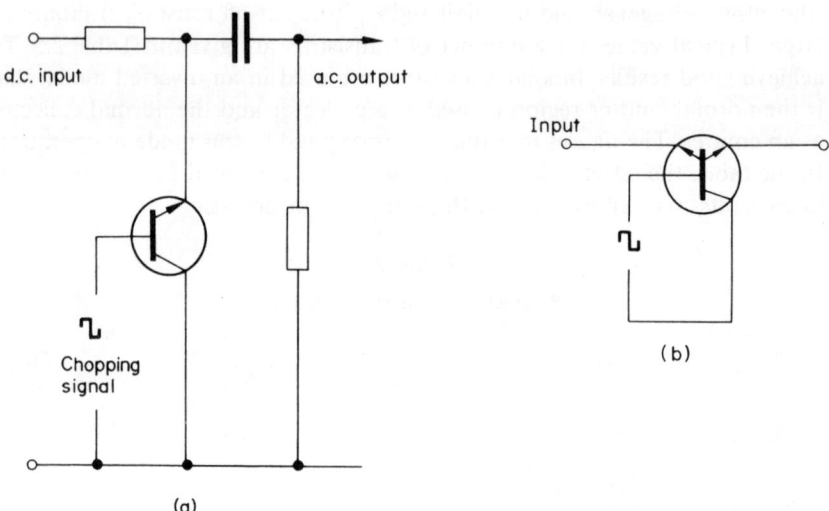

Fig. 2.24(a)   A shunt bipolar chopper and (b) a series twin-emitter bipolar chopper.

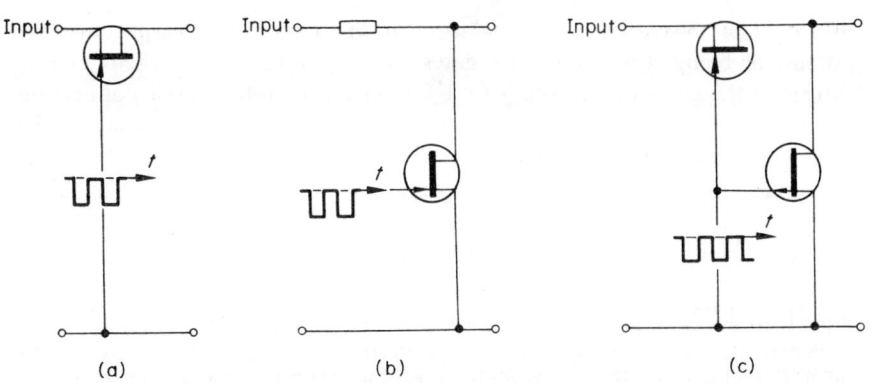

Fig. 2.25   Field-effect transistor chopper circuits.

Figures 2.25(a) and (b) show basic configurations of n-channel FET choppers in the series and shunt modes, respectively. A series-shunt chopper using complementary FETs is illustrated in Fig. 2.25(c). In circuits 2.25(a) and (b) the amplitude of the gate voltage swing must be large enough to pinch the FET off, whatever the input voltage. In the case of circuit 2.25(c), the gate voltage must swing between a positive voltage to pinch off the p-channel device and a negative voltage to pinch off the n-channel device. In a practical version of circuit 2.25(c), diodes would have to be included in series with the gates.

A practical chopper amplifier is shown in Fig. 2.26. Here, complementary voltages are generated at the collectors of TR1 and TR2, the voltages switching between $+V_{CC}$ and $-V_{EE}$ so that complementary voltages are applied to the gates of F1 and F2. When TR1 is saturated, its collector voltage is nearly $-V_{EE}$ so that D2 is forward biased and TR2 is cut off. At this time, the collector voltage of TR2 is nearly $+V_{CC}$ so that D1 is reverse

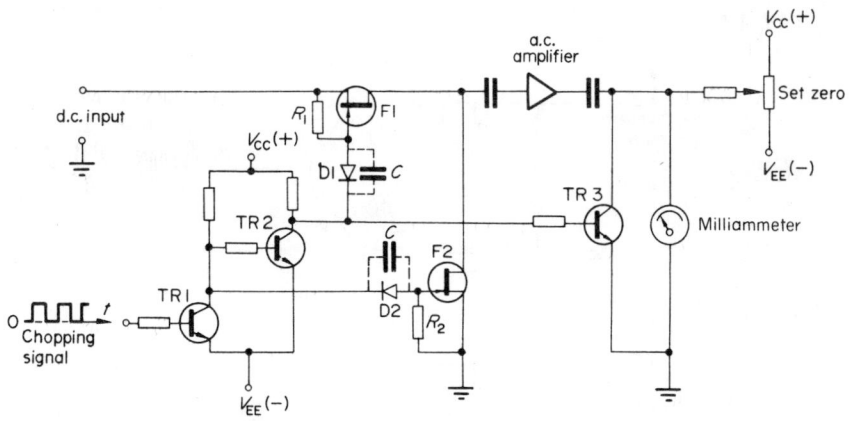

Fig. 2.26   A practical form of chopper amplifier.

biased. The two diodes are included in the circuit to prevent positive potentials being applied to the gates of the FETs. Thus, when TR1 is saturated the gate-source voltage of F1 is zero (due to the ohmic connection through R_1), so that F1 effectively connects the d.c. input signal to the amplifier. Also F2 is cut off since the drain current is pinched off by the large negative voltage applied to its gate, and appears as an open-circuited switch to the input voltage.

The positive potential is also applied to the base of TR3, causing it to short-circuit the output. Thus, F1, F2, and TR3 are equivalent to switches S3, S1, and S2, respectively, in the basic circuit Fig. 2.22(a).

When the base voltage of TR1 is reduced to zero, it causes F1 to be cut off, F2 to short-circuit the amplifier input, and TR3 to be cut off. If diodes D1 and D2 are fast recovery types, they may cut off before the FET capacitances have changed to the new steady values. In this event, the capacitances must charge through the very high leakage resistance of the diodes, so increasing the switching time. This problem can be overcome by shunting the fast recovery diodes by capacitors of capacitance of about 2 pF. Alternatively, the diodes may be replaced by slow recovery types (e.g., Zener diodes).

For the best performance in the circuit described here the input impedance of the a.c. amplifier should be as high as possible, and one with a FET input stage is advantageous.

2.14   Other applications of choppers

An application of the chopper technique applied to *data sampling* or *time division multiplexing* is shown in Fig. 2.27. Here, sampling pulses are applied successively to the three transistors so that inputs A, B, and C are gated

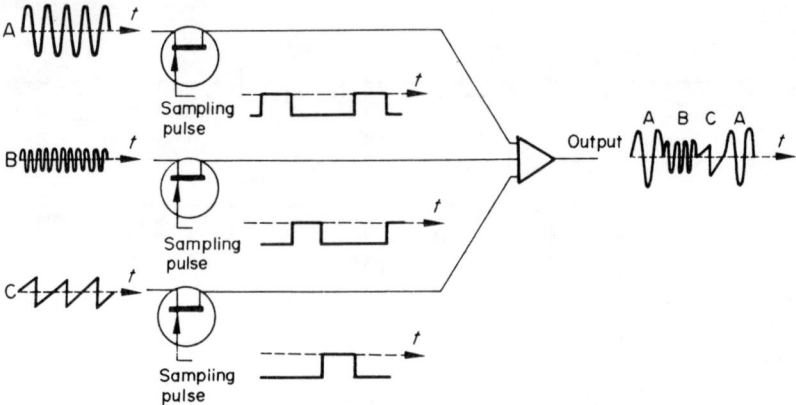

Fig. 2.27   Application of the 'chopping' principle to a data sampling system.

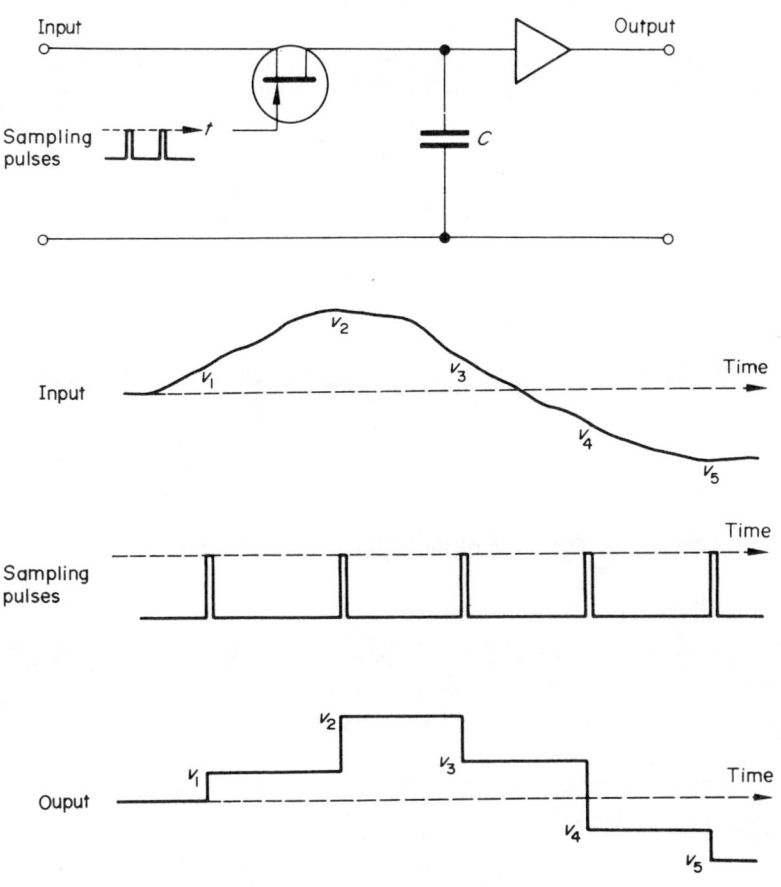

Fig. 2.28   A sample-and-hold circuit.

sequentially to the input of the amplifier. The amplifier output signal is a sequence of bursts of the respective inputs, the length of the burst being equal to the sampling signal duration.

A *sample-and-hold circuit* or *analogue memory* is shown in Fig. 2.28. Here, the input voltage is sampled at intervals of time and is used to charge capacitor C. For accuracy, the sampling pulse must be short and the value of C must be small. Additionally, the input impedance of the amplifier must be very high in order not to discharge C between the sampling pulses.

2.15   Chopper stabilized wide-band d.c. amplifiers

For many applications the drift of conventional d.c. amplifiers is too large, and the bandwidth of a chopper modulated amplifier is much too small. By combining a chopper modulated amplifier with a differential amplifier, as in

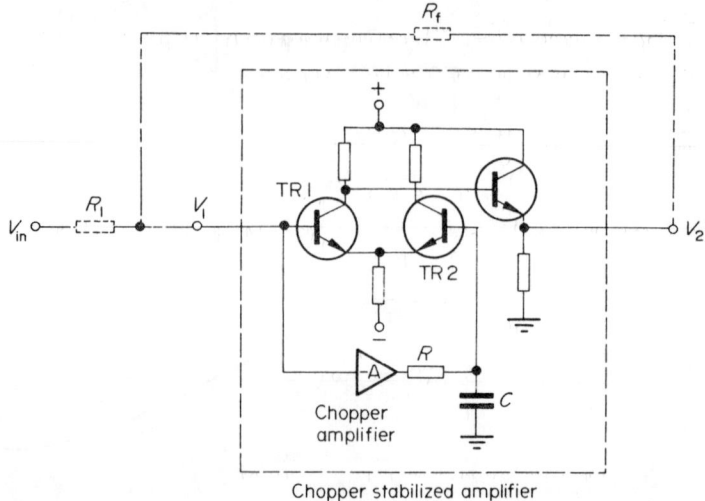

Chopper stabilized amplifier

Fig. 2.29   The basis of the chopper-stabilized d.c. amplifier.

Fig. 2.29, the best features of both types are realized in one amplifier. Such an amplifier is described as a *chopper-stabilized d.c. amplifier* and combines the advantages of high gain and wide bandwidth with low drift.

Since the chopper amplifier used in Fig. 2.29 is phase-inverting, any signal which is applied to the base of TR2 is effectively added to the voltage at the base of TR1. Thus,

$$V_2 \simeq -(KV_1 + KAV_1) = -V_1 K(1 + A)$$

where K is the gain of the differential amplifier and A is the gain of the chopper amplifier. In these applications, the chopper filter RC time constant is usually of the order 20 to 30 s, so that the bandwidth of the chopper amplifier is very low indeed. At very low frequencies (i.e., those associated with drift effects), the overall gain of the amplifier is $-K(1 + A)$, which may be of the order of 10^6 in some amplifiers. As a result, any output drift away from the correct value is fed back via resistor R_f to produce an output which counteracts the drift effect, and reduces it practically to zero.

At normal operating frequencies the gain of the chopper amplifier falls off very rapidly, so that the gain soon falls to $-K$, which is the gain of the direct coupled amplifier. Thus, the chopper amplifier is only effective at drift frequencies. Provided that the amplifier gain is high, then the gain when feedback is applied (see also chapter 3) is

$$\frac{V_2}{V_{in}} = -\frac{R_f}{R_1}$$

2.16 Linear microcircuits

A wide range of low cost linear microcircuits is commercially available, making system construction a relatively easy task. All that is required apart from the microcircuit and power supplies are a few components. One popular circuit, the Fairchild μA709 amplifier is shown in Fig. 2.30.

The input stage of this amplifier comprises the differential pair of transistors TR1 and TR2 with associated load resistors R_1 and R_2 and constant current source TR3 and TR4. The second stage contains transistors TR5 and TR6, which drive the output stage containing transistors TR7, TR8, and TR9. Transistors TR7 and TR8 are complementary transistors which operate in class B, and one must be turned off before the other can conduct. Cross-over distortion in this amplifier is reduced to a very small value by the feedback applied through R3. The differential voltage gain of such an amplifier is 45 000 typically, and the common-mode rejection ratio is 90 dB. Terminations D, E, and F are provided for frequency compensation circuits (series R–C circuits), which are necessary to prevent instability and oscillation. The values of the components used in the frequency compensation circuits are quoted in manufacturers' data sheets.

The negative sign associated with input A means that it is a phase-inverting input, and that any signal applied to that terminal causes the output voltage to be antiphase to it. Input B is non-inverting, and the output is in phase with the signal at input B.

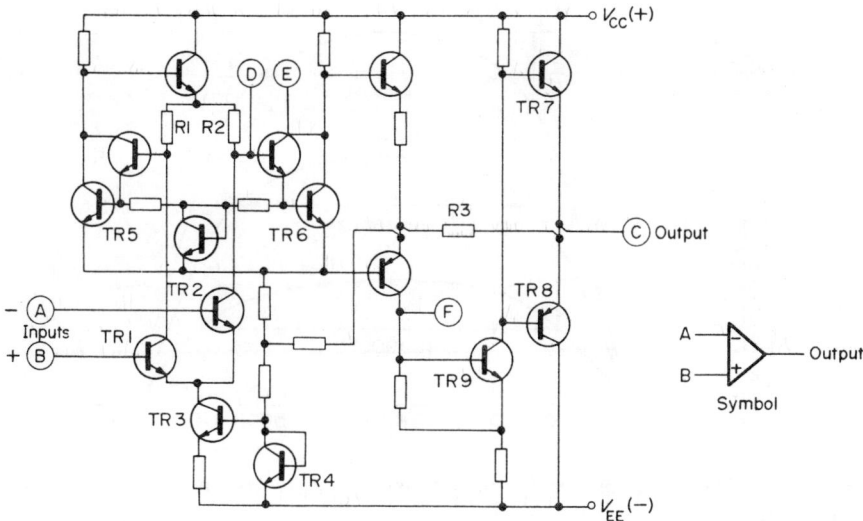

Fig. 2.30 A Fairchild μA709 amplifier.

2.17   Gain and bandwidth of cascaded stages

Voltage gain, current gain, or power gain can be increased by cascading amplifier stages. However, penalties are incurred by using this technique as follows. Firstly, the input resistance of driven stages loads the driver stage, thereby reducing the gain below the product of the gains of the cascaded stages. As a guide to the reduction factor, if the output resistance of the driver stage is R_{out} and the input resistance of the driven stage is R_{in}, then the voltage gain of the cascaded pair is reduced by a factor of $R_{in}/(R_{in}+R_{out})$.

Secondly, the bandwidth of the amplifier is reduced. In Fig. 2.31, curve A is the frequency response diagram of an amplifier with a gain of 8 dB and a bandwidth of 9900 Hz. If two identical amplifiers are cascaded, the mid-band gain is doubled but the bandwidth is reduced to 6278 Hz. The reason for this can be discovered by inspecting curve A. The bandwidth of an amplifier is defined as the band of frequencies between which the gain is 3 dB below the mid-band gain. In the case of curve A, it is between 100 Hz and 10 kHz. When two amplifiers are cascaded the gain is theoretically doubled at all frequencies. This means that for two cascaded amplifiers the bandwidth lies between the frequencies for which the gain of *each amplifier* is 1.5 dB below the mid-band gain. In the case considered, these frequencies are 156 and 6434 Hz, giving a bandwidth of 6278 Hz. The bandwidth is further reduced for three identical cascaded amplifiers, as shown in curve C.

If the lower and upper cut-off frequencies of an amplifier are f_1 and f_2,

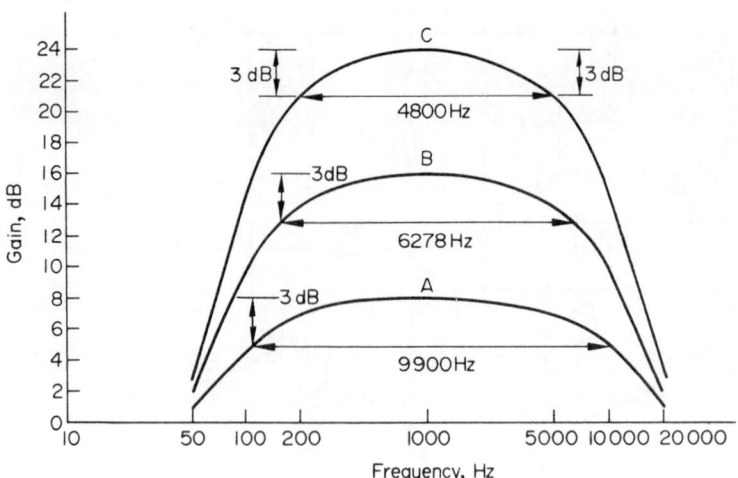

Fig. 2.31   The effect on the frequency response curve of cascading identical amplifiers.

respectively, and there are N identical cascaded stages, then the lower cut-off frequency f_L of the cascaded amplifiers becomes

$$f_L = f_1 / \sqrt{(2^{1/N} - 1)}$$

and the upper cut-off frequency f_H is

$$f_H = f_2 \sqrt{(2^{1/N} - 1)}$$

The curves in Fig. 2.31 take no account of the loading factor mentioned above, so that the actual gain for curves B and C will be less than the values shown.

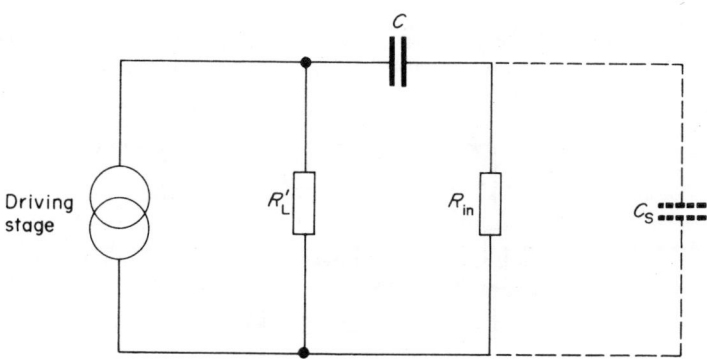

Fig. 2.32 Determination of cut-off frequencies.

2.18 Amplifier cut-off frequencies

Figure 2.32 shows the low frequency equivalent output circuit of an amplifier. Cut-off occurs at the frequency at which the current gain is $1/\sqrt{2}$ of the mid-frequency value, and this occurs at low frequencies at frequency f_1, where

$$f_1 = 1/2\pi C(R'_L + R_{in})$$

where R'_L is the resistance of the collector load in parallel with the output resistance of the driving stage, R_{in} is the input resistance of the driven stage, and C is the capacitance of the inter-stage coupling capacitor.

At high frequencies, cut-off occurs at frequency f_2, where

$$f_2 = 1/2\pi RC_S$$

where $R = R_{in} R'_L / (R_{in} + R'_L)$, and C_S is the total capacitance shunting the driving stage.

Problems

2.1 The circuit shown in Fig. 2.5 is that of a simple transistor amplifier. Estimate the values of the collector load resistor R_L and of the bias resistor R_B if the quiescent collector current and voltage values are 9.2 mA and 4.4 V, respectively. The transistor has a d.c. current gain of 115, V_{BE} is 0.7 V, and $V_{CC}=9$V.

To improve the d.c. stabilization of the circuit the bias is to be obtained by returning the bias resistor to the collector. Draw a circuit diagram to show how this can be done without introducing unwanted a.c. feedback. Calculate suitable values for the components required and compare the relative merits of this method of d.c. stabilization with other methods.

2.2 Calculate the thermal stability factors of each of the circuits in problem 2.1.

2.3 Explain why the circuit in Fig. 2.33 gives better d.c. stability than one in which the emitter is connected directly to the 0 volt line.

The transistor used has negligible leakage current at room temperature and a d.c. current gain (h_{FE}) of 100. Under quiescent conditions the collector current is 4 mA and V_{BE} is 600 mV. Determine a suitable value for R_b.

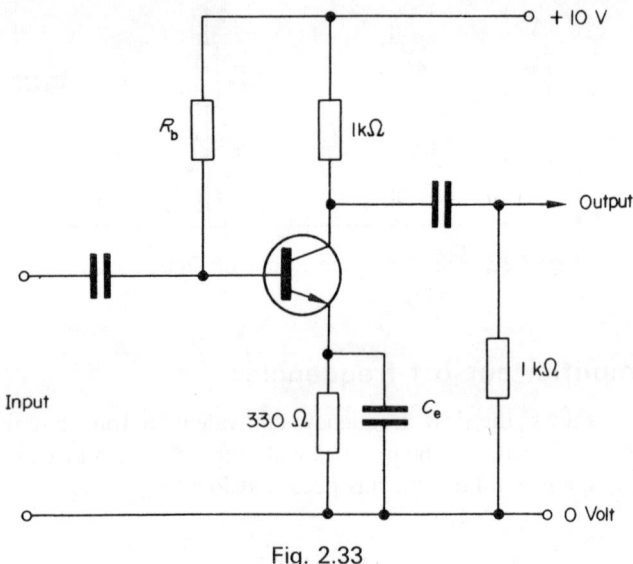

Fig. 2.33

If the amplifier a.c. current gain (A_i) is 90 and the a.c. input resistance is 800 Ω, calculate the voltage gain assuming that all capacitors have negligible reactance at the frequency of operation.

Explain briefly what the effect would be of removing C_e from the circuit.

2.4 The circuit shown in Fig. 2.34 is that of a single stage amplifier. Draw an a.c. equivalent circuit in terms of the transistor h-parameters, incorporating only the labelled components (h_{oe} and h_{re} may be neglected). Using the equivalent circuit, if $R1 = 100 \, k\Omega$, $R2 = 680 \, \Omega$, $h_{ie} = 1 \, k\Omega$, and $h_{fe} = 85$, determine

(a) the a.c. input resistance

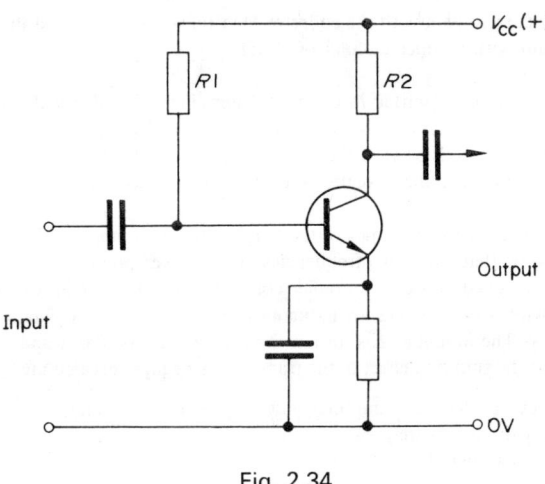

Fig. 2.34

(b) the voltage gain.

Explain why the input resistance will not remain constant if V_{CC} changes.

2.5 Derive an expression for the lower cut-off frequency of a common-emitter R–C coupled amplifier, stating any approximations made.

2.6 A transistor with $h_{fe} = 100$ and $h_{oe} = 25\ \mu S$ is used as the active element in the first stage of a two-stage R–C coupled amplifier. The two stages are coupled by a $2\ \mu F$ capacitor, the input resistance of the second stage being $1\ k\Omega$. If the lower cut-off frequency is 15.9 Hz, determine (a) the effective mid-frequency current gain of the amplifier, and (b) the collector load resistance of the first stage.

2.7 In the case of the amplifier in problem 2.6, compute the effective current gain and its phase shift at 7.95 Hz.

2.8 Explain the meaning of p-n-p and n-p-n as applied to transistors. What is the difference in the circuits appropriate to these two types of transistors?

The characteristics of a transistor in the common-base connection are, for small signals, represented by the following equations

$$V_e = h_{ib}I_e + h_{rb}V_c$$
$$I_c = h_{fb}I_e + h_{ob}V_c$$

In a particular transistor, $h_{fb} = -0.97$, $h_{ob} = 5\ \mu S$, $h_{ib} = 20\ \Omega$, and $h_{rb} = 0.0005$. If the transistor is used in the grounded-base connection with a load resistor of $50\ k\Omega$ in the collector circuit, calculate (a) the voltage amplification, and (b) the input resistance.

2.9 Draw the equivalent circuit of a common-emitter amplifier using a transistor with the following h-parameters:

$$h_{ie} = 1\ k\Omega;\ h_{re} = 2 \times 10^{-4};\ h_{fe} = 50;\ h_{oe} = 20\ \mu S$$

Assuming that the bias conditions remain unchanged throughout, calculate (a) the input resistance and the current gain with the collector load on short-circuit, and (b) the voltage gain and the power gain with a collector load of 50 kΩ.

2.10   A two-stage common-emitter R–C coupled amplifier uses identical transistors in both stages. The relevant parameters of the transistors are $h_{ie} = 2$ kΩ, $h_{fe} = 60$, and $h_{oe} = 40\ \mu S$, and the collector load resistance is 5 kΩ. If the low-frequency cut-off is to be 15.9 Hz, determine the value of the coupling capacitance. Neglect the effect of the bias circuits.

2.11   (a) If the voltage at the extremes of the bandwidth of an amplifier is 3 dB down on the maximum gain, show that these two frequencies are '$\frac{1}{2}$ power points'.
  (b) The load of a tuned collector amplifier consists of a capacitor haivng a capacitance of 2 μF in parallel with a coil having an inductance of 10 mH and a Q-factor of 14.15 at the resonant frequency. The amplifier has an a.c. input resistance of 700 Ω and a current gain of 152. Assuming that the shunting effect of the transistor is negligible, calculate

  (i) the frequency at which the amplifier voltage gain is a maximum
  (ii) the voltage gain at this frequency
  (iii) the amplifier bandwidth.

2.12   An amplifier has an effective output resistance of 20 Ω and an open-circuit voltage gain of unity. It supplies a load consisting of a reed relay coil which has a resistance of 80 Ω and an inductance of 0.1 H. The relay closes when the current in the coil is 40 mA and opens when it falls to 15 mA. Determine

  (a) the time delay before the contacts close.
  (b) the time they remain closed when a pulse of negligible rise time, 5 V in amplitude and 5 ms duration is applied to the input of the amplifier.

Assume that the quiescent current in the relay coil is zero.

2.13   (a) Explain the difference between the terms 'pinchoff region' and 'pinchoff voltage' in a field effect transistor.

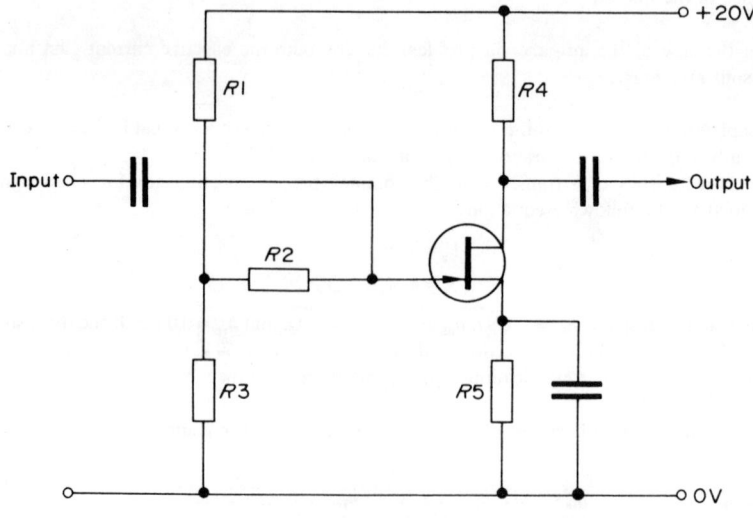

Fig. 2.35

(b) Why does the bias arrangement used in the class A amplifier in Fig. 2.35 give good d.c. stability? What is the purpose of R2?

The FET used has a mutual conductance (transconductance) of 3 mS and the circuit output is connected to an external load of $3\,k\Omega$. If $R1 = 47\,k\Omega$, $R2 = 1\,M\Omega$, $R3 = 5.1\,k\Omega$, $R4 = 1\,k\Omega$, $R5 = 1\,k\Omega$, determine

(i) the voltage gain
(ii) the circuit a.c. input resistance.

State any assumptions made.

2.14 Show that the two-transistor circuit of Fig. 2.15 is equivalent to a single transistor of input resistance

$$h_{IE1} + (1 + h_{FE1})h_{IE2}$$

Parameters h_{RE1} and h_{RE2} can be neglected.

2.15 Give reasons why simple direct coupled amplifiers are not used in practice to amplify small direct voltages.

Sketch the circuit arrangement of a chopper amplifier and explain its operation with particular reference to an electronic method of chopping.

Give the reasons for bandwidth limitations of this type of amplifier.

2.16 Detail the special requirements of the input stage for a balanced-input, direct-coupled amplifier.

With the aid of a schematic diagram, explain the operation of a chopper-stabilized direct-coupled amplifier. What are the advantages of this amplifier over a chopper amplifier?

2.17 The gain-bandwidth product of an amplifier is 50 MHz. An amplifier is to be constructed using (a) two, (b) three, (c) seven of the above amplifiers, which are cascaded. If the upper cut-off frequency of the cascaded amplifiers is to be limited to 10 MHz, determine (i) the bandwidth (f_2) per stage, and (ii) the gain per stage of the individual amplifiers.

3. Feedback amplifiers

3.1 Methods of application of feedback

Feedback amplifiers or *closed-loop amplifiers* are amplifiers in which a proportion of the output signal (which may be a voltage or a current) is fed back to the input; the signal fed back is added to the input signal so that the amplifier output is related to the sum of the two signals.

Where the signal fed back is in phase with the input signal we say that *positive feedback* or *regenerative feedback* is applied. When the input signal and the signal fed back are antiphase to one another then *negative feedback* or *degenerative feedback* is applied. A number of block diagrams of feedback systems are shown in Fig. 3.1, in all of which it is possible to apply either positive or negative feedback. In this chapter, emphasis is placed on negative feedback; positive feedback, which can lead to instability, is dealt with in chapter 4.

The type of feedback employed has its predominant effect on the *gain* of the amplifier. If positive feedback is applied, the output signal is added to the input signal so that the output tends to increase further. In this way, **positive feedback increases the gain** of the closed-loop circuit above that of the amplifier itself. Where negative feedback is applied, the signal fed back is effectively subtracted from the input signal, thereby reducing the output below the value which would obtain if feedback were not applied. In this way, **negative feedback reduces the gain** of the closed-loop circuit below that of the amplifier itself.

The signal fed back is generally proportional either to output current or output voltage. Where the signal fed back is proportional to the amplifier output voltage we say that *voltage feedback* (also known as *parallel-derived* feedback) is applied; where the signal fed back is proportional to output current then *current feedback* (or *series-derived* feedback) is applied. In

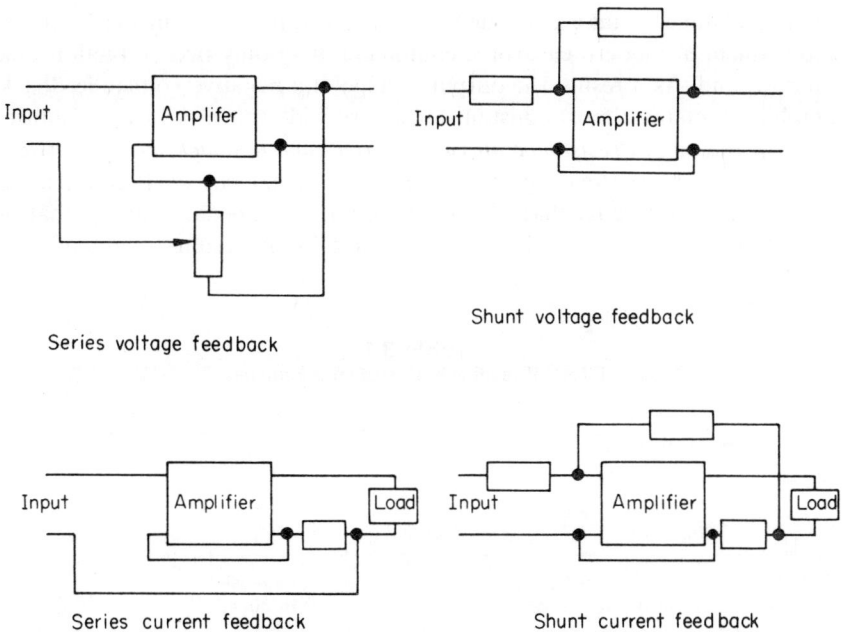

Series voltage feedback

Shunt voltage feedback

Series current feedback

Shunt current feedback

Fig. 3.1   Basic block diagrams of feedback amplifiers.

many amplifiers, including those in Fig. 3.1, current is measured by allowing it to flow in a resistor so that the p.d. across the resistor is proportional to current. The voltage so developed is then used as the feedback signal.

While the way in which the feedback signal is derived has a predominant effect on the *output resistance* of the amplifier, the way in which it is connected at the input affects the amplifier *input resistance*. To explain the above, we will consider individual cases where negative feedback is applied. The results for positive feedback are, in particular cases, opposite to those for negative feedback.

Consider the case of negative voltage feedback. An increase in load current due to, say, a change in load resistance causes the output voltage to drop initially by an amount dependent on the output resistance of the amplifier. The reduced voltage is fed back and, in a negative feedback amplifier,

Net voltage applied to amplifier = Input voltage
                                   − Voltage fed back from output

so that the net voltage applied to the amplifier terminals is increased as a result of the reduced feedback signal. This causes the output voltage to rise to a level which is only slightly less than the original value. The time taken for this change to take place depends on the time delays in the feedback loop which, in modern electronic equipment, may only take a fraction of a microsecond. As a result, the output voltage of a negative voltage feedback amplifier remains almost constant for a very wide range of load resistance. That is, the *output resistance of a negative voltage feedback amplifier is less than the output resistance of the amplifier used in the circuit.* It can also be argued that positive feedback increase the output impedance above that of the amplifier used in the circuit. A complete list of results is tabulated in Table 3.1.

Table 3.1
Effect of feedback on system parameters

Type of feedback		General effect on gain	General effect on R_{out}	General effect on R_{in}
Positive		Increased		
Negative		Reduced		
Voltage	Positive		Increased	
(shunt derived)	Negative		Reduced	
Current	Positive		Reduced	
(series derived)	Negative		Increased	
Shunt injected	Positive			Increased
	Negative			Reduced
Series injected	Positive			Reduced
	Negative			Increased

When negative current feedback is applied, the signal fed back is a voltage developed across a resistor in series with the load. An increase in load current due, say, to a change in load resistance causes the feedback signal to be increased. The equation given above still applies since negative feedback is employed, so that the increased signal fed back reduces the voltage applied to the amplifier terminals. This reduces the output current to a value almost equal to the original value, the time taken for the new steady-state condition to be reached being very short in electronic circuits. Thus, in a negative current feedback amplifier the load current remains almost constant over a wide range of load resistance. That is, the *output resistance of an amplifier employing negative current feedback is greater than the output resistance of the amplifier itself.* Positive current feedback reduces the output resistance below that of the actual amplifier used in the closed-loop (in fact the output resistance can become negative!)

In a negative feedback amplifier in which the signal fed back is shunt-injected, the input signal source must not only provide the input current to the amplifier, but must also provide a current to offset the current in the feedback resistor. As a result, the current drawn from the input source is greater than the input current to the amplifier used within the closed-loop; this is equivalent to a reduction in the input resistance of the feedback amplifier when compared with the input resistance of the amplifier itself. Therefore, *shunt negative feedback reduces the input resistance of the feedback amplifier below that of the amplifier used in the circuit.*

In series negative feedback circuits, the signal fed back opposes the input signal, thereby reducing the current drawn from the input signal source. This is equivalent to an increase in input resistance, hence *series negative feedback increases the input resistance of the feedback amplifier above that of the amplifier used in the circuit.* Positive series feedback has the opposite effect.

A useful equation relating to feedback amplifiers is deduced from eq. (2.6), and is

$$A_{vf} = -A_{if} R_L / R_{inf}$$

where the terms with the subscript f are parameters of the amplifier *after* feedback has been applied.

3.1.1 Series negative voltage feedback amplifier

To allow readers to compare the results of applying series negative voltage feedback to an amplifier, some general results are quoted below. In the following the parameters A_v, R_1, and R_2 are the voltage gain, the input resistance, and the output resistance, respectively, of the amplifier before feedback is applied, whilst A_{vf}, R_{in}, and R_{out} are the parameters after feedback has been applied.

$$A_{vf} = A_v / (1 - A_v \beta)$$
$$R_{in} = R_1 (1 - A_v \beta)$$
$$R_{out} = R_2 / (1 - A_v \beta)$$

In negative feedback amplifiers, the product $A_v \beta$ has a negative value so that

$$A_{vf} < A_v$$
$$R_{in} > R_1$$
$$R_{out} < R_2$$

[**Note:** A detailed analysis is given in *Industrial Electronics* (McGraw-Hill) by N. M. Morris.]

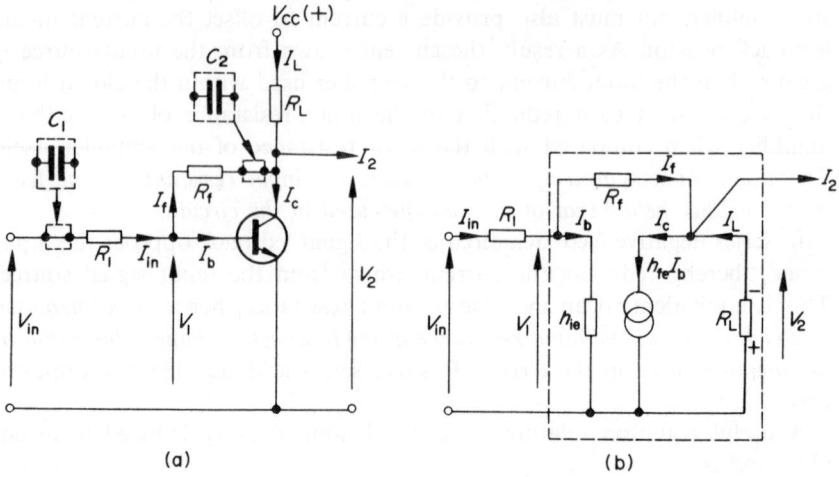

(a)   (b)

Fig. 3.2 (a) Basic transistor shunt negative voltage feedback amplifier and (b) its simplified a.c. equivalent circuit.

3.2 Bipolar transistor shunt negative voltage feedback amplifier

In Fig. 3.2(a) resistor R_f is connected between the collector and base of the transistor, and it is this element which provides the feedback path between the output and input. Since the amplifier is phase-inverting, the signal fed back is in the correct sense to provide negative feedback. In some versions of the circuit, it may be necessary to include capacitors C_1 and C_2 for d.c. blocking purposes.

The simplified h-parameter equivalent circuit of the amplifier is shown in Fig. 3.2(b), in which parameters h_{re} and h_{oe} are assumed to be negligibly small. The current I_2 shown is drawn by a load which is connected externally to the amplifier.

We will analyse the circuit in two parts. First, we consider the section of the amplifier in Fig. 3.2(b) enclosed in broken lines. The collector current is

$$I_c = h_{fe}I_b \tag{3.1}$$

and
$$I_b = V_1/h_{ie} \tag{3.2}$$

Applying Kirchhoff's laws to the collector circuit we deduce

$$I_L = I_c + I_2 - I_f \tag{3.3}$$

where
$$I_f = (V_1 - V_2)/R_f \tag{3.4}$$

also
$$V_2 = -I_L R_L = -(I_C + I_2 - I_f)R_L \tag{3.5}$$

Combining eqs. (3.1) to (3.5) gives the expression

$$V_2\left(1+\frac{R_L}{R_f}\right)=-V_1\left(\frac{h_{fe}R_L}{h_{ie}}-\frac{R_L}{R_f}\right)-R_LI_2 \qquad (3.6)$$

To determine the voltage gain of the section of the amplifier we are considering, let $I_2=0$ in eq. (3.6), when

$$\frac{V_2}{V_1}=-\left(\frac{h_{fe}R_L}{h_{ie}}-\frac{R_L}{R_f}\right)/(1+R_L/R_f) \qquad (3.7)$$

Taking typical values, let h_{fe} be 200, R_L be 4.7 kΩ, R_f be 47 kΩ, and h_{ie} be 1.5 kΩ. This gives a voltage gain of

$$V_2/V_1=-([200\times4.7/1.5]-4.7/47)/(1+4.7/47)$$
$$=-(627-0.1)/(1+0.1)=-570$$

The above calculation shows that eq. (3.7) can be reduced to the approximate expression $-h_{fe}\,R_L/h_{ie}(1+R_L/R_f)$. Equation (3.7) also verifies that the voltage gain of the feedback amplifier is less than the gain of the basic amplifier without feedback, which is $-h_{fe}R_L/h_{ie}$ (see Table 2.4, p. 46).

The output resistance of the amplifier, from eq. (3.6), is computed from the change in V_2 with I_2 at a constant value of input voltage

$$R_{out}=-V_2/I_2=R_L/(1+R_L/R_f)=R_LR_f/(R_L+R_f) \qquad (3.8)$$

That is, the effective output resistance is the parallel combination of R_L and R_f. With the values given above

$$R_{out}=4.7\times47/(4.7+47)=4.27\ \text{kΩ}$$

which is less than the output resistance of the amplifier without feedback,

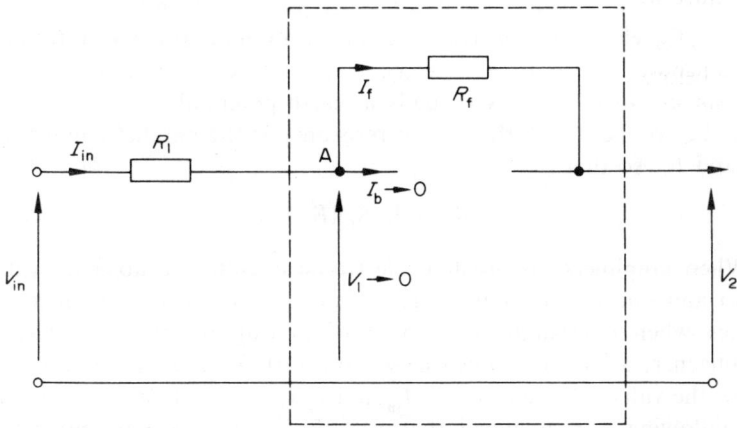

Fig. 3.3 Completion of the analysis of the shunt feedback amplifier.

which is $4.7\,k\Omega$. In cases where $R_L \ll R_f$, the output resistance is approximately equal to R_L.

The effective input resistance viewed between the base and emitter of the transistor is approximately $h_{ie}/(1 + h_{fe}R_L/R_f)$ (see also section 3.10), which is less than the input resistance of the transistor in the absence of feedback ($\simeq h_{ie}$).

The second part of the analysis of the amplifier is completed with the aid of Fig. 3.3, in which the amplifier input is applied via resistor R_1. As was shown above, the gain of a single-stage amplifier is relatively large so that V_1 is small compared with V_2. To simplify our calculations we assume that V_1 is small enough to be neglected; in this case we may also assume that I_b is very small so that all the input current I_{in} flows through R_f, hence

$$I_{in} = I_f$$

but $I_{in} = (V_{in} - V_1)/R_1$ and, since V_1 is very small, then $I_{in} \simeq V_{in}/R_1$. Also, $I_f = (V_1 - V_2)/R_f \simeq -V_2/R_f$, hence

$$\frac{V_{in}}{R_1} = -\frac{V_2}{R_f}$$

Therefore, the voltage gain A_{vf} is

$$A_{vf} = \frac{V_2}{V_{in}} = -\frac{R_f}{R_1} \tag{3.9}$$

If $R_1 = 2.2\,k\Omega$, $R_f = 47\,k\Omega$, then the voltage gain is $-47/2.2 = -21.4$.

Moreover, the input resistance of the feedback amplifier is

$$R_{in} = (V_{in} - V_1)/I_{in} \simeq V_{in}/I_{in} \simeq R_1 \tag{3.10}$$

That is, the effective input resistance of the amplifier is equal to the value of R_1. Since we have assumed V_1 to be practically zero, we say that junction A is a *virtual earth*, i.e., it is virtually at earth potential.

As before (eq. (3.8)), the output resistance is the parallel combination of R_L and R_f, so that

$$R_{out} = R_L R_f/(R_L + R_f) \tag{3.11}$$

When amplifiers are used in electronic circuits the absolute values of input and output impedances are not always significant, but their relative values when compared with those of driving or driven circuits is of significance, as will be shown in section 3.10. As a result we also need to know the values of the ratios I_2/I_{in} and I_2/V_{in} for the feedback amplifier. In the following it is assumed that $R_f \gg R_L$, so that the output resistance simplifies to R_L. Now

$$\frac{I_2}{V_{in}} = \frac{I_2}{V_2} \cdot \frac{V_2}{V_{in}} = \frac{1}{R_L} \left(-\frac{R_f}{R_1} \right) = -\frac{R_f}{R_L R_1} * \qquad (3.12)$$

and since $V_{in} = R_1 I_{in}$, then $I_2/V_{in} = I_2/R_1 I_{in} = -R_f/R_L R_1$, hence

$$\frac{I_2}{I_{in}} = -\frac{R_f}{R_L} \qquad (3.13)$$

and

$$\frac{V_2}{I_{in}} = \frac{V_2}{V_{in}} \cdot \frac{V_{in}}{I_{in}} = \left(-\frac{R_f}{R_1} \right) . R_1 = -R_f \qquad (3.14)$$

Simplified analysis: Assuming that the base-emitter voltage V_1 is small compared with V_{in}, then $I_{in} = V_{in}/R_1$ and the input resistance is

$$R_{in} = V_{in}/I_{in} = R_1$$

Also, if V_1 is very small then $I_{in} = I_f$, and since $I_f = -V_2/R_f$ then $V_{in}/R_1 = -V/R_f$, giving an expression for voltage gain of

$$V_2/V_{in} = -R_f/R_1$$

The output resistance can be computed from the relationship $R_{out} = V_{2(oc)}/I_{2(sc)}$, where $V_{2(oc)}$ is the open-circuit (no-load) output voltage and $I_{2(sc)}$ is the current which flows in a short-circuit applied across the load. From the basic equations of the common-emitter amplifier, we can say that $V_{2(oc)} = h_{fe} R_L V_1/h_{ie}$. When a short-circuit is applied to the output, the current in the short-circuit is $I_{2(sc)} = h_{fe} I_b = h_{fe} V_1/h_{ie}$, hence

$$R_{out} = V_{2(oc)} \big/ I_{2(sc)} = R_L$$

3.3   FET shunt negative voltage feedback amplifier

The basic circuit is shown in Fig. 3.4 for which the equations are

$$I_d = y_{fs} V_1$$
$$I_L = I_d + I_2 - I_f$$
$$I_f = (V_1 - V_2)/R_f$$
$$V_2 = -I_L R_L$$

*Note: The ratio I_2/V_{in} is phase-inverting (has a negative sign) if the external load is connected between the collector and the common line (as is assumed to be the case here). If the load is connected between the collector and the supply line, then I_2 increases with V_{in} and I_2/V_{in} is non-inverting.

Solving for V_2 between the equations gives

$$V_2/V_1 = -(y_{fs}R_L - R_L/R_f)/(1 + R_L/R_f) \tag{3.15}$$

and

$$R_{out} = R_L/(1 + R_L/R_f) \tag{3.16}$$

If, in the circuit used $R_f \gg R_L$, the equations reduce to

$$V_2/V_1 = -y_{fs}R_L \quad \text{and} \quad R_{out} = R_L$$

The above equations are derived from a simplified equivalent circuit of the FET in which parameters y_{is}, y_{rs}, and y_{os} are assumed to be negligibly small. If the gain of the FET amplifier (eq. (3.15)) is large, then V_1 is sufficiently small to be neglected so that the overall expressions for voltage gain (V_2/V_{in}), input resistance, and output resistance are identical to those of the bipolar transistor amplifier, eqs. (3.9), (3.10), and (3.11), respectively. In this circuit, the gate-to-drain capacitance is magnified by the Miller effect, so that the effective gate-drain capacitance is $C_{gd}(1 - V_2/V_1)$.

In all circuits where we can make the general assumption that V_1 is very small in comparison with both V_{in} and V_2, the equations for the feedback amplifier (eqs. (3.10) to (3.16)) hold good.

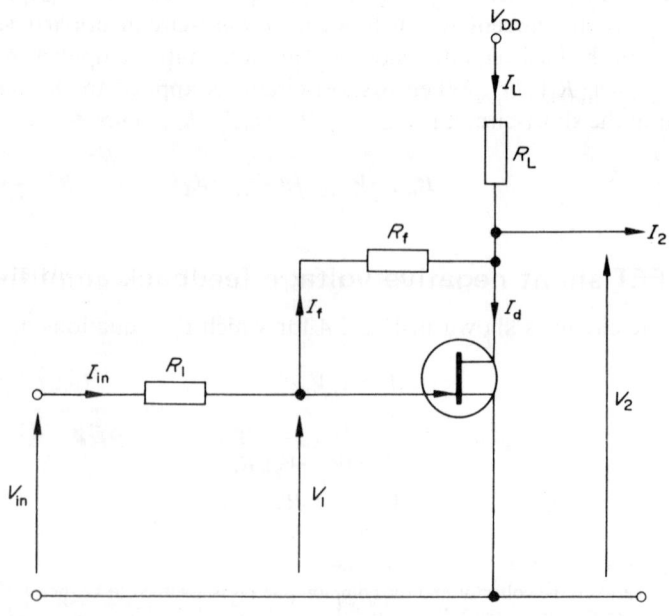

Fig. 3.4   FET shunt voltage feedback amplifier.

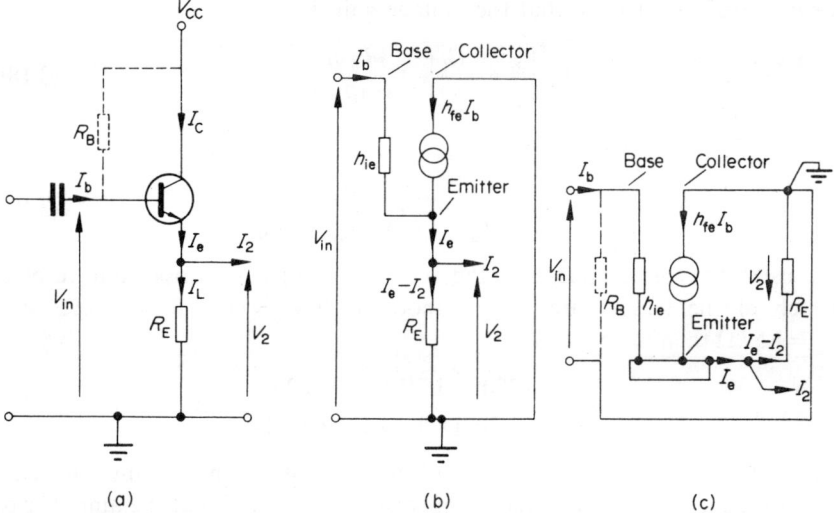

(a)   (b)   (c)

Fig. 3.5 (a) Basic emitter follower circuit, (b) its a.c. equivalent circuit, and (c) the equivalent circuit redrawn in terms of the basic block diagram in Fig. 3.1.

3.4 The emitter follower

The emitter follower is so named because the potential of the emitter terminal 'follows' the input signal potential. The basic circuit of the emitter follower is shown in Fig. 3.5(a), in which bias is provided by resistor R_B. In the analysis which follows, the effect of R_B on the input resistance is neglected but, in practice, the value of R_B may be smaller than the input resistance of the emitter follower.

The simple principle of the emitter follower is as follows. An increase in input signal voltage causes the base current and emitter current to increase. This, in turn, results in an increase in the output voltage V_2, which is in phase with V_{in}. In most circuits, the signal voltage appearing across the emitter junction of the transistor is very small, so that for all practical purposes V_2 and V_{in} are equal in magnitude, giving the circuit a unity voltage gain.

Now
$$I_b = (V_{in} - V_2)/h_{ie}$$
and
$$I_e = I_c + I_b = (1 + h_{fe})I_b$$
also
$$V_2 = (I_e - I_2)R_E$$

Combining the above equations yields

$$V_2 = \frac{V_{in}(1 + h_{fe})R_E/h_{ie}}{1 + (1 + h_{fe})R_E/h_{ie}} - \frac{I_2 R_E}{1 + (1 + h_{fe})R_E/h_{ie}} \qquad (3.17)$$

From which we deduce that the voltage gain is

$$\frac{V_2}{V_{in}} = \frac{(1+h_{fe})R_E/h_{ie}}{1+(1+h_{fe})R_E/h_{ie}} \tag{3.18}$$

and the output resistance is

$$R_{out} = -\frac{V_2}{I_2} = \frac{R_E}{1+(1+h_{fe})R_E/h_{ie}} \tag{3.19}$$

By inserting typical values into the above equations, it is possible to resolve which of the factors are most important. If $h_{fe} = 200$, $R_E = 4.7\,k\Omega$, and $h_{ie} = 1.5\,k\Omega$ then from eq. (3.18)

$$\frac{V_2}{V_{in}} = \frac{(1+200) \times 4.7/1.5}{1+(1+200) \times 4.7/1.5} = \frac{629}{1+629} \simeq 1.0$$

From the above we can see that the factor of unity in the denominator can be ignored so that the amplifier gain is nearly unity and that the amplifier is non-inverting. The output resistance is

$$R_{out} = \frac{4700}{1+629} = 7.46\,\Omega$$

Again, the factor of unity in the denominator can be neglected, and R_{out} can be reduced to

$$R_{out} \simeq R_E/[(1+h_{fe})R_E/h_{ie}] = h_{ie}/(1+h_{fe})$$

$$\simeq h_{ie}/h_{fe} \tag{3.20}$$

Inserting values into eq. (3.20) gives $R_{out} = 1500/200 = 7.5\,\Omega$. During this calculation we have, in fact, neglected the output resistance of the signal source in calculating R_{out}, and to account for it we must add it to h_{ie} so that eq. (3.20) becomes

$$R_{out} = (R_S + h_{ie})/h_{fe} \tag{3.21}$$

In the emitter follower the input capacitance is significantly reduced below that of the common-emitter amplifier since C_{cb} is not magnified by amplifier action and the current drawn by C_{be} is reduced by virtue of the small voltage across it.

The input resistance of the emitter follower is computed as follows. The output voltage is approximately $I_L R_E$ and, for low values of load current,

$$V_2 \simeq I_e R_E = (1+h_{fe})I_b R_E$$

Now, in the emitter follower $V_2 = h_{ie}I_b + V_2$, therefore

$$V_{in} = h_{ie}I_b + (1+h_{fe})I_b R_E = I_b[h_{ie} + (1+h_{fe})R_E]$$

Hence, the input resistance is

$$R_{in} = V_{in}/I_b = h_{ie} + (1+h_{fe})R_E \simeq h_{ie} + h_{fe}R_E \qquad (3.22)$$

$$\simeq h_{fe}R_E \qquad (3.23)$$

With the values given, the exact value of R_{in} is $1.5 \pm (1+200) \times 4.7 = 946.2\,k\Omega$ (or $940\,k\Omega$ by the approximate equation). Unfortunately, the emitter follower input is effectively shunted by R_B so that the actual value of the input resistance is the parallel combination of R_{in} and R_B. If, for instance, R_B is a $220\,k\Omega$ resistor then the input resistance is $179\,k\Omega$.

The current gain of the emitter follower is given by the ratio I_e/I_b which is $(1+h_{fe})$, and has the value 201 in the case considered.

To summarize, the principal features of interest in the emitter follower are:

(a) Its gain is approximately unity
(b) It is non-inverting
(c) Its input impedance is high
(d) Its output impedance is low.

Factors (c) and (d) above make it particularly useful as an impedance level converter.

Simplified analysis:
Current gain $= I_e/I_b = (1+h_{fe})I_b/I_b = 1+h_{fe} \simeq h_{fe}$

If we assume V_{be} to be very small, then

$$\text{Voltage gain} = V_2/V_{in} = (V_{in} - V_{be})/V_{in} \simeq 1$$

and

$$\text{Input resistance} = R_{in} = V_{in}/I_b \simeq V_2/(I_e/h_{fe}) = I_e R_E/(I_e/h_{fe}) = h_{fe}R_E$$

The no-load output voltage $V_{2(oc)} = V_{in}$, and the short-circuit output current is $I_{2(sc)} = h_{fe}V_{in}/h_{ie}$, therefore

$$R_{out} = V_{2(oc)}/I_{2(sc)} = V_{in}/(h_{fe}V_{in}/h_{ie}) = h_{ie}/h_{fe}$$

3.5 The source follower

In the source follower, Fig. 3.6(a), the output voltage is taken from the source electrode. And, as in the case of the emitter follower, the name of the circuit is derived from the fact that the potential of the source electrode follows the gate potential. The simplified equivalent circuit of the FET is used in Fig. 3.6(b) in which parameters y_{is}, y_{rs}, and y_{os} are neglected. In the circuit shown

$$I_s = y_{fs}V_g = y_{fs}(V_{in} - V_2)$$

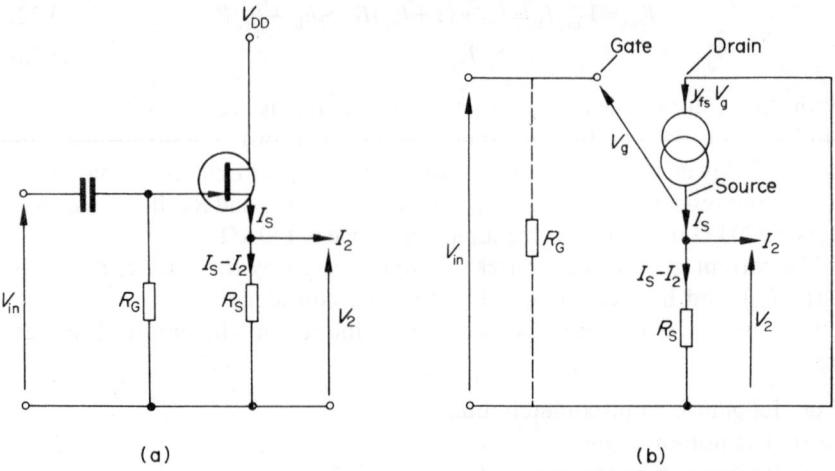

(a)   (b)

Fig. 3.6 (a) Source follower and (b) its a.c. equivalent circuit.

and
$$V_2 = (I_s - I_2)R_S$$

Solving between the equations for V_2 yields

$$V_2 = \frac{y_{fs}R_S}{1 + y_{fs}R_S}V_{in} - \frac{R_S}{1 + y_{fs}R_S}I_2$$

giving a voltage gain of

$$\frac{V_2}{V_{in}} = \frac{y_{fs}R_S}{1 + y_{fs}R_S} \tag{3.24}$$

and an output resistance of

$$R_{out} = R_S/(1 + y_{fs}R_S) \tag{3.25}$$

If we employ a FET with $y_{fs} = 5\,mS$ and R_S of $4.7\,k\Omega$, then $V_2/V_{in} = 5 \times 4.7/[1 + (5 \times 4.7)] = 0.96$, and $R_{out} = 4700/[1 + (5 \times 4.7)] = 192\,\Omega$. Readers will note that in the example given $y_{fs}R_S \gg 1$, so that

$$R_{out} \simeq R_S/y_{fs}R_S = 1/y_{fs} \tag{3.26}$$

This simplification gives an estimated output resistance of $200\,\Omega$.

The input resistance of the source follower itself at low frequency is infinitely large due to the input resistance of the FET, so that the actual input resistance of the amplifier is equal to the value of the bias resistor R_G which shunts the input. There is, however, an upper limit to the value of R_G due to thermal leakage effects of the FET. To obtain a high input resistance, the lower connection of R_G can be taken to a tapping on the source resistor as shown in Fig. 3.7. In this circuit, the bias voltage developed across R_B

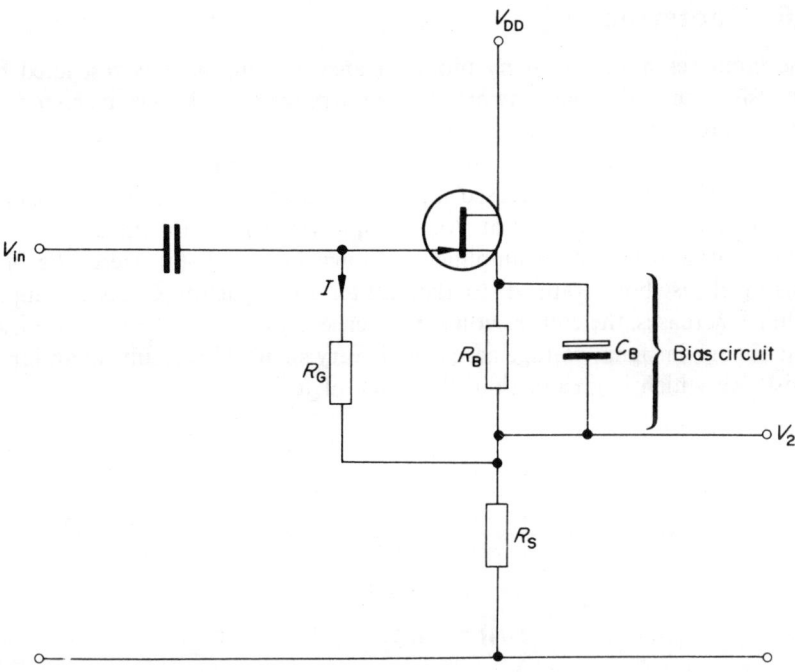

Fig. 3.7   A practical form of source follower circuit.

and C_B is connected to the gate via resistor R_G. If the voltage gain of the source follower is A_{vf}, then current I is

$$I = \frac{V_{in} - V_2}{R_G} = \frac{V_{in} - A_{vf}V_{in}}{R_G} = \frac{V_{in}(1 - A_{vf})}{R_G}$$

Since I is the current drawn from the signal source, then the effective input resistance of the amplifier is

$$R_{in} = V_{in}/I = R_G/(1 - A_{vf}) \qquad (3.27)$$

If we use a FET with $y_{fs} = 5$ mS and a source resistor of 4.7 kΩ, then $A_{vf} = 0.96$ and the input resistance is $R_G/(1 - 0.96) = 25R_G$. A value of $R_G = 1$ MΩ therefore gives an a.c. input resistance of 25 MΩ. The physical reason for this increase in input resistance above that of R_G is as follows. With an input of 1V, the output voltage is 0.96V and the voltage appearing across R_G is only 0.04V. As a result, the input current is $0.04V/1$ MΩ $= 0.04$ μA; so far as the input circuit is concerned the effective input resistance is $1V/0.04$ μA $= 25$ MΩ.

3.6   Bootstrapping

The input resistance of many bipolar transistor amplifiers is restricted by the resistance of the bias network. Circuit arrangements known as *bootstrap circuits* are sometimes adopted to increase the input resistance. In these circuits, the voltage across the bias resistor is 'pulled up by its own bootstraps', so that an increased input voltage does not necessarily mean a large increase in input current. An example was shown in Fig. 3.7.

   One form of bootstrap amplifier is shown in Fig. 3.8(a). Here, the bias resistor R_B is 'bootstrapped' to the emitter by capacitor C. As the input voltage increases, the emitter voltage increases by emitter follower action, so that change in signal voltage across R_B is very small. This results in an input resistance which is greater than the value of R_B.

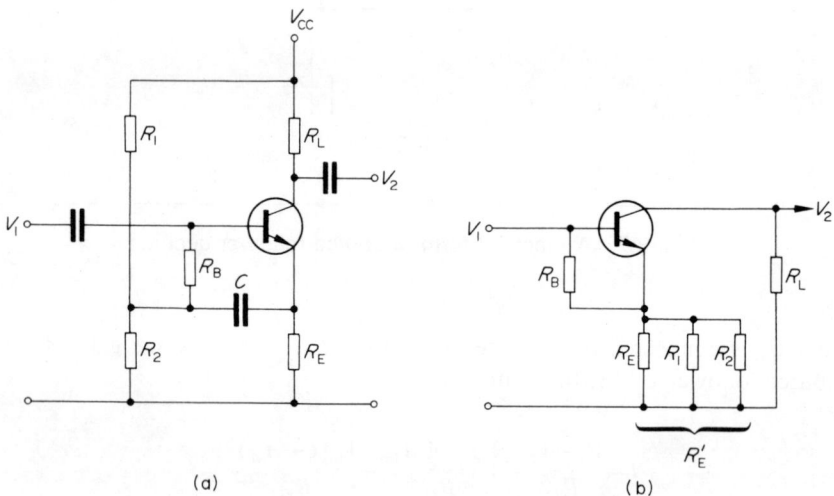

(a)   (b)

Fig. 3.8   One form of bootstrap amplifier.

   The reactance of capacitor C at the operating frequency must be very low; the a.c. equivalent circuit in Fig. 3.8(b) allows us to estimate the input resistance of the amplifier. From the work on the emitter follower, the input resistance is approximately

$$R_{in} \simeq h_{ie} + R'_E h_{fe} \simeq R'_E h_{fe}$$

where R'_E is the resistance of the parallel combination of R_E, R_1, and R_2. We can neglect the effect of R_B since it is shunted by the input resistance of the transistor ($\simeq h_{ie}$), which is small in comparison with R_B. A disadvantage of this circuit is that the voltage gain is reduced to approximately R_L/R'_E (see section 3.7).

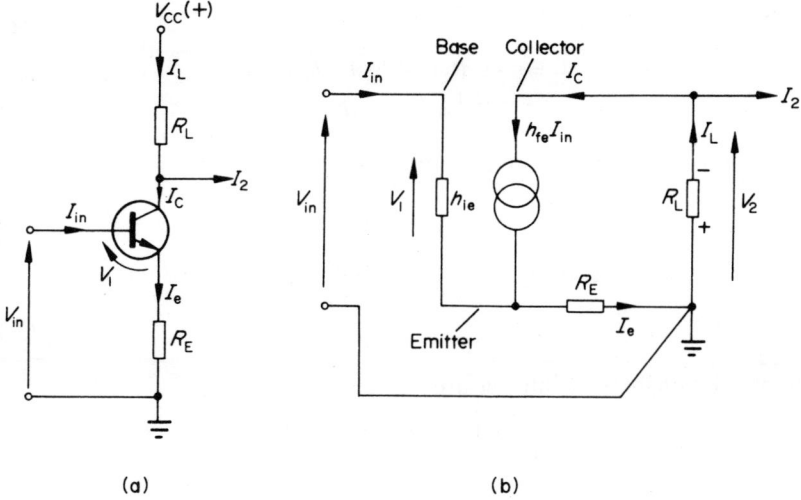

(a)   (b)

Fig. 3.9   (a) An amplifier using series negative current feedback and (b) its a.c. equivalent circuit.

3.7   A series negative current feedback amplifier

In the basic circuit, Fig. 3.9(a), bias and coupling circuits have been omitted for clarity. To provide feedback, an unbypassed emitter resistor is used so that the net signal voltage applied to the transistor is

$$V_1 = V_{in} - I_e R_E \qquad (3.28)$$

From this expression we can see that the signal fed back is $I_e R_E$, which is related to the output current. Clearly,

$$V_{in} = V_1 + I_e R_E \simeq I_{in} h_{ie} + (1 + h_{fe}) I_{in} R_E$$
$$= I_{in} [h_{ie} + (1 + h_{fe}) R_E]$$

which gives the input resistance as

$$R_{in} = V_{in}/I_{in} = h_{ie} + (1 + h_{fe}) R_E \qquad (3.29)$$

As anticipated in Table 3.1, the input resistance is greater than is the case without feedback, i.e., greater than h_{ie}. If h_{fe} has a large value, the equation simplifies to

$$R_{in} \simeq h_{fe} R_E \qquad (3.30)$$

Applying Kirchhoff's laws to the collector node we have

$$I_L = I_c + I_2 \simeq h_{fe} I_{in} + I_2$$

where $$I_{in} = V_{in}/R_{in} \simeq V_{in}/h_{fe}R_E$$

also
$$V_2 = -I_L R_L = -h_{fe}I_{in}R_L - I_2 R_L$$
$$= -R_L V_{in}/R_E - I_2 R_L \tag{3.31}$$

The voltage gain of the circuit, from the first term of eq. (3.31) is

$$A_{vf} = V_2/V_{in} = -R_L/R_E \tag{3.32}$$

and the output resistance, from the second term of the equation is

$$R_{out} = -V_2/I_2 = R_L \tag{3.33}$$

Other relationships of interest are

$$\frac{I_2}{V_{in}} = \frac{I_2}{V_2} \cdot \frac{V_2}{V_{in}} = \frac{1}{R_L} \cdot \left(-\frac{R_L}{R_E}\right) = -\frac{1}{R_E} \tag{3.34}$$

$$\frac{V_2}{I_{in}} = \frac{V_2}{V_{in}} \cdot \frac{V_{in}}{I_{in}} = -\frac{R_L}{R_E} \times h_{fe}R_E = -h_{fe}R_L \tag{3.35}$$

$$\frac{I_2}{I_{in}} = \frac{I_2}{V_{in}} \cdot \frac{V_{in}}{I_{in}} = -\frac{1}{R_E} \times h_{fe}R_E = -h_{fe}$$

[**Note:** This ratio in eq. (3.34) has a negative sign (phase-inverting) if the external load is connected between the collector and the common line, as is assumed to be the case here.]

Simplified analysis: By emitter follower action $V_{in} \simeq I_e R_E$, hence $I_e = V_{in}/R_E$. But, if $I_c \simeq I_e$, then $I_c \simeq V_{in}/R_E$. The output voltage is given by the expression $V_2 = -I_c R_L$, hence $V_2 = -V_{in}R_L/R_E$ from which the expression for voltage gain is

$$V_2/V_{in} = -R_L/R_E$$

The input resistance is

$$R_{in} = V_{in}/I_b \simeq I_e R_E/(I_e/h_{fe}) = h_{fe}R_E$$

The output resistance, so far as an external load is concerned, consists of R_L in parallel with a transistor which has series negative current feedback applied to it (see Fig. 3.9(b)). The output resistance of the transistor with this form of feedback is very high, so that the approximate value of the output resistance is

$$R_{out} = R_L$$

Example 3.1: Design an amplifier circuit of the type in Fig. 3.10 in which

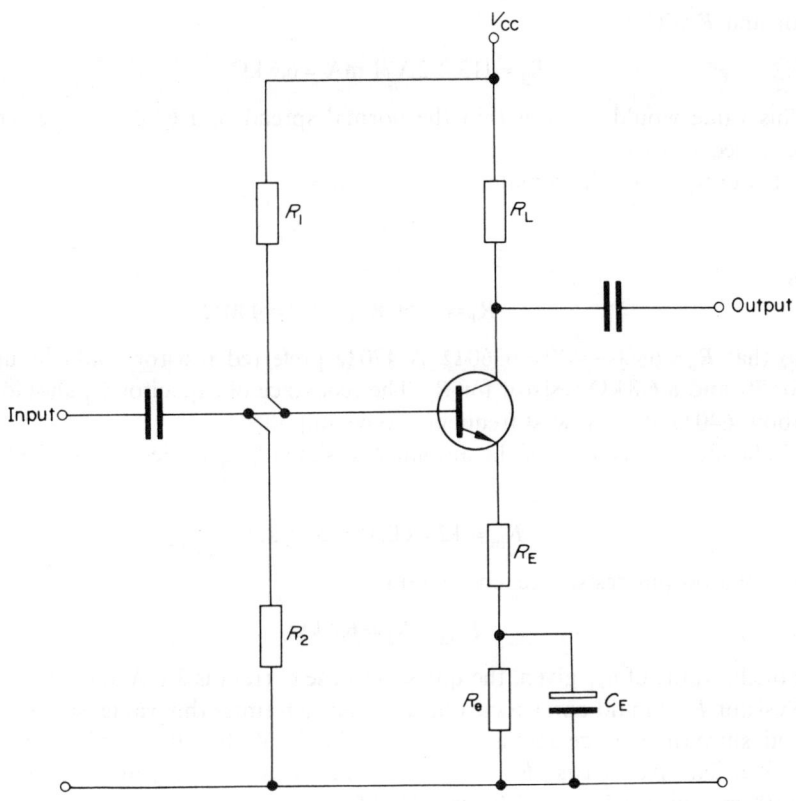

Fig. 3.10 A practical form of current feedback amplifier.

the supply voltage is 20 V, the mean collector current is to be 1 mA, and the transistor is a silicon device with an h_{fe} and h_{FE} of 120. The voltage gain of the amplifier is to be -15.

Solution: The circuit in the figure is a practical version of the basic circuit, in which $R_e C_E$ is part of the bias and thermal stability circuit. To establish a design for adequate thermal stability, the *total* resistance $(R_E + R_e)$ in the emitter lead should be of the order of one-tenth the resistance of R_2 (see section 2.5). Also, the p.d. across $(R_E + R_e)$ when carrying the quiescent current should be significantly greater than the quiescent base-emitter voltage of the transistor, which is about 0.7 V in silicon devices. Let us assume that this p.d. is 6.8 V, so that

$$R_E + R_e = 6.8 \text{ V}/1 \text{ mA} = 6.8 \text{ k}\Omega$$

In order to give the largest possible output voltage swing the remaining voltage, i.e. $20 - 6.8 = 13.2$ V, should be divided equally between the transistor and R_L. Thus,

$$R_L = (13.2/2)V/1\ mA = 6.6\ k\Omega$$

This value would come within the normal spread of a 6.8 kΩ 10 per cent preferred resistor.

From eq. (3.32) the signal frequency gain is

$$-15 = -R_L/R_E$$

or

$$R_E = -6600/(-15) = 440\ \Omega$$

so that $R_e = 6800 - 440 = 6360\ \Omega$. A 470 Ω preferred resistor could be used for R_E and a 6.8 kΩ resistor for R_e. The reactance of capacitor C_E should be about 640 Ω at the lowest frequency to be amplified.

The input resistance of the amplifier to signal frequencies, from eq. (3.30) is

$$R_{in} = 120 \times 0.44 = 52.8\ k\Omega$$

and the output resistance, eq. (3.33) is

$$R_{out} = R_L = 6.6\ k\Omega$$

For the value of h_{FE} given, the quiescent base current is 1 mA/120 = 8.33 μA. Resistor R_2 should carry a current of about ten times this value, say 0.1 mA and, since the voltage across R_2 is 6.8 + 0.7 = 7.5 V, then $R_2 = 7.5\,V/0.1\ mA = 75$ kΩ. The p.d. across R_1 is 20 - 7.5 = 12.5 V when carrying a current of 0.108 mA, giving $R_1 = 12.5/0.108 = 116$ kΩ.

3.8 A FET negative current feedback amplifier

Field-effect transistors can be used in a configuration similar to that in Fig. 3.9, for which the relevant equations are

$$R_{in} \simeq \infty$$
$$A_{vf} = -R_L/R_S$$
$$R_{out} = R_L$$

where R_S is the resistance in series with the source electrode.

3.9 A phase splitting amplifier

If, in Fig. 3.9(a), the emitter and collector load resistors have equal values, then the voltage gain between the input signal and the collector signal is

$$A_{vf} = -R_L R_E = -1$$

Also, by emitter follower action the signal voltage developed at the emitter is approximately equal to the input signal voltage, and is in phase with the input signal. As a result, the signal voltages developed at the collector and emitter of the transistor are equal in magnitude to one another but of opposite phase.

This type of circuit has been given various names including *concertina* and *split-load circuit*, and can be used as a driver stage for push-pull and other circuits.

A disadvantage of this circuit is that the output resistance at the collector is high (approximately R_L) when compared with the output resistance at the emitter (approximately h_{ie}/h_{fe}).

3.10 Multi-stage feedback amplifiers

Amplifiers with specific parameters can be constructed by applying feedback to a number of stages. For example, a basic form of two-stage current amplifier is shown in Fig. 3.11(a). Here, the first stage employs negative shunt voltage feedback to reduce the input impedance so that it can be current driven, and the second stage is a series negative current feedback amplifier to provide a high output resistance. Both features are desirable in a current amplifier.

For the first stage of Fig. 3.11(a), from eq. (3.14)

$$V_2'/I_{in} = -R_f$$

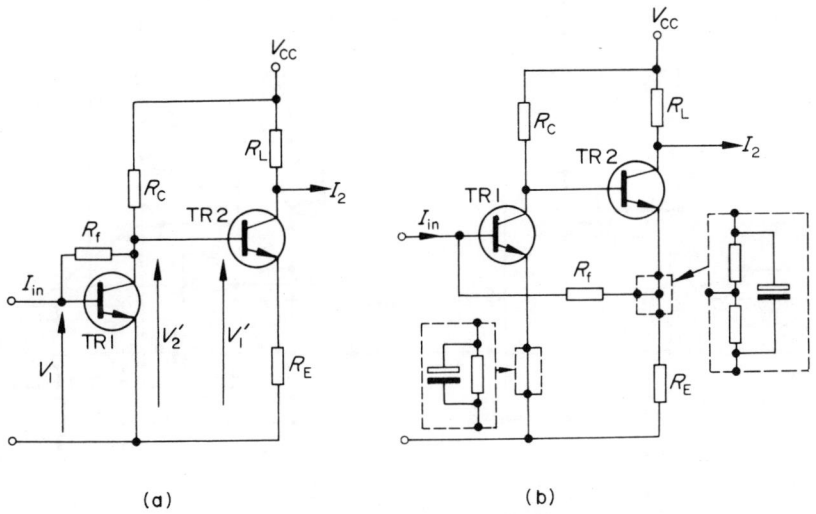

(a)   (b)

Fig. 3.11   A two-stage current amplifier.

and for the second stage, from eq. (3.34)

$$I_2/V_1' = -1/R_E$$

But, since $V_2' = V_1'$, the amplifier current gain is

$$\frac{I_2}{I_{in}} = \frac{V_2'}{I_{in}} \cdot \frac{I_2}{V_1'} = \frac{R_f}{R_E}$$

The output resistance of the amplifier is R_L and the input resistance is evaluated as follows. The voltage gain of the first stage is $V_2'/V_1 = -h_{fe1}R_C/h_{ie1}$, where h_{fe1} and h_{fe1} are parameters of TR1, while eq. (3.14) applied to this stage gives $V_2'/I_{in} = -R_f$, hence

$$R_{in} = \frac{V_2'}{I_{in}} \cdot \frac{V_1}{V_2'} = -R_f/(-h_{fe1}R_C/h_{ie1}) = h_{ie1}R_f/h_{fe1}R_C$$

A practical version of the circuit is shown in Fig. 3.11(b) in which the bias arrangements are shown in the insets. In this version, the feedback resistor is returned to the emitter of TR2, since for all practical purposes the emitter of TR2 is the same equivalent point in the circuit as the collector of TR1. If R_f has a value of 3.9 kΩ and R_E is a 100 Ω resistor, then the current gain is -39. Now, if $h_{ie1} = 2$ kΩ, $h_{fe1} = 100$, and $R_C = 22$ kΩ, then $R_{in} = 2 \times 3.9/(100 \times 22)$kΩ or about 3.5 Ω. If R_L has a value of 18 kΩ, the output resistance is 18 kΩ.

A basic two-stage voltage amplifier is shown in Fig. 3.12(a). The first stage employs series negative current feedback to give a high input re-

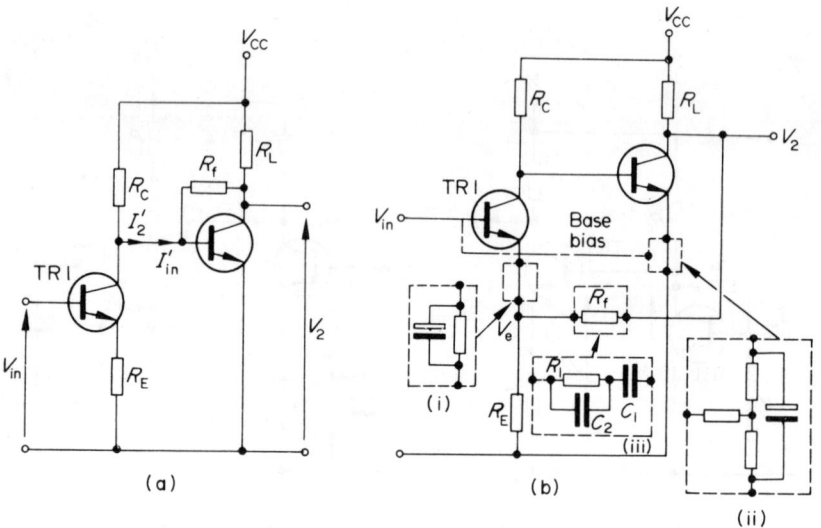

Fig. 3.12   Two-stage voltage amplifiers.

sistance, and the second stage employs shunt negative voltage feedback to give a low output resistance.

The voltage gain is determined as in the previous case using eqs. (3.14) and (3.34) as follows:

$$\frac{V_2}{V_{in}} = \frac{V_2}{I'_{in}} \cdot \frac{I'_2}{V_{in}} = \frac{R_f}{R_E}$$

The input resistance of the circuit is $h_{fe1} R_E$, where h_{fe1} is the current gain of the TR1 and the output resistance is $R_L R_f/(R_L + R_f)$.

A practical version of the circuit is shown in Fig. 3.12(b) in which the feedback resistor R_f is returned to the emitter of TR1. That this arrangement is equivalent to the circuit in Fig. 3.12(a) is not immediately apparent, but can be understood from the following. By potential divider action in Fig. 3.12(b), the signal voltage at V_e is $V_2 R_E/(R_f + R_E)$. As a general rule the value of R_E is much less than that of R_f, so that $V_e \simeq V_2 R_E/R_f$. Also, by emitter follower action, V_e is very nearly equal to the input signal voltage V_{in}, so that $V_{in} \simeq V_2 R_E/R_f$, giving an expression for voltage gain of

$$\frac{V_2}{V_{in}} = \frac{R_f}{R_E}$$

which is identical to the expression for the gain of Fig. 3.12(a). When constructing a circuit of this kind it will be necessary to adjust the value of R_f to obtain the required gain, since R_E carries two interdependent currents, and the equation above provides a starting point for the circuit design.

Bias arrangements for the circuit are shown in insets (i) and (ii). Inset (iii) shows a modified feedback network which gives increased voltage gain at low frequencies (bass boost) when compared with the gain at high frequencies. At low frequencies the reactance of capacitor C_1 is high, thereby limiting the amount of negative feedback. This results in a high gain at low frequencies. In its 'mid-band' frequency range, the reactance of C_1 is less than the resistance of R_1, and the amplifier gain is approximately R_1/R_E. At high frequencies, capacitor C_2 shunts R_1 to increase the amount of feedback and to reduce the gain further. This type of characteristic is desirable in some forms of tape replay amplifier.

Yet another multi-stage amplifier is shown in Fig. 3.13. For clarity, bias and thermal stability arrangements are omitted. In this circuit, by emitter follower action, we see that $V_s \simeq I_{e2} R_E$. Also, if the current gain of the final transistor is reasonably large, then $V_2 \simeq -I_{e2} R_L$. Solving for the voltage gain between these relationships gives

$$\frac{V_2}{V_s} = \frac{R_L}{R_E}$$

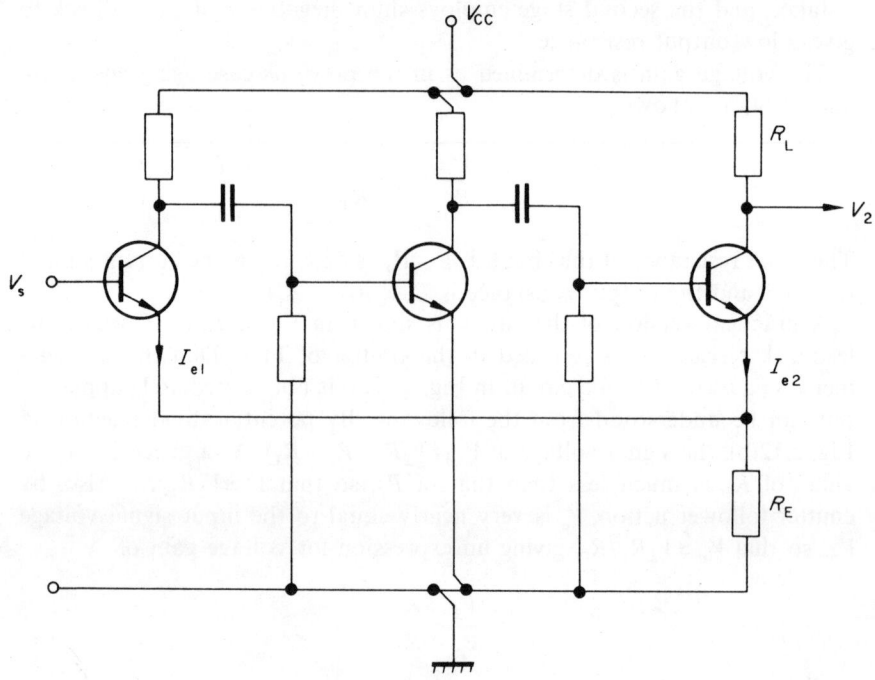

Fig. 3.13   A three-stage feedback amplifier.

3.11   Operational amplifier circuits

High-gain wide-band direct-coupled amplifiers, described as *operational amplifiers*, were originally designed for use in analogue computers. They are now used in integrated circuit form in applications which cover a wide range of industrial, commercial, and domestic circuits. One form of integrated circuit linear operational amplifier was described in section 2.16.

Ideally, the gain of the operational amplifier is infinity so that the voltage at the *summing junction* in Fig. 3.14 is practically zero. So small is it that the summing junction is a virtual earth point, and for the purpose of analysis it is taken to be at earth potential. Operational amplifiers which are commercially available have gains in the range 20 000 to 10^6.

The complete circuit in Fig. 3.14 is that of an *inverting amplifier* with a single-ended output. The operational amplifier in the figure is shown as having two input lines; when a signal is applied to the input marked with a negative sign, it causes the output signal V_2 to be 180 degrees out of phase with the input. The '−' input line is therefore known as the *inverting input*. When a signal is applied to the line marked '+', the output signal is in phase with the input signal. The '+' input is therefore described as the *non-*

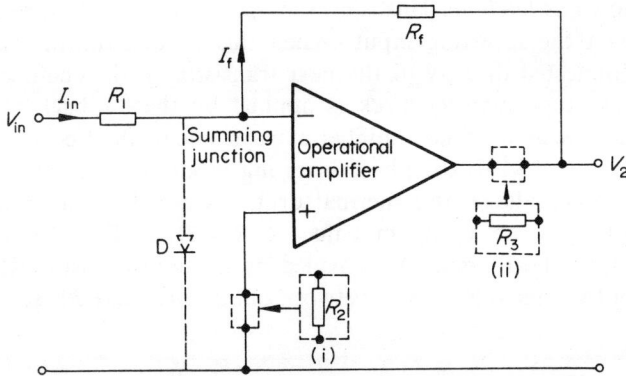

Fig. 3.14 An inverting amplifier using an integrated circuit operational amplifier.

inverting input line. The '+' and '−' signs have nothing whatever to do with the polarity of the voltage, since either positive or negative polarity voltages can be applied to either input line. In the circuit shown, the non-inverting input line is earthed, so that the output voltage is antiphase to the input signal.

Let us suppose for the moment that the voltage at the summing junction is V_x. The currents flowing in the input and feedback resistors are

$$I_{in} = (V_{in} - V_x)/R_1 \qquad I_f = (V_x - V_2)/R_f$$

Since the gain of the amplifier is very large, then V_x is practically zero so that

$$I_{in} \simeq V_{in}/R_1 \quad \text{and} \quad I_f \simeq -V_2/R_f$$

Since V_x is very small then, whatever the input resistance of the amplifier, the input current must also be very small and for all practical purposes we may assume that $I_{in} = I_f$, or $V_{in}/R_1 = -V_2/R_f$, giving a voltage gain of

$$\frac{V_2}{V_{in}} = -\frac{R_f}{R_1} \tag{3.36}$$

and the input resistance of the amplifier, from the above, is

$$R_{in} = V_{in}/I_{in} \simeq R_1$$

The output resistance of the circuit is the parallel combination of R_f and the output resistance of the operational amplifier. It is normal practice in operational amplifiers to use an emitter follower output stage (see Fig. 2.30), so that the output resistance is low.

At this point it is necessary to say a few words about some circuit precautions which may be necessary. Some circuits are liable to a phenomenon known as *latch-up*. This is erratic operation due to saturation of the transistor on the inverting input. When this occurs, the inverting input signal is connected directly to the next transistor in the chain of amplification, so that positive feedback is applied by the feedback resistor R_f. Latch-up can be avoided in a number of ways, one method being to connect diode D in Fig. 3.14 to the phase inverting input. To minimize the worst effects of offset voltage and thermal drift, resistor R_2 can optionally be included and has an optimum value of $R_2 = R_1 R_f/(R_1 + R_f)$. Protection against output short-circuits is provided by including resistor R_3 in series with the output, its value being between about $50\,\Omega$ and $200\,\Omega$.

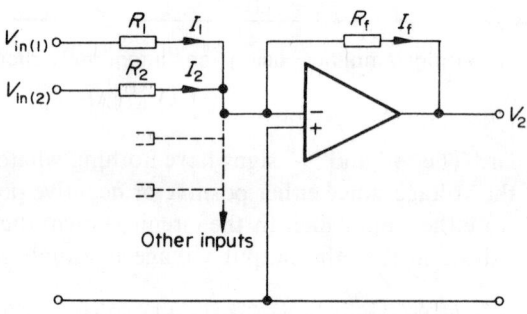

Fig. 3.15   A summing amplifier.

A variant of the inverting amplifier is the *summing amplifier* in Fig. 3.15. In the case shown

$$I_f = I_1 + I_2 + \ldots$$

or

$$-\frac{V_2}{R_f} = \frac{V_{in(1)}}{R_1} + \frac{V_{in(2)}}{R_2} + \ldots$$

therefore

$$V_2 = -\left(\frac{R_f}{R_1} V_{in(1)} + \frac{R_f}{R_2} V_{in(2)} + \ldots\right)$$

That is, the output voltage is of the opposite polarity (or phase) to the sum of the input voltages when each input voltage has been multiplied by the appropriate factor. For a three input summing amplifier with $R_1 = R_2 = 10\,\text{k}\Omega$, $R_3 = 100\,\text{k}\Omega$, and $R_f = 100\,\text{k}\Omega$, the output voltage for $V_{in(1)} = 0.2$, $V_{in(2)} = 0.1\,V$, and $V_{in(3)} = -0.3\,V$ is

$$V_2 = -\left(\frac{100 \times 0.2}{10} + \frac{100 \times 0.1}{10} + \frac{100 \times (-0.3)}{100}\right) = -2.7\,\text{V}$$

The circuits described above can be used as a.c. amplifiers by incorporating blocking capacitors in series with each lead. Additionally, it is possible to tailor the frequency response curve to suit individual requirements by using frequency-dependent elements in the circuit, as follows. If R_1 in Fig. 3.14 is replaced by an impedance (an R–C network) Z_1, and R_f is replaced by Z_f then the voltage gain is $-Z_f/Z_1$. The gain-frequency characteristic of the amplifier is then dependent on the relative frequency response curves of Z_f and Z_1.

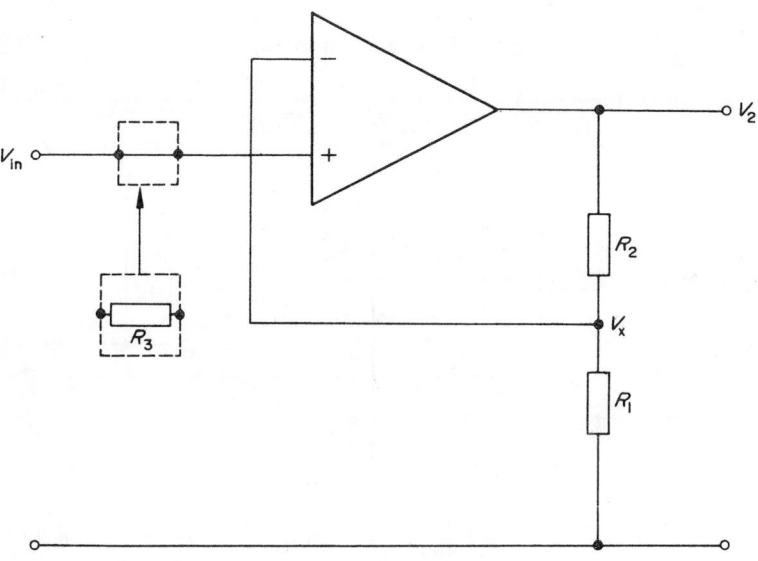

Fig. 3.16   A non-inverting amplifier.

Figure 3.16 shows a basic form of *non-inverting amplifier* in which V_2 is in phase with V_{in}. If we assume that the gain of the operational amplifier is very high, then the inverting and non-inverting input terminals are at the same potential so that $V_{in}=V_x$. But, since $V_x=V_2R_1/(R_1+R_2)$, then

$$V_2=V_{in}(R_1+R_2)/R_1$$

which, with $R_1=1\,k\Omega$ and $R_2=47\,k\Omega$ gives a voltage gain of 48. Resistor R_3 can optionally be included in the input line to minimize the effects of offset voltage and, if used, should have a value equal to $R_1R_2/(R_1+R_2)$.

A variant of Fig. 3.16, known as a *voltage follower*, is shown in Fig. 3.17. If we apply the reasoning outlined above we see that for this circuit $V_2=V_{in}$. The input resistance of this circuit is very high and its output resistance is low.

So far we have described only single-ended input amplifiers. A *differential input amplifier* is shown in Fig. 3.18 in which inputs $V_{in(1)}$ and $V_{in(2)}$ are both

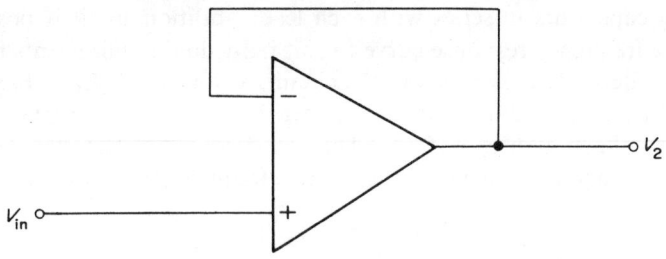

Fig. 3.17   A voltage follower.

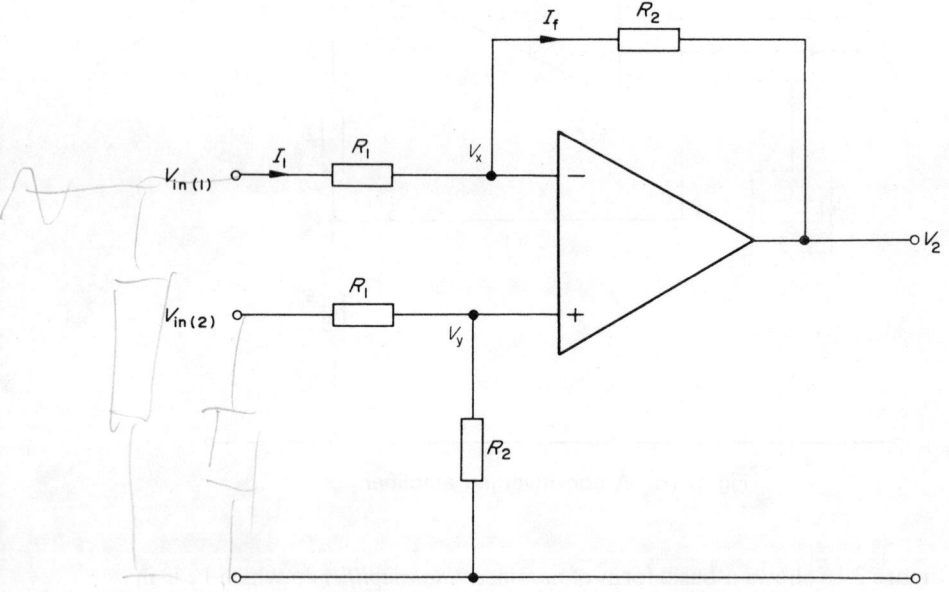

Fig. 3.18   A differential input amplifier.

at some potential relative to the common line. Now

$$V_x = \frac{V_{in(2)}R_2}{R_1 + R_2}$$

and, since the amplifier has a high gain, the voltage at the inverting input of the operational amplifier is practically equal to V_x so that

$$I_1 = (V_{in(1)} - V_x)/R_1 \quad \text{and} \quad I_f = (V_x - V_2)/R_2$$

But $I_f = I_1$, hence

$$(V_x - V_2)/R_2 = (V_{in(1)} - V_x)/R_1$$

also

$$V_y = \frac{R_2}{R_1 + R_2} V_{in(2)}$$

Solving for V_2 gives

$$V_2 = (V_{in(1)} - V_{in(2)})R_2/R_1$$

so that if $R_2 = 100\,\text{k}\Omega$ and $R_1 = 1.8\,\text{k}\Omega$, then the differential voltage gain is 55.5.

Applications of operational amplifiers are legion and include high input impedance amplifiers, band-pass and band-stop amplifiers, filters, thermocouple and resistance bridge amplifiers, transducer amplifiers, photo-electric effect amplifiers, voltage level detectors, voltage-to-frequency convertors, multivibrators, sine-wave oscillators, audio-frequency mixing circuits and many others.

3.12 Electronic integrator circuits

The current which flows through a capacitor is given by the relationship

$I = C \times$ the rate of change of voltage across the capacitor $= C \times dV/dt$

In the circuit in Fig. 3.19

$$I = V_{in}/R = C \times (-dV_2/dt) = -CdV_2/dt$$

or
$$\frac{dV_2}{dt} = -\frac{V_{in}}{CR} \qquad (3.37)$$

$$dV_2 = -\frac{V_{in}}{CR}\,dt$$

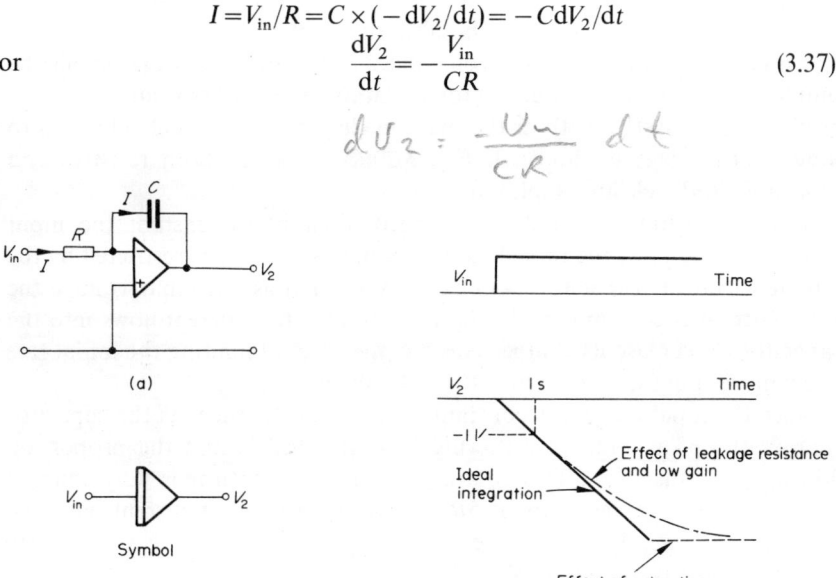

(a)

Symbol

(b)

Fig. 3.19 An electronic integrator circuit.

That is, the *rate of change of output voltage* is dependent on the product CR, so that if $R=0.1\,\text{M}\Omega$ and $C=1\,\mu\text{F}$ then, with a constant input voltage of 0.1 V,

$$dV_2/dt = -0.1/1 \times 0.1 = -1 \text{ V/s}$$

and the output voltage increases with negative polarity at the rate of 1 V/s. For a *step change* in input voltage (shown in Fig. 3.19), the output voltage is proportional to the time integral of the input voltage. By solving eq. (3.37) for V_2 we deduce that

$$V_2 = -\frac{1}{RC}\int V_{\text{in}}\, dt$$

which shows that the output of the circuit in Fig. 3.19 is proportional to the time integral of the input voltage whatever its waveform.

In practical versions of the circuit there are, unfortunately, factors which limit the accuracy of integration over long periods of time, the principal factors being:

(a) The finite open-loop gain of the operational amplifier
(b) The leakage resistance of the capacitor
(c) Saturation of the amplifier.

For a step input voltage change the effects of these factors are shown in Fig. 3.19(b).

It is sometimes necessary to be able to construct an electronic circuit which has the characteristics of some piece of apparatus. An example is the simulation of an electrical generator or motor with its associated time lag, or of a heating system with its thermal time lag. A circuit which allows us to simulate time lags is shown in Fig. 3.20(a), in which both resistive and capacitive feedback are employed.

If the capacitor is initially uncharged, then at the instant the input voltage V_{in} is applied the rate of change of output voltage is restricted by the rate at which the capacitor can charge. With a constant input voltage the input current is constant $(I=V_{\text{in}}/R)$ and, initially, this current flows into the capacitor to increase its charge. This has the effect of limiting the initial rate of change of output voltage to $-V_{\text{in}}/CR$ volts per second.

Since the input current is constant for constant V_{in} then, as the capacitor charges the proportion of I flowing in C diminishes and the proportion flowing in R_{f} increases. This continues until the capacitor is fully charged (after a period of time of about $5RC$), when all the input current has been transferred to R_{f}. When this occurs $I=V_{\text{in}}/R=-V_2/R_{\text{f}}$. The steady-state output voltage is, therefore, $V_2 = -V_{\text{in}}R_{\text{f}}/R$.

Applications of circuits of this kind are described in more detail in specialized texts on analogue computing and control systems (see for example, *Control Engineering* by N. M. Morris).

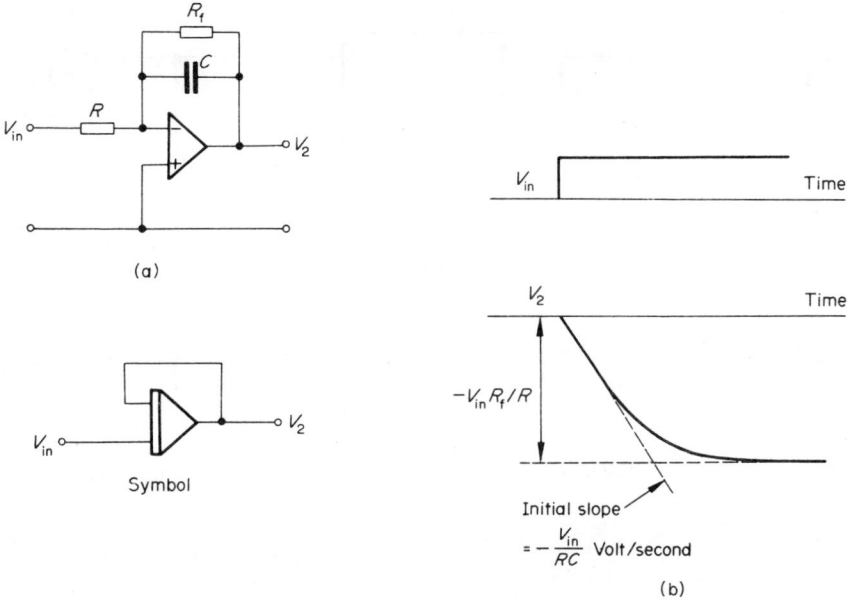

Fig. 3.20 Simulation of a time lag.

3.13 Effect of negative feedback on distortion

As we have already shown, the effects of feedback are largely dependent on the type of feedback, on how the feedback signal is derived, and on how it is injected. In the remaining sections of this chapter we will concentrate on the effects of series negative voltage feedback on various parameters.

Firstly, we will derive a general expression for the voltage gain of such an amplifier. [A detailed treatment of basic feedback amplifier theory is given in *Industrial Electronics* by N. M. Morris, McGraw-Hill, London.] A block diagram of the rudimentary amplifier is given in Fig. 3.21, in which A_v is the forward gain of the amplifier and β is the constant of the feedback network. The instantaneous polarities around the circuit are as shown, so that

$$V_1 = V_{in} + \beta V_2$$

and

$$V_2 = A_v V_1$$

Combining the two equations yields a closed-loop voltage gain A_{vf} of

$$A_{vf} = V_2/V_{in} = A_v/(1 - A_v\beta)^*$$

so that if $A_v = -900$ and $\beta = 0.01$, then the closed-loop gain is

$$A_{vf} = -900/[1 - (-900 \times 0.01)] = -90$$

*In the great majority of electronic amplifiers A_v has a negative value, i.e., they are phase-inverting, so that the voltage fed back is antiphase to the input signal.

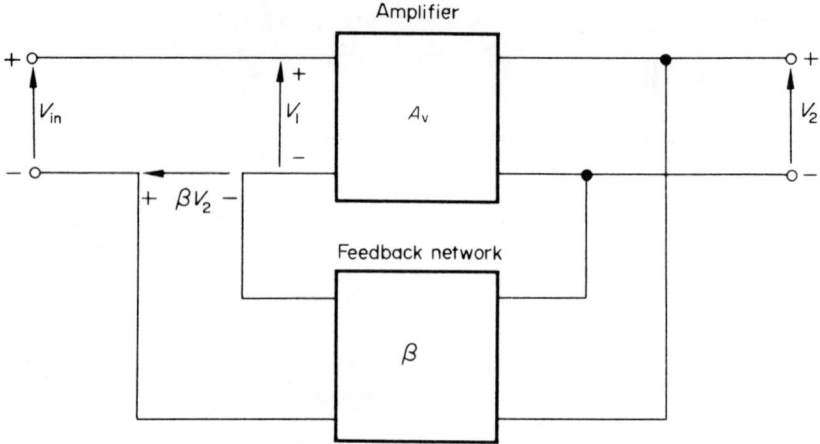

Fig. 3.21   A block diagram of a series voltage feedback amplifier.

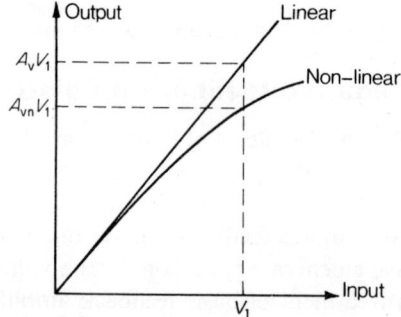

Fig. 3.22   Linear and non-linear characteristics.

Practical amplifiers do not have perfect characteristics, so that the output signal is not a true reproduction of the input signal. This is particularly true for very small or very large input signals. The characteristics in Fig. 3.22 illustrate the effect of a saturating characteristic. With an input voltage V_1 the amplifier with the linear characteristic provides an output A_vV_1, while the output from the non-linear amplifier is $A_{vn}V_1$, where A_v and A_{vn} are the respective gains of the linear and non-linear amplifiers. We can define the *fractional distortion* due to the non-linear characteristic as follows

$$\frac{\text{Linear output} - \text{Non-linear output}}{\text{Linear output}} = \frac{A_vV_1 - A_{vn}V_1}{A_vV_1} = 1 - \frac{A_{vn}}{A_v}$$

Now, when negative series voltage feedback is applied to both amplifiers, the respective closed-loop gains are

$$Linear\ amplifier\ output = A_v V_1/(1 - A_v\beta)$$
$$Non\text{-}linear\ amplifier\ output = A_{vn} V_1/(1 - A_{vn}\beta)$$

so that the fractional distortion with feedback is

$$\left\{\frac{A_v V_1}{1 - A_v\beta} - \frac{A_{vn} V_1}{1 - A_{vn}\beta}\right\} \Big/ \frac{A_v V_1}{1 - A_v\beta} = 1 - \frac{A_{vn}}{A_v} \frac{(1 - A_v\beta)}{(1 - A_{vn}\beta)}$$

Since $A_{vn} < A_v$, the ratio $(1 - A_v\beta)/(1 - A_{vn}\beta) > 1$ so that the fractional distortion with feedback is less than it is without feedback, as illustrated in the following example.

Example 3.2: If $A_v = -1000$, $\beta = 0.01$, and $A_{vn} = -800$, calculate the fractional distortion both without and with feedback.

Solution: Without feedback the fractional distortion is

$$1 - (-800/-1000) = 0.2$$

and with feedback it is

$$1 - [-800(1 + 1000/100)]/[-1000(1 + 800/100)] = 0.022$$

3.14 Effect of negative feedback on noise

All amplifiers produce spurious voltages at their outputs in addition to the amplified input signal. The spurious signals are known as electrical *noise*, and are due to several causes. In transistors, the causes of noise include *surface leakage effects*, *recombination effects*, and *thermal effects*. Other sources of noise include periodic variations at mains frequency or a harmonic of mains frequency and thermal effects in circuit components.

An important parameter in defining the noise content in the output of an amplifier is the *signal-to-noise ratio*, that is the ratio of the information signal at the output to the noise at the output.

An important parameter in defining the noise content in the output of an amplifier is the *signal-to-noise ratio*, that is the ratio of the information signal at the output of the noise at the output.

If the noise is produced by the first stage of the amplifier, then negative feedback has practically no effect on the signal-to-noise ratio since the feedback reduces the gain by equal amounts to both the noise and signal frequencies. If the noise is generated at a late stage in the amplifier, it can be shown that the signal-to-noise ratio is reduced by negative voltage feedback.

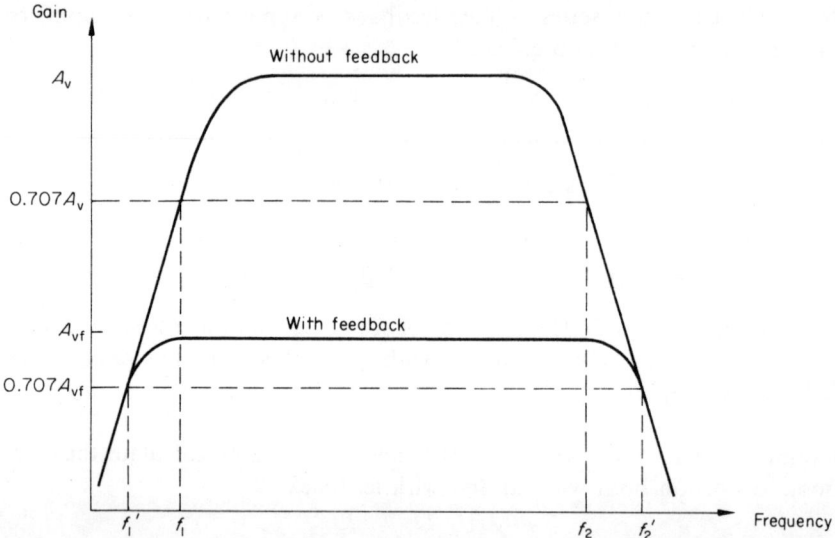

Fig. 3.23   Frequency response curves of an a.c. amplifier both with and without feedback.

3.15   Effect of negative feedback on bandwidth

A typical frequency response diagram of an a.c. coupled amplifier is shown in Fig. 3.23. When negative feedback is applied the mid-band gain is reduced in the manner described in section 3.13. Also, as we have already described, series injected negative feedback increases the input resistance of the amplifier. Now, the lower cut-off frequency f_1 (in the absence of feedback) is determined by the frequency at which the reactance of the input coupling capacitor C_1 is equal to the amplifier input resistance, that is when $R_{in} = 1/2\pi f_1 C_1$, giving

$$f_1 = 1/2\pi R_{in} C_1$$

The effect of the form of feedback used here is to increase the input resistance by a factor $(1 - A_v \beta)$ (**Note:** in a phase-inverting amplifier the product $-A_v \beta$ has a positive value), so that the lower cut-off frequency is reduced to

$$f_1' = 1/2\pi R_{in}(1 - A_v \beta)C_1 = f_1/(1 - A_v \beta)$$

If the output resistance of the amplifier in the absence of feedback is R_{out} then, with the form of feedback used, the closed-loop output resistance is reduced to $R_{out}/(1 - A_v \beta)$. From this, it follows that the upper corner frequency of the closed-loop amplifier is increased to

$$f_2' = f_2(1 - A_v \beta)$$

As a general rule, both f_1 and f_1' are small when compared with f_2 and f_2', respectively, so that the respective bandwidths of the amplifier with and without feedback are approximately f_2 and f_2'. The gain-bandwidth product of the amplifier itself is therefore $A_v f_2$, and that of the closed-loop amplifier is

$$\frac{A_v}{1 - A_v\beta} \times f_2(1 - A_v\beta) = A_v f_2$$

That is, the gain-bandwidth product of the amplifier remains constant irrespective of the amount of feedback applied, so that negative feedback increases the bandwidth to compensate for the reduction in gain.

3.16   Stability margins of feedback amplifiers

Under certain operating conditions, a feedback amplifier can become unstable, that is the output can be oscillatory for a constant (or even zero) input. We have already seen that the gain of a series voltage feedback amplifier is

$$A_{vf} = A_v/(1 - A_v\beta)$$

so that if $A_v\beta = 1$, then the gain becomes infinitely large. This implies that however minute the input signal, an output signal is generated. While the conditions for instability in a series voltage feedback amplifier have been quoted, similar conditions can also exist in all forms of feedback amplifier.

Let us investigate the conditions for instability further. The statement that $A_v\beta = 1$ implies that for instability the *loop gain* must be unity and that the *loop phase shift* must be zero (or 360 degrees). To determine if a system will be stable when the feedback loop is connected, it is necessary to test the system in some way, one popular method being to determine the frequency response characteristic with the feedback loop broken, as shown in Fig. 3.24. A sinusoidal signal V_x is injected into the amplifier, and the magnitude of the signal V_y is measured together with its phase shift with respect to V_x. The loop gain $A_v\beta$ is then evaluated from the relationship

$$A_v\beta = \frac{V_y}{V_x} \angle \phi$$

where ϕ is the phase shift of V_y with respect to V_x, and the ratio V_y/V_x is the ratio of the magnitudes of the two voltages. The loop gain is evaluated at a large number of frequencies and is plotted in the form of a polar curve, known as a *Nyquist diagram* (after its inventor). The relationship between the polar curve and the $+1$ point on the diagram gives an indication of the degree of stability of the amplifier.

A typical Nyquist diagram is shown in Fig. 3.25. If the first test frequency

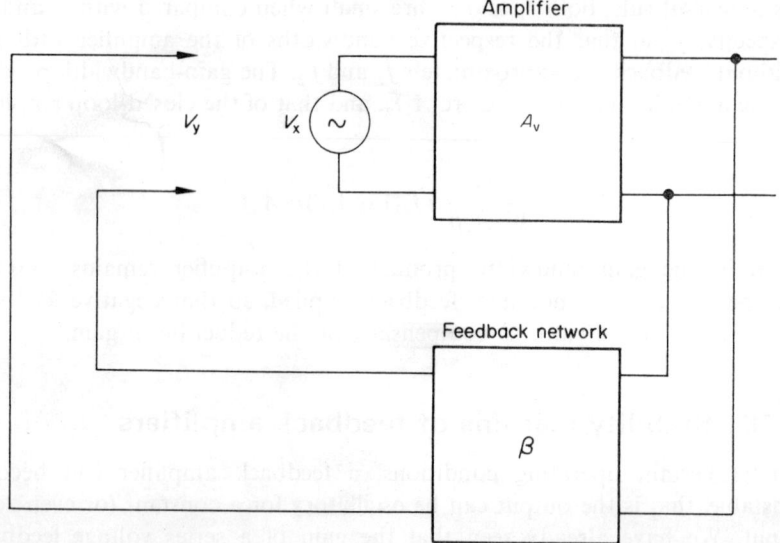

Fig. 3.24 Test circuit to determine the Nyquist diagram.

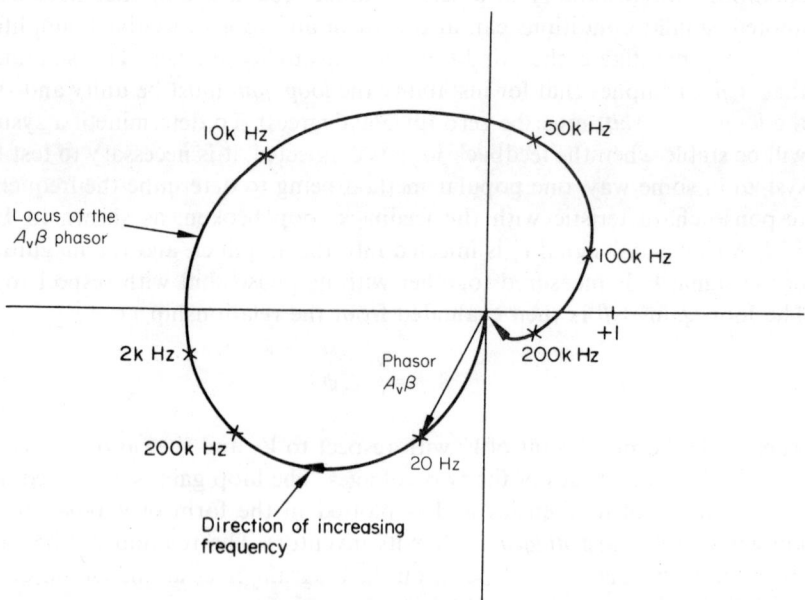

Fig. 3.25 A typical Nyquist diagram for an a.c. coupled amplifier.

is 20 Hz, the relative magnitudes of V_x and V_y are measured (possibly with electornic voltmeters) and the phase shift between them is determined (say, on an X–Y C.R.O.). Suppose these values are $V_x = 1V$, $V_y = 2V$, $\phi = -100$ degrees. The length of the phasor $A_v\beta$ is 2 (dimensionless in this case) and is plotted at an angle of -100 degrees from the reference axis, as shown in the figure. Other points are then plotted after tests at the frequencies shown.

[It is possible to approach feedback amplifier theory on the basis that the signal fed back is *subtracted from* the input signal and not added to it, as is assumed in the theory here. In this event, due to the addition of phase inversion at the input, it is found that instability occurs if $A_v\beta = -1$. This approach is generally adopted in control system theory and leads to a slightly different form of Nyquist diagram. (See, for example, *Control Engineering* by N. M. Morris.)]

The locus of the $A_v\beta$ phasor in the diagram commences at the origin (indicating zero gain at zero frequency) and progresses in a clockwise direction until it falls to zero again at a high value of frequency. This type of Nyquist plot is in general agreement with the frequency response curve in Fig. 3.23. For the system to be stable when the loop is closed, the $+1$ point must lie outside the $A_v\beta$ locus, and a simplified statement of Nyquist's stability criterion is

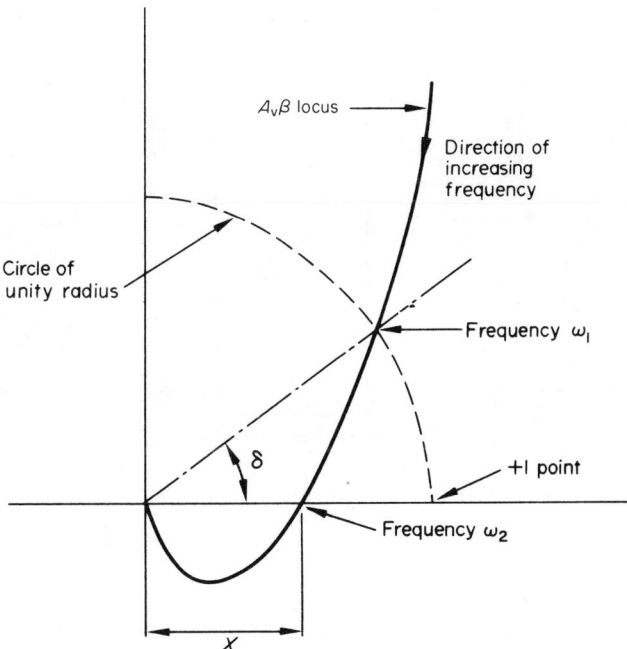

Fig. 3.26 Method of evaluating gain and phase margins.

The point $(+1, 0)$ must lie on the left of the locus as it is traversed in the direction of increasing frequency.

The locus may be more complex than that shown in Fig. 3.25, and to give an idea of the degree of system stability certain margins are defined at specified points on the diagram. The points usually used for definition purposes are shown in Fig. 3.26 and are the *gain crossover frequency* ω_1 and the *phase crossover frequency* ω_2.

The first stability margin we define is the *phase margin* δ, which is the angle in degrees by which the phase lag can be increased (at the gain crossover frequency ω_1) without change in gain before the locus passes through the $+1$ point. The second stability margin is the *gain margin*, which is the amount by which the gain can be increased (at the phase crossover frequency ω_2) without change in phase shift before the locus passes through the $+1$ point. In Fig. 3.26 the loop gain at ω_2 is X so that the numerical value of the gain margin is $1/X$; the gain margin is usually given in decibels so that its value is $20 \lg 1/X$. For satisfactory operation the phase margin should be at least 30 degrees and the gain margin at least 10 dB.

Problems

3.1 A negative feedback amplifier is shown in Fig. 3.27. When the 330 Ω feedback resistor is short-circuited, the magnitude of the voltage gain (V_{out}/V_{in}) is 1000. What is the voltage gain when the short-circuit is removed?

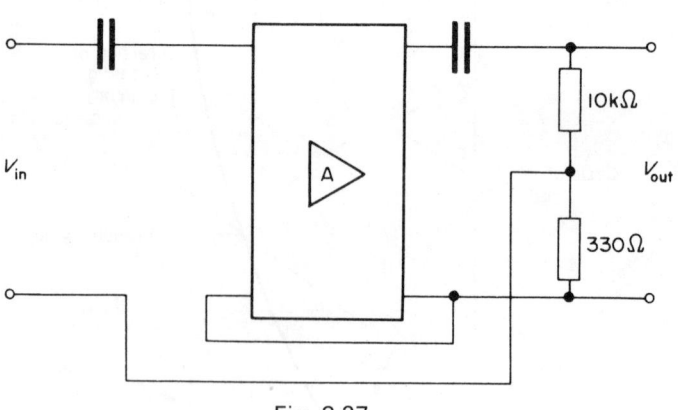

Fig. 3.27

If the resistors used in the feedback network have a tolerance of $\pm 5\%$, calculate the maximum and minimum values of voltage gain with feedback. (The effects of the input and output impedances of the amplifier may be assumed to be negligible.)

State how negative feedback affects the following properties of the amplifier shown:

(i) harmonic distortion
(ii) bandwidth
(iii) d.c. and a.c. stabilities.

3.2 Prove from first principles that

(a) Series negative voltage feedback reduces the output impedance of an amplifier.
(b) Series negative current feedback increases the output impedance of an amplifier.

3.3 Prove that shunt negative voltage feedback reduces the input impedance of an amplifier.

3.4 Sketch circuit diagrams of transistor amplifiers using (i) voltage feedback, and (ii) current feedback. For each circuit state how the equivalent feedback fraction is determined from the values of the circuit components.

3.5 An extract from the specifications of an integrated circuit amplifier is as follows:

Nominal gain 86 dB
95% sample range 80–89 dB
input impedance 100 kΩ

The frequency response characteristic is shown in Fig. 3.28. This amplifier is to be used with voltage negative·feedback in series with the input to give a nominal gain of 1000. For the feedback amplifier determine

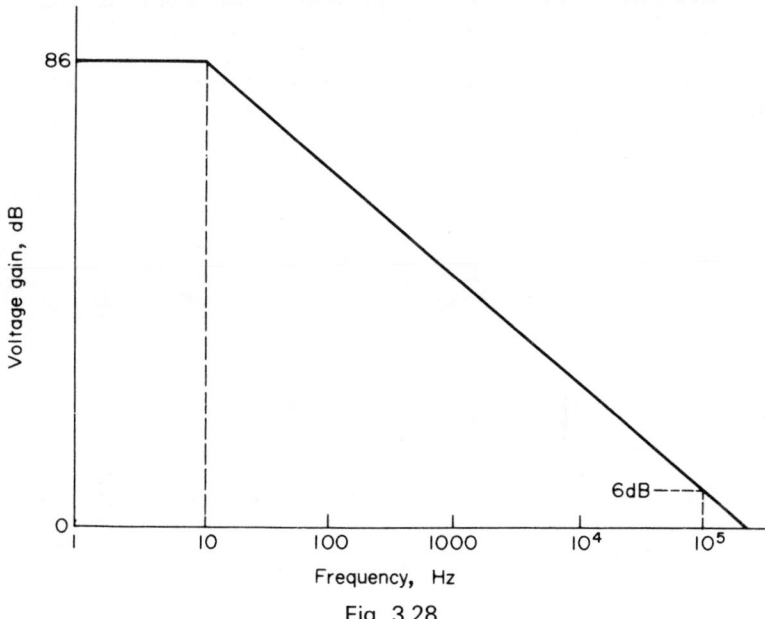

Fig. 3.28

(a) the gain spread of the system
(b) the input impedance of the system
(c) the bandwidth of the system.

3.6 If the gain of an amplifier without feedback is represented by A, derive an expression for the gain when a fraction β of the output voltage is fed back in opposition to the input.

In an amplifier with a constant input of 1 V, the output falls from 50 V to 25 V when feedback is applied. Calculate the fraction of the output which is fed back. If due to ageing, the amplifier gain fell to 40, find the percentage reduction in stage gain:

(i) without feedback
(ii) with the feedback connection.

3.7 Discuss the effects of (i) negative and (ii) positive series current feedback on the steady-state output impedance of amplifiers.

A d.c. amplifier with an output impedance of 1 kΩ supplies a current of 3 mA at a terminal voltage of 7 V when the input to the the amplifier is 0.2 V. Calculate the effective output impedance of the amplifier if (a) series negative current feedback, and (b) series positive current feedback is used. The signal fed back is developed across a 10 Ω resistance in series with the output. The input impedance of the amplifier is infinite.

3.8 If overall series negative voltage feedback is applied in conjunction with positive current feedback to the amplifier in problem 3.7, determine the amount of voltage feedback required to reduce the net output impedance to 50 Ω.

3.9 Explain how the ripple voltage at the output of an amplifier is reduced by the application of negative series voltage feedback.

An amplifier has a gain of −90, and the ripple voltage at the output (introduced in the final stage of amplification) is 1 V r.m.s. If 10% series negative voltage feedback is applied to the amplifier, calculate the r.m.s. ripple at the output of the feedback amplifier.

3.10 Sketch circuit diagrams of (a) an emitter follower, and (b) a source follower. Explain why the circuits may be used as matching stages between a source and a load.

Determine the approximate power gain (in decibels) of an emitter follower which has an a.c. input resistance of 50 kΩ and a load resistance of 500 Ω.

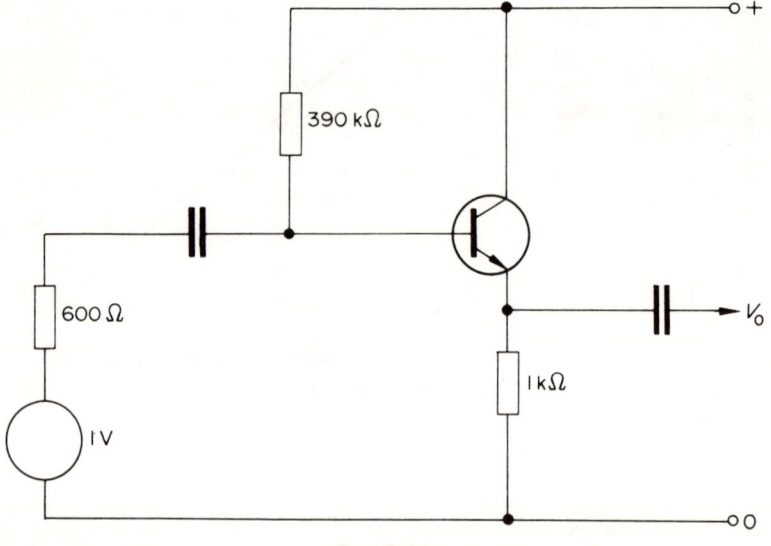

Fig. 3.29

3.11 The circuit diagram of an emitter follower is shown in Fig. 3.29, and the h-parameters of the transistor are as listed below:

COMMON EMITTER: $h_{ie} = 600\,\Omega$; $h_{re} = 0$; $h_{fe} = 99$; $h_{oe} = 0$
COMMON COLLECTOR: $h_{ic} = 600\,\Omega$; $h_{rc} = 1$; $h_{fc} = -100$; $h_{oc} = 0$

Draw *either* (a) the common-emitter, *or* (b) the common-collector equivalent circuit diagram, and hence find approximately

(i) the input resistance
(ii) the output resistance of the stage.

3.12 A resistor summing network has three input lines A, B, and C connected through resistors of $100\,\text{k}\Omega$, $200\,\text{k}\Omega$, and $10\,\text{k}\Omega$, respectively, to the summing junction S. A resistor R is connected between S and earth. The voltages at the input lines are $-2.5\,\text{V}$, $5\,\text{V}$, and $0.5\,\text{V}$, respectively. If the value of R is (i) $1\,\text{k}\Omega$, (ii) $1\,\text{M}\Omega$, determine the voltage at the summing junction.

3.13 An operational amplifier has three input lines W, X, Y, having input resistors of $100\,\text{k}\Omega$, $1000\,\text{k}\Omega$, and $200\,\text{k}\Omega$, respectively, connected to the summing junction. When the voltages on lines W, X, and Y are $5\,\text{V}$, $-2\,\text{V}$, and $-1\,\text{V}$, respectively, the output voltage is $-2.5\,\text{V}$. What value of feedback resistor is used?

3.14 An operational amplifier has two input lines A and B, and a $100\,\text{k}\Omega$ feedback resistor is used. When the input voltages applied to lines A and B are 1 and 2 V, respectively, the output voltage is $-2\,\text{V}$. When the inputs to A and B are $-1\,\text{V}$ and 2 V, the output voltage is zero. Determine the values of input resistance used.

3.15 Explain the conditions which can give rise to oscillation in a multistage amplifier employing overall negative feedback.

A three-stage amplifier has an open-loop voltage gain of 5×10^5, an input impedance of $100\,\text{k}\Omega$, and an output impedance of $100\,\Omega$. Overall negative feedback is applied in series with the input to reduce the gain to 10^3. Determine

(a) the feedback factor required
(b) the input impedance
(c) the output impedance.

4. Sinusoidal oscillators and active filters

4.1 Instability in feedback amplifiers

An oscillator is a circuit which converts power from a unidirectional source into alternating power, the frequency of the oscillations being a function of the constants of the system. The general conditions for maintaining oscillations in feedback amplifiers, outlined in chapter 3, are that the loop gain must be unity and the loop phase shift must be zero (or a multiple of 360 degrees).

These are the steady-state conditions which must exist during operation and if, for example, the loop gain momentarily falls below unity then oscillations cease. For oscillations to commence in a circuit at the instant of switch-on, it is necessary for the initial loop gain to be greater than unity. After the first few cycles of operation, the circuit settles down so that the *average* loop gain taken over the cycle is unity.

Electronic oscillators employ active devices which introduce a phase shift of 180 degrees, so that when an odd number of stages is cascaded the overall phase shift of the amplifier is 180 degrees (or an odd multiple of 180 degrees). For oscillations to take place in this case, the feedback circuit must introduce a further 180 degrees of phase shift at the oscillatory frequency to maintain a loop phase shift of zero. When an even number of phase-inverting stages is used, the amplifier phase shift is zero (or a multiple of 360 degrees), so that for instability in this case the feedback network must not introduce any phase shift at the oscillatory frequency.

Oscillators can be treated analytically either as *negative resistance circuits* or as *feedback circuits*. In this chapter, we shall regard oscillators as feedback circuits, but before proceeding to a discussion of circuit principles a few words about negative resistance oscillators would not be out of place.

The classification of negative resistance oscillators includes circuits incorporating devices which have an effective negative resistance region in their characteristics, e.g., tunnel diodes, etc. The negative resistance device is connected to a tuned circuit in such a way that the negative resistance of the element compensates for the dynamic resistance of the tuned circuit. In this way, the circuit damping is reduced to zero so that sustained oscillations occur. From an analytical viewpoint, negative resistance oscillators and feedback oscillators are identical, since the latter can be regarded as having introduced negative resistance into the feedback circuit.

Below a frequency of about 0.1 Hz, integrator-type circuits are popular (see section 4.11), but in the range from 1 Hz to 1 MHz circuits employing R–C feedback are almost universally used since high quality components are readily obtainable. Moreover, high-Q L–C circuits at these frequencies are difficult to manufacture and are both large and costly. Also, in L–C oscillators the frequency of oscillation is usually proportional to $1/\sqrt{LC}$, whereas in R–C oscillators the frequency is often proportional to $1/RC$. For a given change in C, a greater frequency range can be obtained from R–C oscillators than is the case in L–C oscillators.

Certain types of oscillators can also be subdivided into *series-fed* and *shunt-fed* circuits. In series-fed circuits, the steady (quiescent) component of the amplifier current flows through part of the frequency-determining network. In shunt-fed versions, the frequency-determining network does not carry the quiescent current.

4.2 R–C ladder feedback oscillators

R–C oscillators can be broadly divided into those using R–C *ladder feedback networks* and those with *Wien-Bridge* (or half-bridge) *networks*. The former are dealt with in this section.

Types of ladder network in common usage are shown in Fig. 4.1, the circuit in Fig. 4.1(a) being used in circuits with a high input impedance, e.g.,

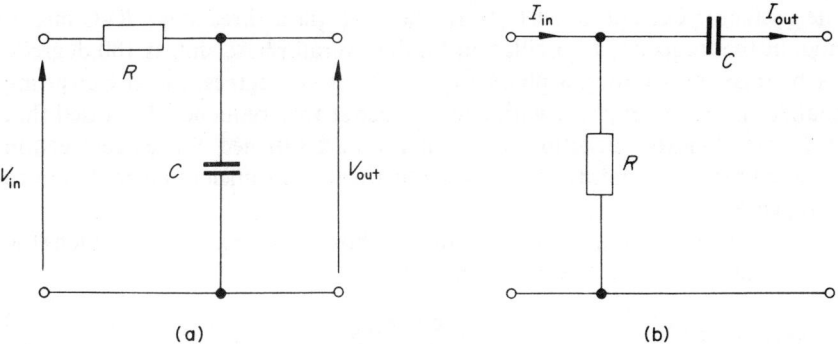

(a)   (b)

Fig. 4.1   (a) A voltage transfer ladder section and (b) a current transfer section.

FET, and certain BJT circuits. The circuit in Fig. 4.1(b) is appropriate for use with circuits with a low input impedance, e.g., most types of BJT circuits. The voltage transfer network imparts a phase lag to the output voltage with respect to the input voltage, while the current transfer section imparts a phase lead to the output current with respect to the input current. Ideally, the voltage transfer network is terminated by a load of infinite impedance and the current transfer network is terminated by a short-circuit.

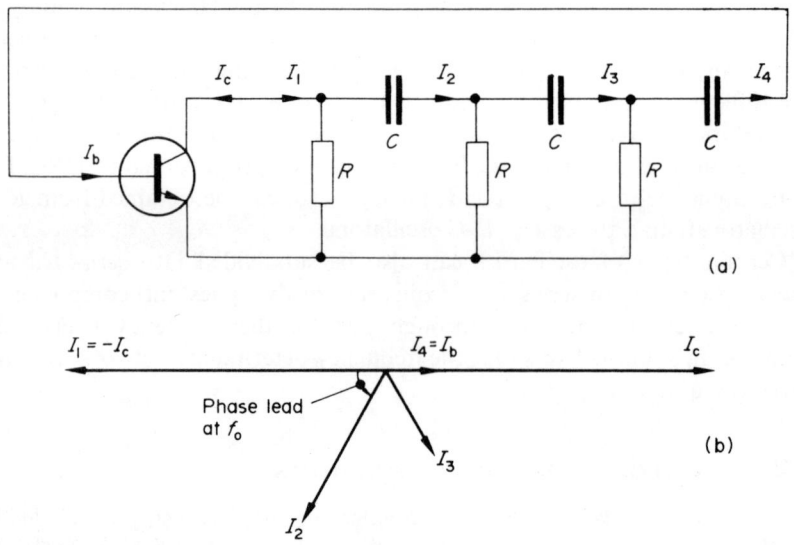

Fig. 4.2   (a) Simplified circuit of a BJT R–C oscillator and (b) its phasor diagram.

The principle of one form of R–C oscillator is shown in Fig. 4.2(a). Here, the oscillator current is fed back to the base via a three-stage R–C circuit and, at the frequency of oscillation f_0, the overall phase shift is 180 degrees, each stage contributing a phase lead of about 60 degrees, thereby ensuring that I_b and I_c are in phase with one another at this frequency. Provided that $h_{fe}I_b \gg I_c$, then the conditions for oscillation are satisfied. Since each section attenuates the current, the transistor must provide sufficient current gain to compensate for this.

A practical version of the circuit is shown in Fig. 4.3 in which the theoretical frequency of oscillation is

$$f_0 = 1/2\pi R_1 C_1 \sqrt{6}$$

and for oscillations to occur the current gain of the transistor must be at

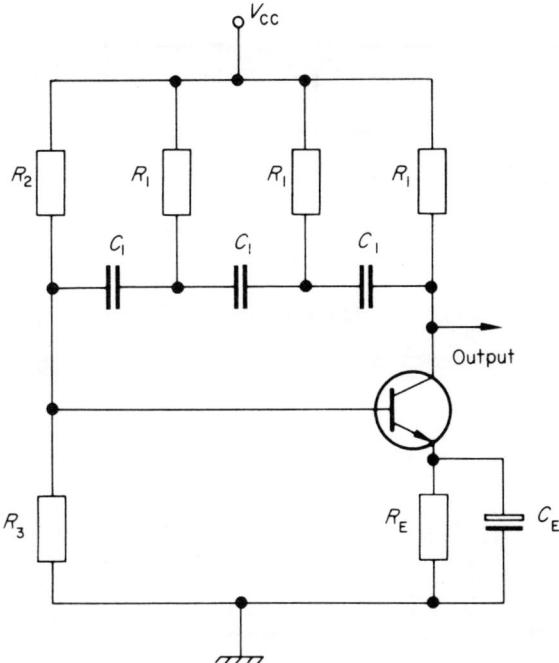

Fig. 4.3   A simple R–C oscillator.

least 29. In fact, to ensure self-starting the current gain should be twice this value, say 60. The net effect of the bias circuit resistors and the finite value of the transistor parameters is to cause the oscillatory frequency to be higher than the theoretical frequency by possibly as much as 25 per cent. Changing the values of either R or C in Fig. 4.3 causes both the frequency and the gain to change, thereby changing the amplitude of the oscillations. Also, if the gain is changed by bypassing only part of R_E, the frequency of oscillation is also altered. As a result, the circuit in Fig. 4.3 is best suited to applications which require a fixed oscillatory frequency.

4.3   Wien bridge oscillators

Some oscillators use an R–C combination in the feedback path similar to the reactive half of the Wien bridge, two of the most popular circuits being shown in Fig. 4.4. Figure 4.4(a) shows a voltage transfer type of network used to drive a circuit with a high input impedance, and Fig. 4.4(b) is a current transfer network used to drive a low (ideally zero) impedance circuit. An analysis of Fig. 4.4(a) yields

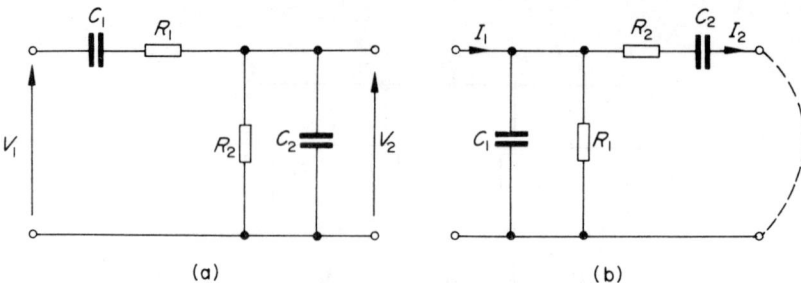

Fig. 4.4   Half-Wien bridge circuits.

$$\frac{V_2}{V_1} = \frac{\dfrac{R_2}{1+j\omega C_2 R_2}}{R_1 + \dfrac{1}{j\omega C_1} + \dfrac{R_2}{1+j\omega C_2 R_2}} \tag{4.1}$$

$$= \frac{R_2}{j\omega C_2\left(R_1 R_2 - \dfrac{1}{\omega^2 C_1 C_2}\right) + \left(R_1 + \dfrac{R_2 C_2}{C_1} + R_2\right)}$$

In the case of Fig. 4.4(b) the relationship is

$$\frac{I_2}{I_2} = \frac{R_1}{\left\{R_1\left(1+\dfrac{C_1}{C_2}\right) + R_2\right\} + j\omega\left(C_1 R_1 R_2 - \dfrac{1}{\omega^2 C_2}\right)} \tag{4.2}$$

One form of oscillator using a current transfer network is shown in Fig. 4.5, the feedback network comprising R_1, C_1, C_2 and a resistor having the value $(R_2 - R_{in})$. This value is chosen so that the net series feedback resistance including the input resistance is R_2. The overall voltage gain is controlled by potentiometer RV, and the function of the unbypassed resistor R_E is to ensure that the drive to the feedback network appears as a current source. Comparing Figs. 4.4(b) and 4.5, we see that $I_1 = -I_{c2}$ and $I_2 = I_{b1}$, hence

$$\frac{I_{b1}}{I_{c2}} = \frac{-R_1}{\left\{R_1\left(1+\dfrac{C_1}{C_2}\right) + R_2\right\} + j\omega\left(C_1 R_1 R_2 - \dfrac{1}{\omega^2 C_2}\right)}$$

If the overall current gain is $A_1 = I_{c2}/I_{b1}$, then

$$R_1\left(1+\frac{C_1}{C_2}\right) + R_2 + j\omega\left(C_1 R_1 R_2 - \frac{1}{\omega^2 C_2}\right) = -A_i R_1$$

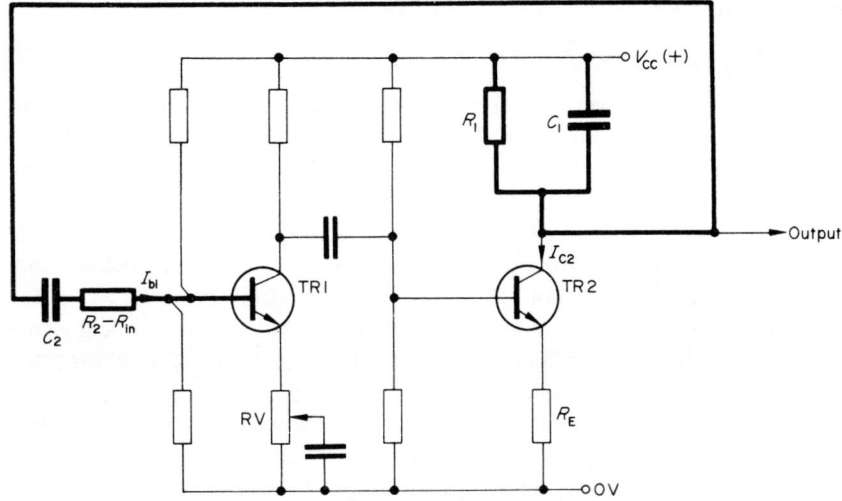

Fig. 4.5   One form of Wien bridge oscillator.

Since the quadrature term on the right-hand side of the equation is zero, then

$$\omega_0\left(C_1 R_1 R_2 - \frac{1}{\omega_0{}^2 C_2}\right) = 0$$

where ω_0 is the oscillatory frequency, but as $\omega_0 \neq 0$ then

$$\omega_0{}^2 = 1/R_1 R_2 C_1 C_2 \tag{4.3}$$

hence

$$\omega_0 = 1/\sqrt{R_1 R_2 C_1 C_2} \tag{4.4}$$

and

$$f_0 = 1/2\pi\sqrt{R_1 R_2 C_1 C_2} \tag{4.5}$$

The minimum current gain for oscillations to occur is determined by equating the 'real' terms in the equation as follows:

$$R_1\left(1 + \frac{C_1}{C_2}\right) + R_2 = -A_i R_1$$

hence

$$A_i = -\left(1 + \frac{C_1}{C_2} + \frac{R_2}{R_1}\right) \tag{4.6}$$

The circuit elements are generally chosen so that $R_1 = R_2 = R$, say, and $C_1 = C_2 = C$. Substituting these values into eq. (4.5) yields

$$f_0 = 1/2\pi RC \tag{4.7}$$

and eq. (4.6) gives

$$A_i = -3$$

That is to say, the overall current gain required to maintain oscillations need only have a value of three. The negative sign results from the fact that I_{c2} is assumed to flow *into* the collector of TR2 rather than into the feedback network.

In the case of circuits using the feedback arrangement of Fig. 4.4(a), an analysis shows that the frequency of oscillation is $1/2\pi RC$, as before, and that the overall voltage gain of the amplifier need only be three.

Oscillators using the full Wien bridge circuit provide better frequency and amplitude stability than half-bridge circuits. The basis of one form of full-bridge oscillator is illustrated in Fig. 4.6, and uses a differential input amplifier. Here the 'plus' sign at the upper input terminal indicates that any signal applied at this terminal causes the output voltage to be in phase with it. A signal applied to the 'minus' (lower) input terminal causes the output to be antiphase to that signal. Clearly, the reactive arm of the bridge provides positive feedback and the resistive network (R_3 and R_4) provides negative feedback; if the loop gain (including the effect of negative feedback) is equal to or greater than three, then the correct conditions for oscillations exist. If β is the attentuation provided by the branch $R_3 R_4$, and the voltage gain of the amplifier itself is A_v, then for oscillations to be maintained

$$3 = A_v/(1 + A_v\beta) \tag{4.8}$$

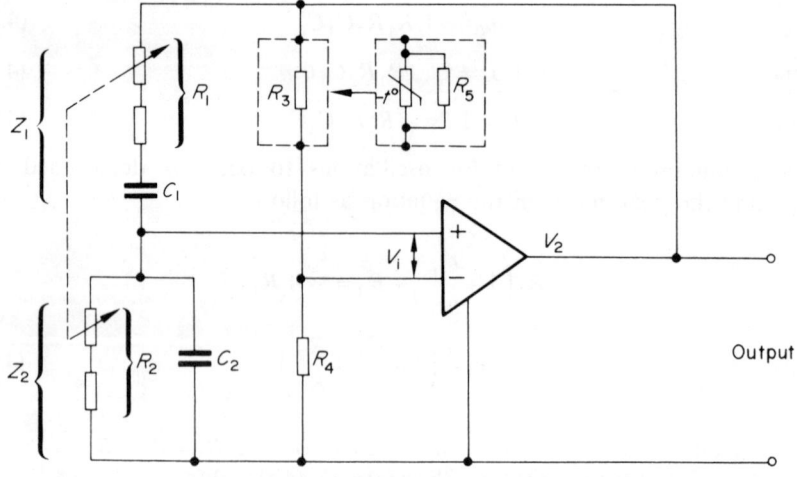

Fig. 4.6   A block diagram of an oscillator using a full-Wien bridge feedback circuit.

so that if $A_v = 5$, then for oscillations to be maintained $\beta = 0.133$. That is, if $R_4 = 200\,\Omega$, then $R_3 = 1304\,\Omega$.

Fine adjustment of frequency in Fig. 4.6 can be obtained by using the double-ganged variable resistor shown. Coarse frequency changes can then be introduced by switching different values of capacitance into circuit. Alternatively, fine frequency control can be obtained by using twin-ganged capacitors and coarse control by switching different values of resistor into the circuit.

The range of frequencies produced by one complete sweep of the potentiometer in Fig. 4.6 has its upper limit fixed by the value of the fixed resistor and the lower limit by the total resistance of the fixed and variable resistors. If the fixed and variable resistors have values of 1.8 and 20 kΩ, respectively, then the frequency range covered by one sweep of the potentiometer is $(20 + 1.8):1.8 = 12.1:1$, i.e., slightly greater than one decade.

If no precautions are taken in Fig. 4.6, the amplitude of the output voltage changes significantly when either R or C is altered. In transistor circuits this is usually compensated for by replacing R_3 with a thermistor as shown in the inset in Fig. 4.6. When the thermistor is cold its resistance is high so that only a small amount of negative feedback is applied. As the oscillation amplitude increases, the mean temperature of the thermistor increases and its resistance reduces, thereby increasing the amount of negative feedback and reducing the amplifier gain. The amplitude of the oscillations is stabilized when eq. (4.8) is satisfied. The thermistor may be shunted by resistor R_5 to give the desired response over a range of operating temperature and supply voltage.

In some circuits, including valve circuits, amplitude stabilization is achieved by replacing R_4 by a resistor with a positive temperature-resistance coefficient in the form of a tungsten filament lamp, R_3 then being a fixed resistance.

4.4 L–C oscillators

An equivalent circuit which satisfies a range of L–C oscillators is shown in Fig. 4.7. The circuit may be used in any of the basic configurations, e.g., common-emitter, common-base, etc. without affecting the results of the analysis which follows.

The effort involved in the analysis is reduced if we consider the feedback loop to be broken at point N, so that current I_b flows into the base and I'_b flows out of the feedback network. When $I'_b \geqslant I_b$ we have the correct conditions for oscillation. The resulting equations are also much simplified if we ignore parameter h_{oe} in the equivalent circuit, as shown in Fig. 4.8(a). The feedback circuit is terminated in a load equal to h_{ie} to satisfy the loading conditions of the circuit. Converting the current generator circuit

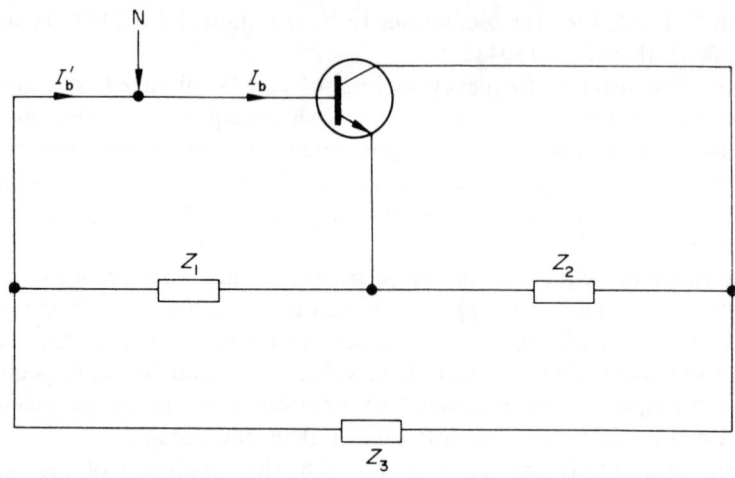

Fig. 4.7   One form of L–C oscillator circuit.

into a voltage generator circuit using the notation of section 1.14, we get the equivalent circuit of Fig. 4.8(b). The relevant equations for this circuit are

$$I'_b = -Z_1 I/(Z_1 + h_{ie}) \tag{4.9}$$

$$I = h_{fe} I_b Z_2/[Z_2 + Z_3 + Z_1 h_{ie}/(Z_1 + h_{ie})] \tag{4.10}$$

Substituting eq. (4.9) into eq. (4.10) and equating I'_b to I_b for instability leads to the equation

$$Z_1\{Z_2(1 + h_{fe}) + Z_3\} + h_{ie}(Z_1 + Z_2 + Z_3) = 0 \tag{4.11}$$

In the circuit the impedances Z_1, Z_2, Z_3 are ideally pure reactances jX_1, jX_2, jX_3. Substituting these values into eq. (4.11) gives

$$-X_1\{X_2(1 + h_{fe}) + X_3\} + jh_{fe}(X_1 + X_2 + X_3) = 0$$

Equating the quadrature terms to zero gives

$$h_{ie}(X_1 + X_2 + X_3) = 0$$

and since $h_{ie} \neq 0$, then

$$X_1 + X_2 + X_3 = 0 \tag{4.12}$$

Equating the 'real' terms to zero yields

$$X_1\{X_2(1 + h_{fe}) + X_3\} = 0$$

and as $X_1 \neq 0$, then

$$X_3 = -X_2(1 + h_{fe}) \tag{4.13}$$

From eq. (4.13) it follows that X_2 and X_3 are of opposite type to one

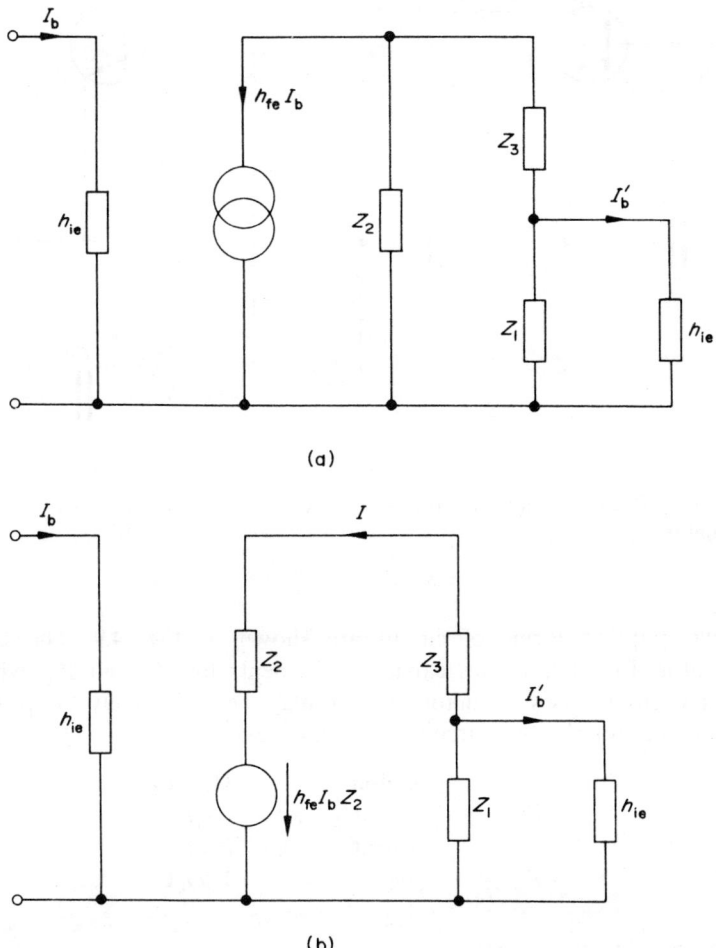

Fig. 4.8   Equivalent circuits of Fig. 4.7 using (a) a current generator equivalent circuit and (b) a voltage generator equivalent circuit.

another, and if one is inductive then the other is capacitive. Also, from eq. (4.13) the minimum value of h_{fe} for oscillations to be maintained is

$$h_{fe} = -\left(\frac{X_3}{X_2} + 1\right) \tag{4.14}$$

The actual value of h_{fe} would need to be greater than this for oscillations to commence.

Substituting for X_3 from eq. (4.13) into eq. (4.12) gives $X_1 = h_{fe} X_2$, so that X_1 and X_2 are reactances of the same type.

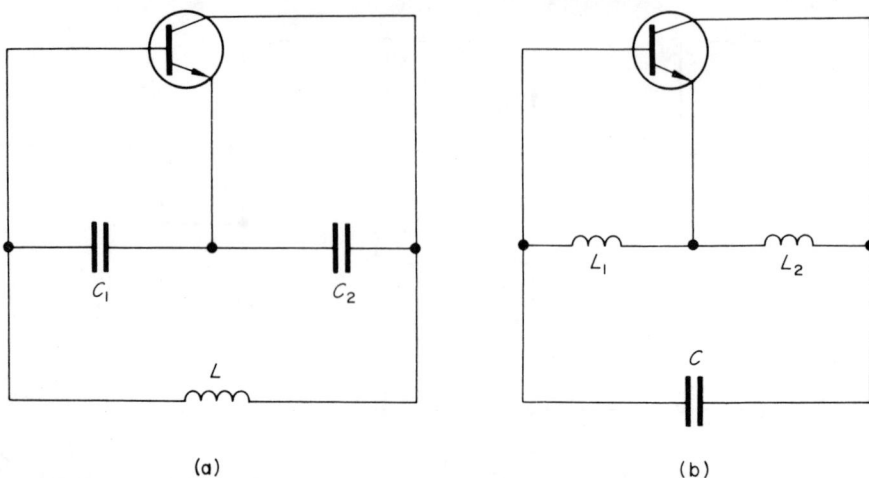

(a)   (b)

Fig. 4.9 Basic circuits for (a) the Colpitts oscillator and (b) the Hartley oscillator.

Two popular forms of circuit are shown in Fig. 4.9. The Colpitts oscillator, Fig. 4.9(a), uses capacitive elements for Z_1 and Z_2, while the Hartley circuit uses inductors which may be inductively coupled. The impedances for the two circuits at resonance are

	Colpitts	Hartley
Z_1	$1/j\omega_0 C_1$	$j\omega_0 L_1$
Z_2	$1/j\omega_0 C_2$	$j\omega_0 L_2$
Z_3	$j\omega_0 L$	$1/j\omega_0 C$

4.5 Colpitts oscillators

Two forms of Colpitts oscillator are shown in Fig. 4.10, the components C_1, C_2, and L being the reactive components of the feedback network, other elements being required for bias purposes. In the series-fed version, Fig. 4.10(a), the quiescent collector voltage is equal to the supply voltage, so that the maximum output voltage swing is from about 1 to 17 V. In the case of the shunt-fed circuit, the quiescent collector voltage is about 4.5 V, and the peak-to-peak output voltage swing is about 8 V. The output voltage swing of the common-emitter circuit can be increased by replacing the resistive load with the radiofrequency choke shown in the inset in Fig. 4.10(b).

For these circuits, from eq. (4.12)

$$\frac{1}{j\omega_0 C_1} + \frac{1}{j\omega_0 C_2} + j\omega_0 L = 0$$

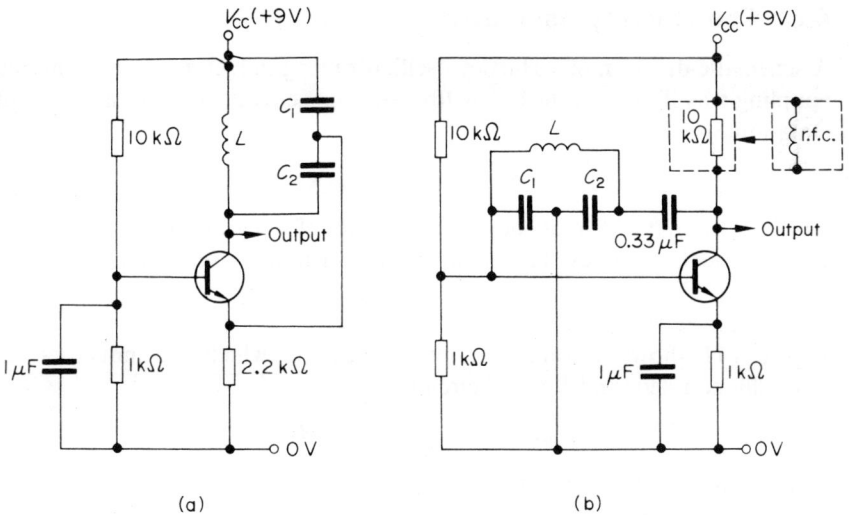

Fig. 4.10   Colpitts oscillators; (a) series-fed and (b) shunt-fed.

or

$$\omega_0 = \sqrt{\frac{1}{L}\left(\frac{1}{C_1} + \frac{1}{C_2}\right)} \qquad (4.15)$$

That is to say, the frequency of oscillations is that of a tuned circuit consisting of inductor L and two capacitors C_1 and C_2 all connected in series.

Also, from eq. (4.14) the current gain for maintenance of oscillations is

$$h_{fe} = -\left(\frac{j\omega_0 L}{1/j\omega_0 C_2} + 1\right) = \frac{C_2}{C_1} \qquad (4.16)$$

Normally the reactance of C_1 at ω_0 is kept small so that it is not shunted by the transistor input impedance. That is $C_1 \gg C_2$ (C_1 may be 10 to 30 times greater than C_2), hence

$$\omega_0 \simeq 1/\sqrt{LC_2}$$

Capacitor C_1 effectively controls the feedback proportion so that any change in its value has more effect on the magnitude of the output voltage than on the frequency. For example, increasing the value of C_1 reduces the output voltage and has little significant effect on the frequency. On the other hand, doubling the value of C_2 reduces the frequency by a factor of nearly $1/\sqrt{2}$ and has only a small effect on the output amplitude.

This type of circuit can be used to generate frequencies between audio-frequencies and a few gigahertz.

4.6   The Hartley oscillator

A schematic diagram of a Hartley oscillator using inductors without mutual coupling was illustrated in Fig. 4.9(b), and the general solution of its circuit equations yields

$$\omega_0 = 1/\sqrt{(L_1 + L_2)C}$$

which is the natural frequency of oscillation of a series circuit comprising L_1, L_2, and C in series. The condition for sustained oscillations is

$$h_{fe} \geqslant L_1/L_2$$

Figure 4.11 shows a common-emitter circuit in which the inductors are mutually coupled, and for this circuit

$$\omega_0 = 1/\sqrt{(L_x + L_y + 2M)C}$$

and for sustained oscillations

$$h_{fe} \geqslant (L_x + M)/(L_y + M)$$

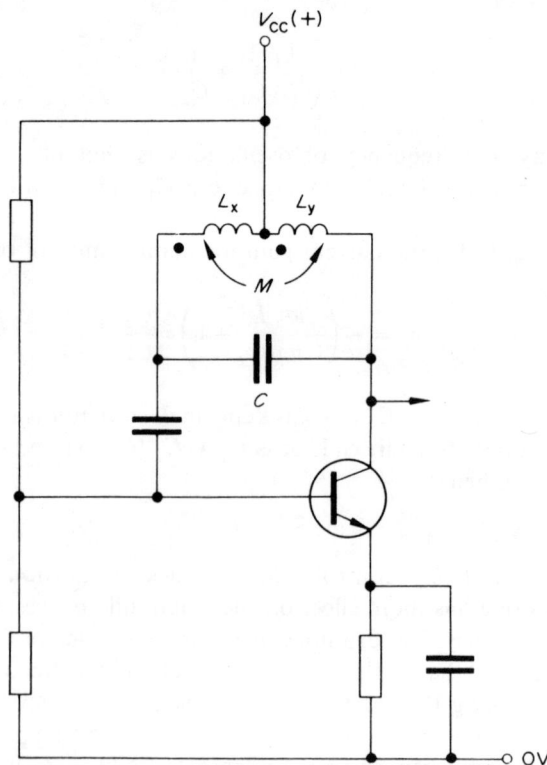

Fig. 4.11   A common-emitter Hartley oscillator.

4.7   The Clapp oscillator

One of the many variants of the basic circuit in Fig. 4.7 is the Clapp oscillator, shown in schematic form in Fig. 4.12. The frequency of oscillations is given by

$$\omega_0 = \sqrt{\frac{1}{L}\left(\frac{1}{C}+\frac{1}{C_1}+\frac{1}{C_2}\right)}$$

By making both C_1 and C_2 large compared with C, the L–C branch becomes the main frequency-determining element, so that $\omega_0 \simeq 1/\sqrt{LC}$. The frequency stability of this circuit is better than that of the Colpitts oscillator, and is further improved by replacing the L–C branch with a piezoelectric crystal (see section 4.9).

4.8   Tuned collector oscillator

A practical form of common-emitter circuit is shown in Fig. 4.13 in which the L and C elements in the collector circuit largely determine the frequency of oscillation. The signal fed back is injected into the base circuit by a winding which is closely coupled to the primary (collector) winding, and the output is delivered by a third winding. Providing that the ratio L/C is large and that the base circuit does not impose a load on the collector circuit,

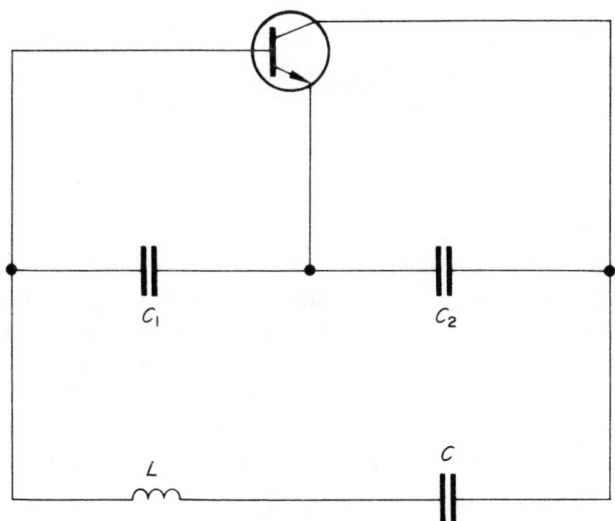

Fig. 4.12   Schematic diagram of a Clapp oscillator.

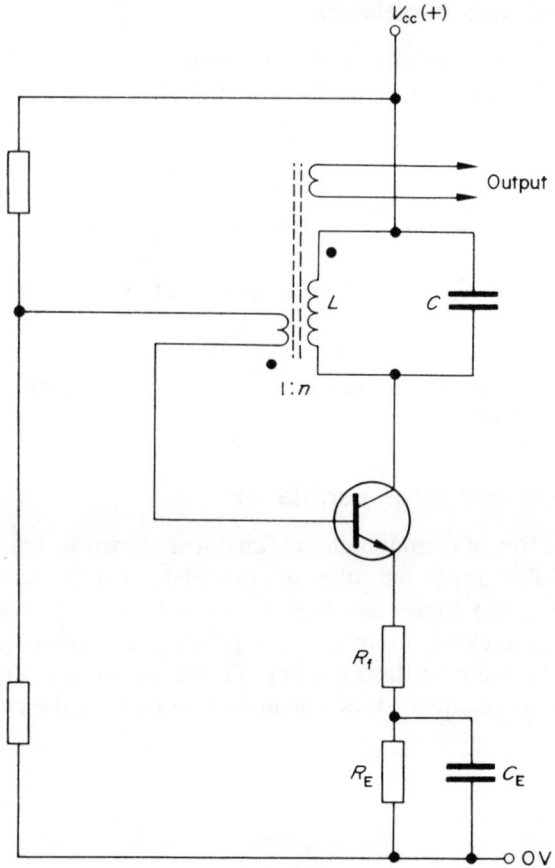

Fig. 4.13   A tuned collector oscillator.

then

$$\omega_0 \simeq 1/\sqrt{LC}$$

For oscillations to commence *at the instant of switch-on*, the *loop-gain* should lie between about 2 and 5. Since it is difficult to determine the exact parameters of components prior to building the circuit, the initial loop gain may be in excess of the value above and may be adjusted by including the unbypassed resistor R_f in the emitter lead.

If the loading imposed by the base circuit is small, then the voltage feedback factor β is approximately equal to $1/n$. Now, for a voltage amplifier

$$\text{Voltage gain} = -\text{Current gain} \times R_L/R_{in} \simeq -h_{fe}R_L/R_{in}$$

where R_L is the effective collector load imposed by the external circuit and

R_{in} is the base circuit input resistance.
Now

$$\text{Loop gain} = \beta \times \text{Voltage gain} = -h_{fe}R_L/nR_{in}$$

For oscillations to commence, the modulus of the loop gain is greater than unity, so that

$$n \geqslant h_{fe}R_L/R_{in}$$

the value of n lying in the range from about 5 to about 40.

With a loop gain which is just sufficient to allow oscillations to commence, the oscillator works in class A and provides a sinusoidal output of small amplitude and low distortion. Increasing the loop gain either by increasing the mutual coupling between the collector and base windings or by reducing R_f causes the circuit to operate in class AB, then class B and, finally, class C. The reason for this change in operating condition as the loop gain is progressively increased is as follows. The *average* emitter current is controlled by the voltage developed across R_E, and the current is equal to the sum of the average values of current in the collector and base circuits. Since the average emitter current remains fairly constant, then with a large loop gain the current in the collector and base are in the nature of pulses. This means that the collector current is cut off for part of the cycle.

A condition known as *squegging* can occur if the value of the time constant $C_E R_E$ is made too large. Squegging is a condition of self-blocking, and the output waveform in this mode of operation is in the form of bursts of oscillations which build up rapidly and then die away slowly in the manner shown in Fig. 4.14. During normal operation, capacitor C_E is charged via the relatively low resistance of the transistor and discharges through R_E. Over the complete cycle the *average* emitter voltage remains

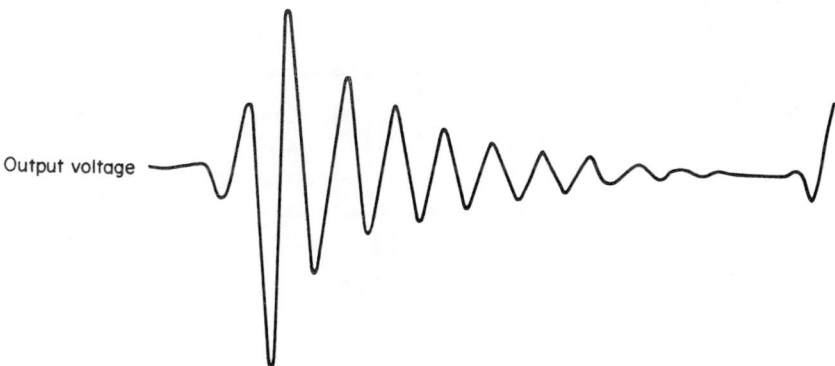

Fig. 4.14   Output waveform from a squegging oscillator.

constant. As the product $C_E R_E$ is increased a condition is reached when C_E is charged more quickly by the transistor during its conduction period than it can discharge through R_E. This causes the average emitter voltage to increase during each cycle, causing a shift in the operating point. This finally causes the loop gain to fall below unity and the oscillator stops operating. The collector voltage oscillations (still at a frequency of $\omega_0 = 1/\sqrt{LC}$) die away slowly, the rate of decay depending on the circuit loading. Only when the base-emitter junction again becomes forward biased (after C_E has discharged sufficiently) can oscillations commence again.

In some instances, large values of capacitance are required to give resonance at a particular frequency. By using a tapped winding in the manner shown in Fig. 4.15 it is possible to reduce the capacitor size required by transformer action. Here the effective tuned circuit capacitance

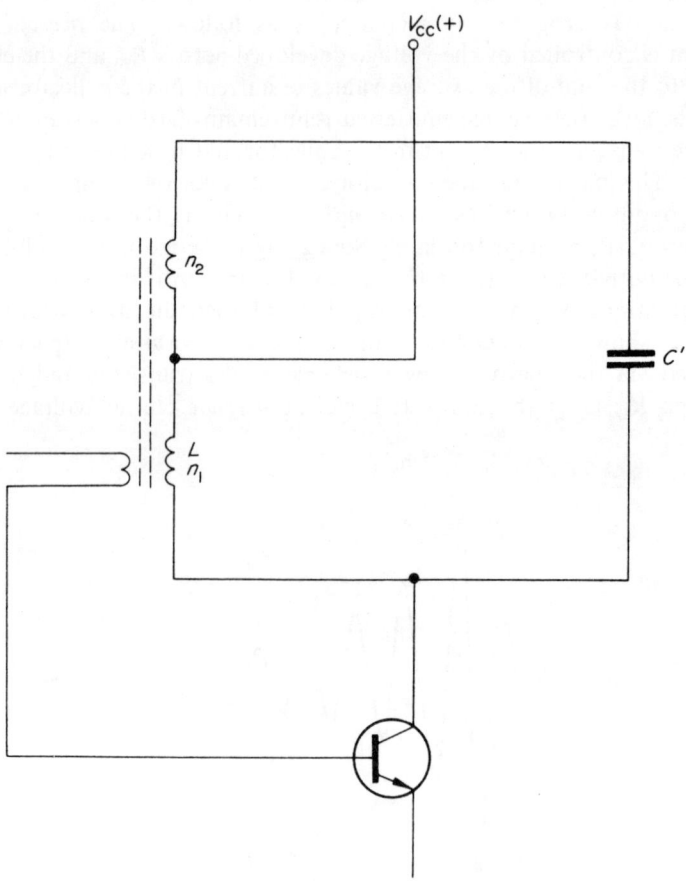

Fig. 4.15 A method of increasing the effective value of tuning capacitance.

C is

$$C = C'(n_2/n_1)^2$$

so that if $C' = 2000$ pF, $n_1 = 80$ turns, and $n_2 = 800$ turns, then $C = (800/80)^2 \times 2000 \times 10^{-12} = 0.2\ \mu$F.

4.9   Crystal-controlled oscillators

Certain crystalline substances, notably quartz, exhibit a *piezoelectric effect* which results either in a p.d. appearing between opposite faces of the crystal

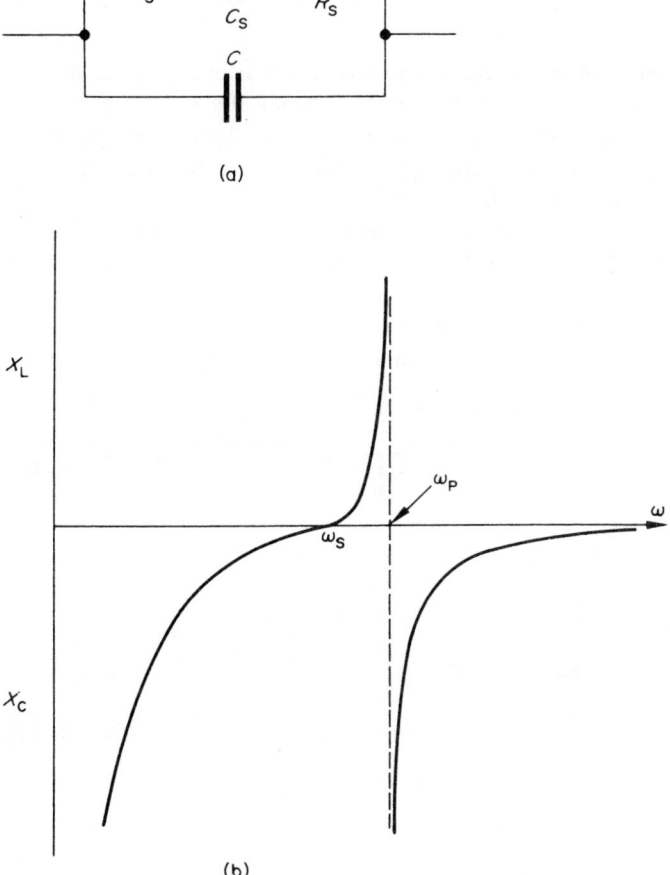

(a)

(b)

Fig. 4.16   (a) The equivalent circuit of a piezoelectric crystal and (b) its reactance-frequency diagram.

when it is mechanically deformed, or in a mechanical strain if it is subjected to an electric field. The electrical behaviour of the crystal can be represented by the equivalent circuit in Fig. 4.16(a) in which L_S, C_S, and R_S represent the inertia, friction loss, and stiffness, respectively, fo the crystal and C is its self-capacitance. Resistance R_S has a value of about $100\,\Omega$, so that it can be neglected without significant loss of accuracy. The crystal has two resonant frequencies, a series-type resonance ω_S and a parallel-type resonance ω_P, where

$$\omega_s = 1/\sqrt{L_s C_s} \quad \text{and} \quad \omega_P = \sqrt{\frac{1}{L}\left(\frac{1}{C_s} + \frac{1}{C}\right)}$$

The lower of the two resonant frequencies is ω_S, shown in Fig. 4.16(b). With crystals it is possible to obtain Q-factors of the order of 10^5, which is about 100 times better than the figure for a high-Q tuned circuit. In most types of crystal used for this purpose, $C \gg C_S$, so that the two oscillatory frequencies are very close to one another, the separation being typically a few hundred hertz per megahertz of oscillatory frequency. Between the two oscillatory frequencies the crystal appears as an inductor. By maintaining the crystal at a constant temperature, oscillators with a frequency stability of 1 part in 10^{10} can be constructed.

A schematic diagram of a crystal-controlled version of the Colpitts oscillator is shown in Fig. 4.17, in which the crystal is used in its inductive mode. The crystal can also be used in Hartley circuit by using it, once more, in the narrow frequency band between ω_S and ω_P as an inductor. Other oscillator circuits have been designed which use either ω_S or ω_P.

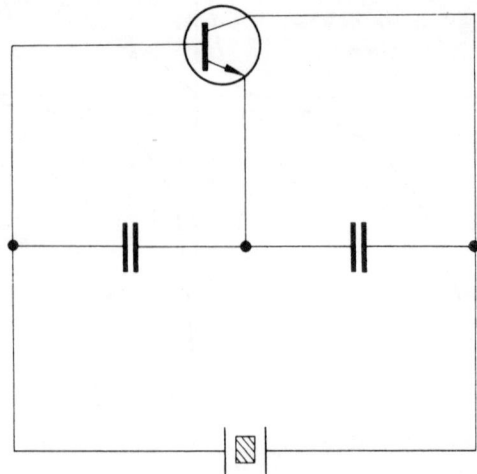

Fig. 4.17   A block diagram of one form of crystal-controlled oscillator.

4.10   Frequency stability of oscillators

A change in the oscillator output frequency can result from changes in many factors including variation of

(a) Parameters of the active device
(b) Values of passive elements (R, L, M, or C)
(c) Supply voltage
(d) Output load.

The most important transistor parameter which affects frequency stability is the collector depletion capacitance C_C. This capacitance varies inversely as the square root of the collector voltage, so that its value varies over the cycle. Since this capacitance is effectively in parallel with the tuning capacitance, it is desirable to use a tuning capacitance which is large in comparison with C_C so as to swamp the effects of variations in its value.

The phase angle of the current gain of BJTs (or voltage gain of FETs) also affects the frequency of oscillation since the loop phase shift must be zero. If the phase shift introduced by the transistor differs from the ideal, then the tuned circuit (or feedback circuit) must operate at a frequency away from resonance in order to satisfy the requirement of zero loop phase shift. The effect can be minimized by using a high-Q circuit, shown in Fig. 4.18. If the active device introduces a phase lag of Φ, then it is necessary for

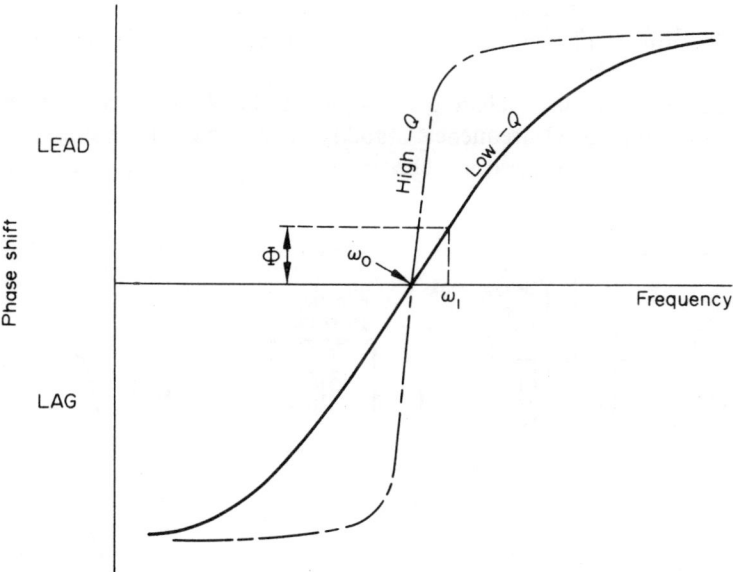

Fig. 4.18   Phase shift characteristics of a parallel tuned circuit.

the tuned circuit to introduce a phase lead of Φ. With a low-Q circuit the operating frequency must change from ω_0 to ω_1 before this condition is satisfied. In the case of a high-Q tuned circuit, the rate of change of phase-shift with frequency $(d\phi/d\omega)$ at ω_0 is high, and the operating frequency is displaced by only a small amount from ω_0. For this reason it is desirable to use high-Q tuned circuits in L–C oscillators.

Since piezoelectric crystals act as high-Q circuits, oscillators using them as the frequency controlling element have a very good frequency stability.

4.11   Ultra-low frequency oscillators

The oscillators so far described cover the frequency range from about 1 Hz to several gigahertz, but their design is such that they do not provide sinusoidal output waveforms of good amplitude stability and frequency stability in the ultra-low frequency region. Ultra-low frequency oscillators include signal generators capable of producing frequencies in the range one cycle in several days to about 1 Hz.

The block diagram of a popular type of circuit for the generation of frequencies above 0.001 Hz is shown in Fig. 4.19. In this circuit, the output from a voltage comparator, which comprises a voltage comparison circuit and a flip-flop (see chapter 8), is applied to an integrator. At any given instant of time the output voltage from the comparator is constant, so that the output voltage from the integrator changes at a constant rate. The integrator output is then applied to a function generator which 'shapes' the output from the integrator into a linearized approximation of a sinewave. Provided that sufficient straight-line segments are used, the distortion introduced is quite small.

Frequency control is achieved by adjusting the RC time constant of the integrator, and good frequency stability is obtained by using low-loss capacitors in the integrator. The circuit has good amplitude stability

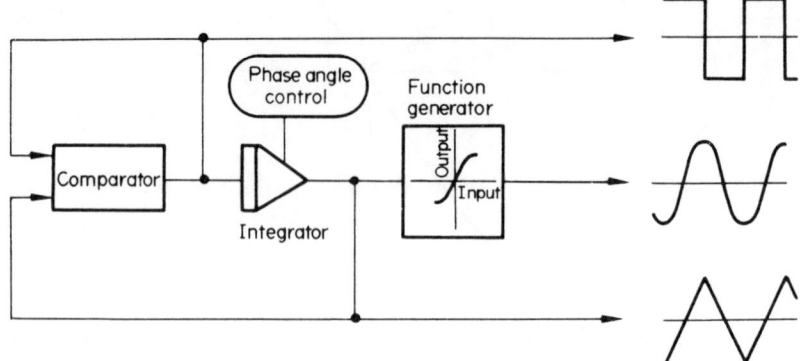

Fig. 4.19   An ultra-low frequency signal generator.

resulting from the comparison of the comparator output with the integrator output. When the integrator output voltage is equal to the comparator output, it causes the comparator output voltage polarity to be reversed, so causing the slope of the integrator output voltage to change sign. In this way, the comparator output is a square wave, the integrator generates a triangular wave and the function generator provides a sinewave, all of good amplitude and frequency stability. The phase angle at which the output sinewave commences operation can be controlled by having an 'initial condition' circuit which charges the integrator capacitor to a pre-determined voltage prior to triggering.

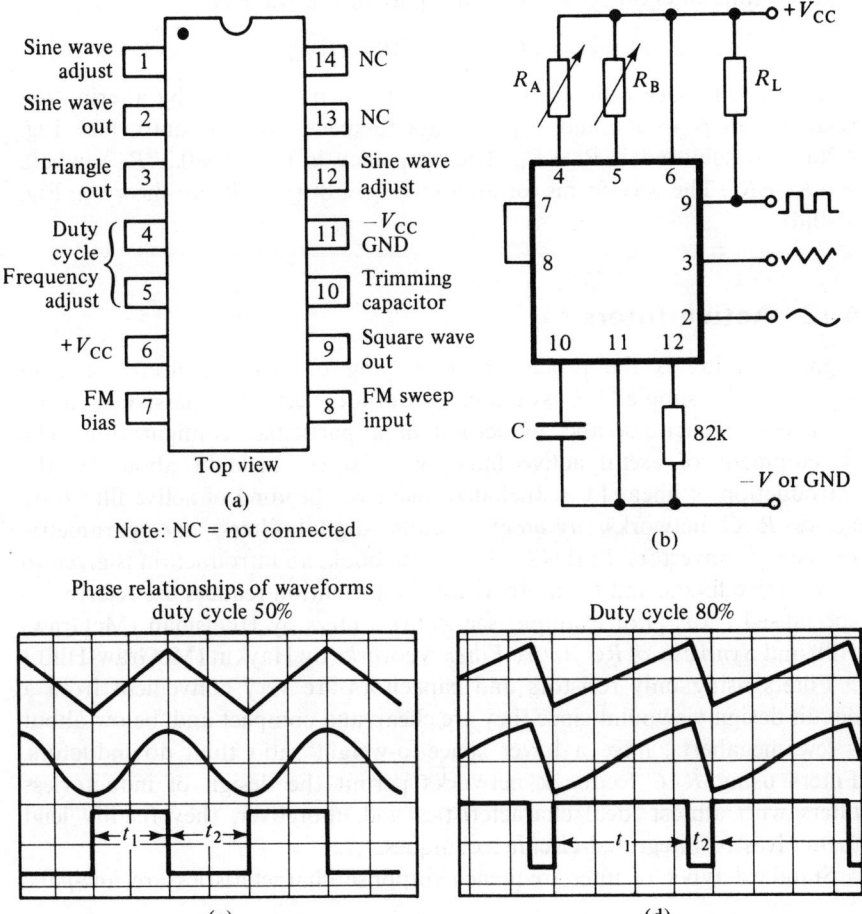

Fig. 4.20 The 8038 waveform generator: (a) top view of IC, (b) basic con-nection diagram, (c) and (d) output waveforms.

(Courtesy of RS Components Ltd)

An integrated circuit which combines the features of Fig. 4.19 is the 8038 IC, which is housed in a 14-pin DIP (see Fig. 4.20(a)) and is capable of producing a frequency in the range 0.001 Hz to 100 kHz with a distortion of not more than one per cent. The basic connections to the circuit are shown in Fig. 4.20(b), its rectangular output waveform appearing at pin 9, its triangular waveform at pin 3, and its sinewave at pin 2, all waveforms being generated simultaneously. The magnitude of the triangular waveform signal is $V_{CC}/3$ and the time interval t_1 of the rising part of the triangular wave is (diagrams (c) and (d))

$$t_1 = 5CR_A/3$$

and the time interval t_2 of the falling part of the wave is

$$t_2 = 5CR_AR_B/[3(2R_A - R_B)]$$

The mark-to space ratio of the waveform can be modified by altering the relative values of R_A and R_B. A mark-to-space ratio of unity (see Fig. 4.20(c)) is achieved if $R_A = R_B$. The frequency is then $f = 0.3/(RC)$, where $R = R_A = R_B$. The waveforms for an 80 per cent duty cycle are shown in Fig. 4.20(d).

4.12 Active filters

Signal filtering is the process of modifying or shaping electronic data signals, and a simple filter would consist of a collection of passive elements (resistors, capacitors, and inductors) in a particular configuration. The development of useful active filters was largely brought about by the introduction of linear I.C.s. Included under the heading of active filters are active *R–C* networks, *gyrator** circuits, digital filters, and parametric frequency convertors. In this section of the book, an introduction is given to *R–C* active filters, and for more detailed information readers are referred to specialised texts. [For example, see *Active Filters* by Huelsman (McGraw-Hill) and *Synthesis of RC Active Filter Networks* by Haykin (McGraw-Hill).]

Filters using only resistors and capacitors are very convenient from a circuit design viewpoint since they are cheap and compact and, below about a few megahertz, have a lower space-to-weight ratio than do inductors. Filters using *R–C* feedback networks permit the design of inductorless filters with almost ideal characteristics and, moreover, they readily lend themselves to integrated circuit techniques.

Standard types of filter frequency response characteristics are *low-pass*,

---

*The gyrator, conceived in a paper by Tellegen in 1948, has impedance inverting qualities which allow a capacitor connected to its output terminals to appear as an inductor at the input. The name gyrator indicates that the impedance has 'gyrated'.

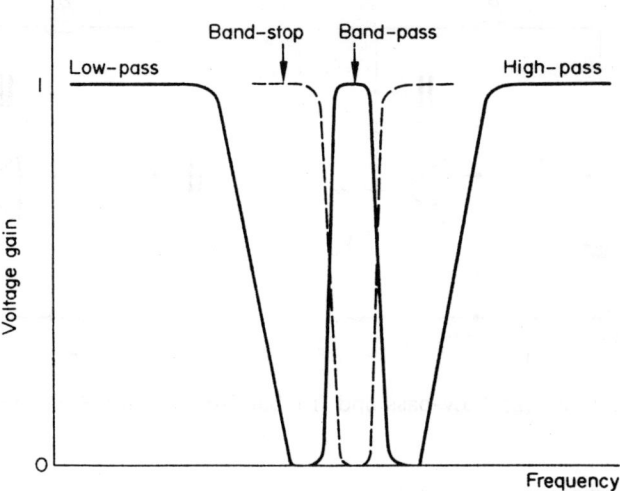

Fig. 4.21   Filter characteristics.

band-pass, band-stop (or *band-reject*), and *high-pass*, shown in Fig. 4.21. Additionally, a variety of characteristics can be obtained for each of the above types. For instance, it is possible to design different types of low-pass filters including those which have characteristics described as Butterworth, Bessel, and Tchebyscheff, each having advantages for particular applications.

Low-pass *Butterworth filters* have a flat ripple-free gain characteristic up to about $f_c/2$, where f_c is the cut-off frequency, and the attenuation at f_c is 3 dB. This type of filter is suitable for general purpose low-pass filtering and for the filtering of signals to be applied to A/D convertors and data acquisition systems. *Bessel filters* have a linear phase-frequency characteristic and are suited to the filtering of rectangular waveforms since the phase-frequency distortion (delay distortion) introduced by the filter is minimal. The high-frequency roll-off of Bessel filters is not as sharp as in the case of Butterworth and Tchebyscheff filters.

Tchebyscheff filters provide a sharper high-frequency cut-off than either of the other two types but only at the expense of variation in gain with frequency (*pass-band ripple*) at the low frequency end of a low-pass filter. The greater the pass-band ripple that can be tolerated, the more rapid the high frequency gain roll-off. The pass-band ripple can be anything from about ± 0.5 dB to several dB, depending on the design. The Tchebyscheff characteristic lends itself to applications where a very sharp cut-off is required and where the pass-band ripple is unimportant. This type of filter is not well suited to filtering rectangular waves and has a poor transient response. For instance, the application of a rectangular wave to a low-pass

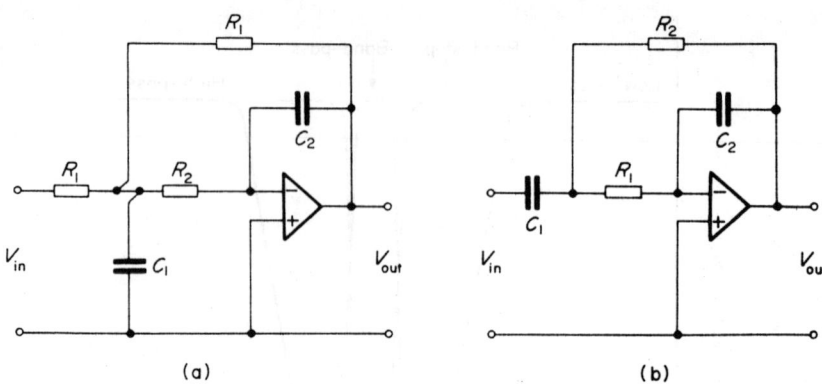

Fig. 4.22   (a) Low-pass and (b) band-pass active R–C filters.

Tchebyscheff filter may result in an overshoot at the output of more than 25 per cent.

Typical active filter circuits using single amplifiers are shown in Fig. 4.22. Figure 4.22(a) illustrates a second-order Butterworth low-pass filter with a cut-off frequency of $(R_1 R_2 C_1 C_2)^{-1/2}$ rad/s and a high-frequency gain roll-off of 12 dB/octave (40 dB/decade). If, in Fig. 4.22(a) the resistors are interchanged with capacitors and capacitors with resistors, then a high-pass filter is realized. The band-pass filter in Fig. 4.22(b) is centered on a frequency of $1/R_1 R_2 C_1 C_2$ rad/s.

It is interesting to see how a lossy inductor may be simulated by an active R–C circuit. If we consider the circuit in Fig. 4.23(a), and apply a sinusoidal

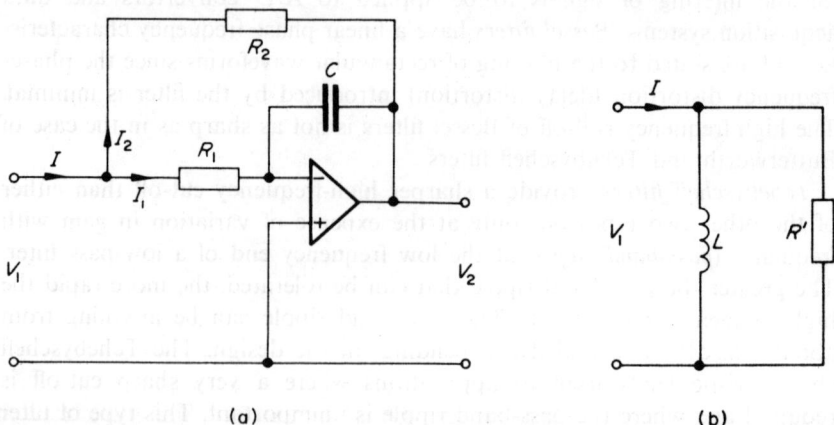

Fig. 4.23   The active R–C circuit in (a) is equivalent to the passive circuit in (b).

voltage V_1 to the input, then

$$V_2 = -\frac{1}{R_1 C}\int V_1 dt = j\frac{V_1}{\omega R_1 C}$$

and
$$I = I_1 + I_2 = \frac{V_1}{R_1} + \frac{V_1 - V_2}{R_2}$$

$$= V_1\left(\frac{1}{R_1} + \frac{1}{R_2} - \frac{j}{\omega R_1 R_2 C}\right)$$

From the above equation, the input admittance of the active circuit is

$$Y_1 = \frac{I}{V_1} = \left(\frac{1}{R_1} + \frac{1}{R_2}\right) - \frac{j}{\omega R_1 R_2 C} \tag{4.17}$$

An inspection of the admittance of the circuit in Fig. 4.23(b) gives

$$\frac{1}{R'} - \frac{j}{\omega L} \tag{4.18}$$

For the two circuits to be equivalent to one another, the coefficients of eqs. (4.17) and (4.18) must be identical to one another, that is

$$R' = R_1 R_2/(R_1 + R_2) \tag{4.19}$$

$$L = R_1 R_2 C \tag{4.20}$$

A negative resistance can be generated using the circuit in Fig. 4.24. Here

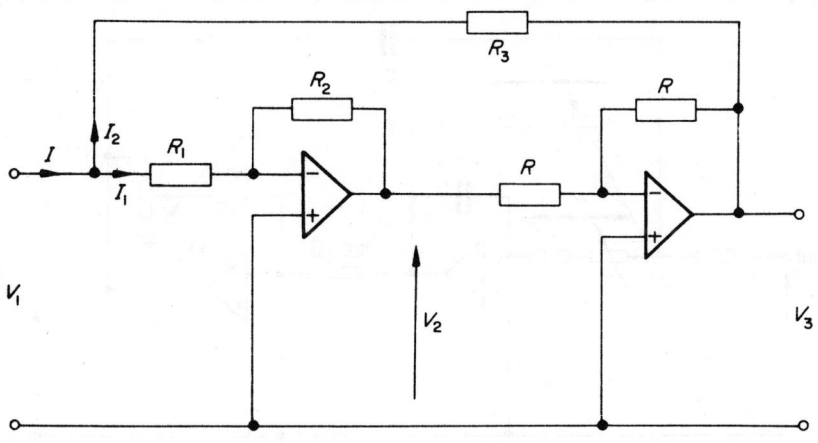

Fig. 4.24   A circuit which can have a negative input resistance.

the circuit relationships are

$$V_2 = -R_2 V_1 / R_1$$

$$V_3 = -V_2 = R_2 V_1 / R_1$$

Now,

$$I = I_1 + I_2 = \frac{V_1}{R_1} + \frac{V_1 - V_3}{R_3}$$

$$= V_1 \left(\frac{1}{R_1} + \frac{1}{R_3} - \frac{R_2}{R_1 R_3} \right)$$

The input conductance is

$$\frac{I}{V_1} = \frac{1}{R_1} + \frac{1}{R_3} - \frac{R_2}{R_1 R_3} = \frac{R_1 + R_3}{R_1 R_3} - \frac{R_2}{R_1 R_3} = \frac{R_1 + R_3 - R_2}{R_1 R_3}$$

Clearly, when $R_2 > (R_1 + R_3)$ then the input resistance of the circuit is negative.

By connecting the circuits in Figs. 4.23 and 4.24 in parallel, the electrical equivalent of a perfect lossless inductor is formed.

Combining the ideas developed above, a high-Q tuneable filter with an adjustable Q-factor can be designed, and is illustrated in Fig. 4.25. The reasonant frequency of the circuit is

$$\omega_0 = 1/RC \text{ rad/s} \qquad (4.21)$$

and its Q-factor is

$$Q = k/(2k - 1)$$

where k is the setting of the potentiometer in the second stage of the amplifier. Clearly, if $k = 0.5$ then the value of Q is infinity! For stability, the

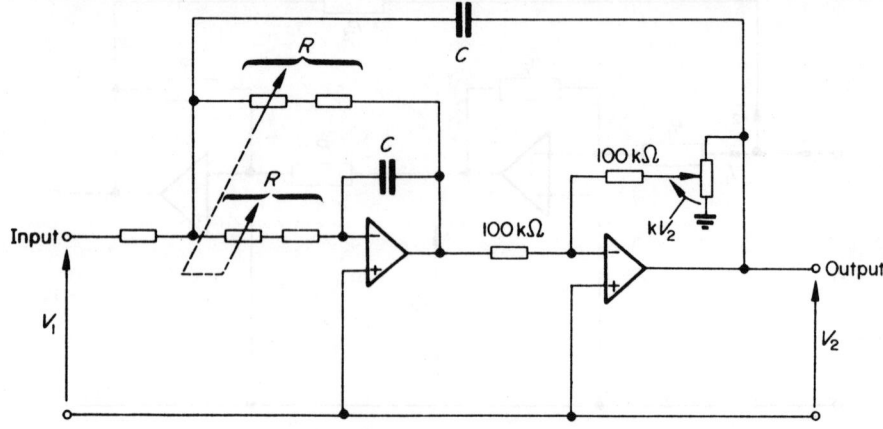

Fig. 4.25 A tuneable filter.

value of k should be greater than 0.5, otherwise overall positive feedback is applied and the circuit operates as an oscillator at the frequency given in eq. (4.21). Comparing the resonant frequency in eq. (4.21) with that of a parallel L–C circuit we have

$$\frac{1}{\sqrt{LC}} \equiv \frac{1}{RC}$$

or $$L \equiv R^2C$$

Resonance occurs at 0.96 kHz if $R=50\,\text{k}\Omega$ and $C=3300\,\text{pF}$, to give an equivalent circuit of $L=8.25\,\text{H}$ and $C=3300\,\text{pF}$. The circuit is capable of providing a high Q-factor down to very low frequencies, and with $R=100\,\text{k}\Omega$, $C=1\,\mu\text{F}$ the resonant frequency is 1.6 Hz, corresponding to the use of a parallel tuned circuit with $L=10\,000\,\text{H}$! Fine frequency control is obtained by using ganged resistors for part of R, and coarse control is obtained if different values of C are switched into circuit.

Problems

4.1   For the Colpitts oscillator shown in Fig. 4.9(a), the value of L is 20 μH, C_2 has a value of 250 pF, and C_1 has a value of 10 nF. Determine the frequency of oscillation and the minimum value of h_{fe} required for the transistor to maintain oscillations.

4.2   A Hartley oscillator uses $L_1 = 400\,\mu\text{H}$, $L_2 = 1000\,\mu\text{H}$, and $C=500$ pF (see Fig. 4.9(b)), and

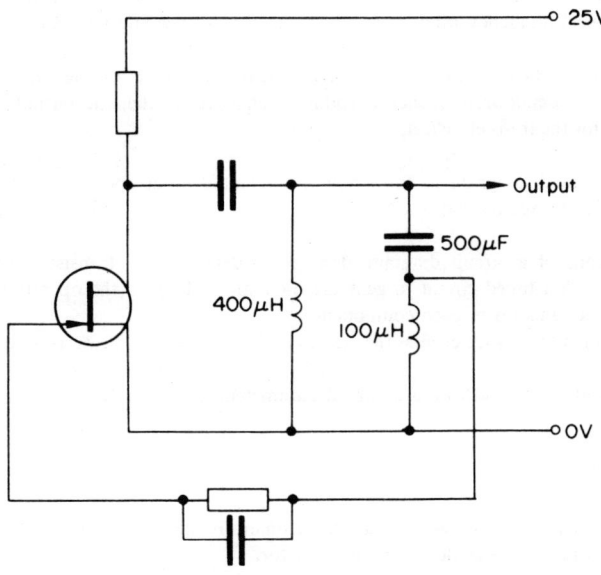

Fig. 4.26

no mutual coupling exists between the coils. Calculate the frequency of oscillations, and also calculate the minimum value of h_{fe} for the transistor.

4.3 The circuit diagram of a Hartley oscillator is shown in Fig. 4.26.

(a) Explain its principle of operation.
(b) Show how the circuit can be modified to form a Colpitts oscillator.
(c) Upon what factors does the frequency stability of an LC oscillator mainly depend?
(d) For the circuit given, estimate the frequency of oscillation assuming that the coils have no mutual inductance and negligible resistance.

4.4 Describe, with the aid of a circuit diagram, the principle of operation of a phase advance, phase shift oscillator. What controls (i) the frequency, and (ii) the amplitude of the oscillations?
State why the amplifier must have a certain minimum gain in order to sustain oscillations. Explain the effect on the output waveform if the open-loop gain is appreciably greater than this minimum value.

4.5 Draw circuit diagrams of *two* different types of resistance-capacitance oscillators and describe the basic action in each case. Compare and contrast the two circuits particularly from the point of view of convenience of control. Explain why $R-C$ oscillators have generally displaced $L-C$ oscillators for the generation of audio-frequency test signals.

4.6 Draw the circuit diagram of a sinusoidal $R-C$ oscillator employing a non-inverting amplifier and, with the aid of a phasor diagram, explain its operation. Determine the minimum amplifier gain required to maintain oscillations. Give reasons why this type of oscillator is preferred to an $L-C$ oscillator for the generation of low frequencies.

4.7 Draw a circuit diagram of a Wien bridge oscillator using transistors, and show how its frequency and its output amplitude are controlled. State the reason for each of the components in the circuit, and describe how amplitude stability is achieved.
Determine the frequency of oscillations in terms of the circuit components.
State a typical frequency range over which the oscillator would be used.

4.8 Describe, for the oscillator in problem 4.7, the transient effect on the output voltage waveform when the oscillator frequency is suddenly changed. Sketch the output waveform, and give reasons for the transient effect.

4.9 Draw a circuit diagram to show how a differential input operational amplifier may be used in a Wien bridge oscillator.

4.10 By means of a circuit diagram, describe in detail how a transistor may be used in conjunction with a tuned circuit to generate oscillations. Explain the operation of the circuit chosen and the function of each component.
Explain which factors determine the stability of the frequency of the oscillations.

4.11 A crystal has the following electrical parameters:

$L_S = 3$ H
$C_S = 0.05$ pF
$C = 5$ pF

Determine the equivalent series and parallel resonant frequencies of the crystal. If the value of R_S for the crystal is 3.5 kΩ, determine its Q-factor.

4.12   The circuit shown in Fig. 4.27 is that of an oscillator. Describe the operation of the circuit and sketch the waveforms appearing at points A and B.

What is the effect of reducing the value of R_1 (a) on the frequency, and (b) on the amplitude of the output signals.

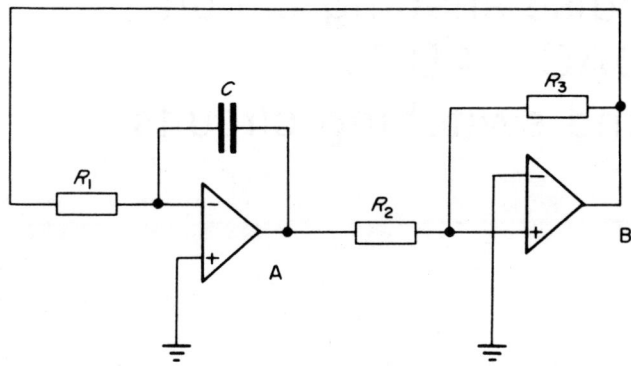

Fig. 4.27

4.13   If the voltage swing at output B in problem 4.12 is limited to ± 5 V, and if $R_1 = R_2 = R_3 = 100$ kΩ, and $C = 0.01$ μF, determine the frequency of oscillation of the circuit.

4.14   A balanced rectangular wave is applied to a network which has a cut-off frequency of 4 kHz, and has a linear phase/frequency characteristic below 4 kHz. The frequency of the wave is varied from 500 Hz to 3 kHz. Describe the waveform which appears at the output of the filter for input frequencies of (i) 500 Hz, (ii) 1 kHz, (iii) 2 kHz, (iv) 3 kHz.

5. Pulse shaping circuits, DACs, ADCs and switching circuits

5.1 Pulse forming and shaping

A wide variety of electronic circuits depend for their operation on pulses, the principal pulse forming and shaping networks being:

(a) Linear passive circuits (containing R, L, and C)
(b) Non-linear passive circuits (diode circuits)
(c) Active circuits (transistor circuits and integrated circuits).

A number of popular circuits are described in this chapter.

5.2 Passive high-pass filter (approximate differentiator)

Two popular versions of this network are shown in Fig. 5.1, both of which can be described under certain circumstances as approximate differentiator networks, but more will be said about this later. In the following analysis it is assumed that the output is unloaded. For the circuit in Fig. 5.1(a),

$$\frac{V_2}{V_1} = \frac{j\omega CR}{1+j\omega CR} = \frac{j\omega \tau}{1+j\omega \tau} = \frac{\omega \tau}{\sqrt{[1+(\omega \tau)^2]}} \angle 90° - \tan^{-1} \omega \tau \qquad (5.1)$$

where $\tau = CR$ and is the circuit time constant having dimensions of time. At very low frequencies ($\omega \to 0$), the voltage gain is approximately equal to $\omega \tau$, i.e., it is very small, but it increases with frequency. At high frequencies ($\omega \to \infty$), the term $(\omega \tau)^2$ in the denominator is much greater than unity so that the gain is approximately $\omega \tau / \sqrt{(\omega \tau)^2} \simeq 1$. Between the two extremes of frequency we reach a condition when $\omega \tau = 1$, when the gain is $1/\sqrt{(1+1)} = 0.707$, i.e., the gain is 3 dB below the high-frequency value of unity. This frequency is known as the *corner frequency* or *cut-off frequency* ω_c. The resulting gain curve is shown in Fig. 5.1(c). Thus, at the corner

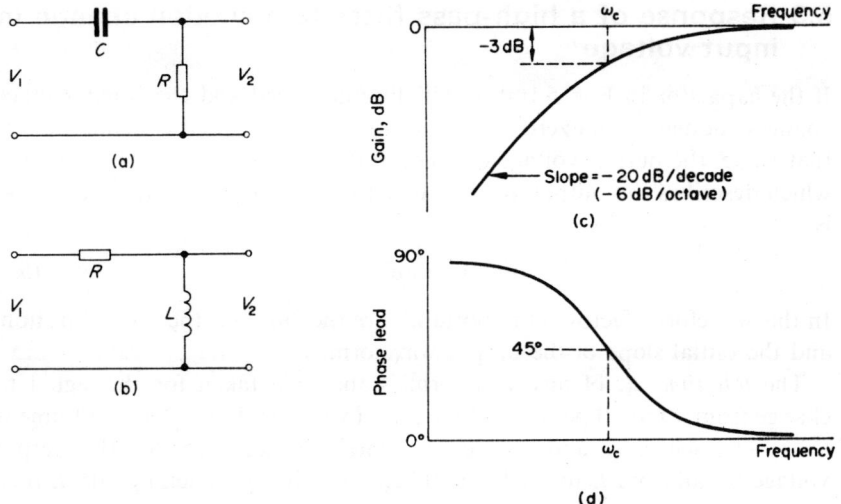

Fig. 5.1   Passive high-pass filters.

frequency $\omega_c \tau = 1$, hence

$$\omega_c = 1/\tau \qquad \text{rad/s} \tag{5.2}$$

and $$f_c = 1/2\pi\tau$$

Also, at very low frequency the phase shift is

$$90° - \tan^{-1} \omega\tau \simeq 90° - \tan^{-1} 0° = 90°$$

that is the phase lead is 90 degrees. At high frequencies (when $\omega\tau \to \infty$), the phase shift is

$$90° - \tan^{-1} \infty = 90° - 90° = 0$$

that is, the output and input are in phase with one another at these frequencies. At frequency ω_c, when $\omega\tau = 1$, the phase shift is

$$90° - \tan^{-1} 1 = 45°$$

The resulting phase curve is shown in Fig. 5.1(d).

Turning now to the circuit in Fig. 5.1(b), the *transfer function* or equation relating the output to the input for sinusoidal signals is

$$\frac{V_2}{V_1} = \frac{j\omega L/R}{1 + j\omega L/R} = \frac{j\omega\tau}{1 + j\omega\tau}$$

where $\tau = L/R$. The form of the equation is seen to be generally similar to eq. (5.1), so that the gain and phase curves for the circuit are the same as those for Fig. 5.1(a).

5.3 Response of a high-pass filter to a sudden change in input voltage

If the capacitor in Fig. 5.1(a) is initially uncharged and the input voltage changes suddenly from zero to $+E$, shown in Fig. 5.2, and is maintained at that value, the output voltage changes in the manner shown. The equation which describes the output voltage after the application of the input signal is

$$V_2 = Ee^{-t/\tau} \tag{5.3}$$

In this waveform, factors of importance are the fall time, the pulse duration, and the initial slope of the output waveform.

The *fall time*, t_f, of any waveform is the time taken for the signal to change from 90 to 10 per cent of the initial value. In Fig. 5.2, the fall time is $t_f = t_2 - t_1$, where t_1 and t_2 are time intervals necessary for the output voltage to fall from E to $0.9\,E$ and $0.1\,E$, respectively. When $V_2 = 0.9\,E$ then

$$0.9\,E = Ee^{-t_1/\tau} \tag{5.4}$$

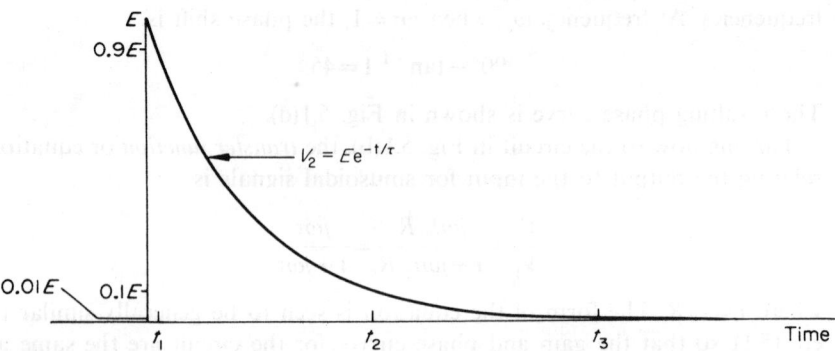

Fig. 5.2 Response to a sudden change in input voltage.

and when $V_2 = 0.1 \, E$ then

$$0.1 \, E = E e^{-t_2/\tau} \tag{5.5}$$

Solving between eqs. (5.4) and (5.5) for t_1 and t_2 gives $t_1 \simeq 0.1 \, \tau$ and $t_2 \simeq 2.3 \, \tau$, so that

$$t_f = t_2 - t_1 = 2.2 \, \tau \tag{5.6}$$

In a circuit using $R = 10 \, \text{k}\Omega$ and $C = 0.01 \, \mu\text{F}$, then $\tau = 100 \, \mu\text{s}$ and the fall time for a step input wave is $220 \, \mu\text{s}$. Also, from eq. (5.6)

$$t_f = 2.2 \, \tau = 2.2/2\pi f_c \simeq 0.35/f_c \tag{5.7}$$

In theory, it takes an infinite time for an exponential curve to die away, and to give a convenient measure of the *pulse duration* we calculate the time taken for the output voltage to drop to 1 per cent of its initial value. This is equal to time t_3 in Fig. 5.2. At this instant $V_2 = 0.01 \, E$, so that

$$0.01 \, E = E e^{-t_3/\tau}$$

or
$$t_3 = 4.6 \, \tau$$

Using the figures above, the pulse width is $4.6 \times 100 = 460 \, \mu\text{s}$.

The slope of the output voltage waveform is obtained by differentiating the expression for V_2 with respect to time as follows:

$$\mathrm{d}(E e^{-t/\tau})/\mathrm{d}t = -E e^{-t/\tau}/\tau \tag{5.8}$$

The *initial slope* of the curve is obtained by equating t in eq. (5.8) to zero as follows

$$\text{Initial slope} = -E/\tau \qquad \text{volts/s} \tag{5.9}$$

Using the values given above once more and assuming a step input voltage change of 1 V, the initial slope of the output voltage waveform is

$$-1/100 \times 10^{-6} = -0.01 \, \text{V}/\mu\text{s} \quad \text{or} \quad -10 \, \text{kV/s}$$

5.3.1   Rectangular impulse wave response of the high-pass filter

For practical purposes, a rectangular wave can be considered as two changes in voltage, one being a sudden change from, say, zero volts to $+E$ followed some time later by a step of $-E$, taking the input to zero once more, as shown in Fig. 5.3. The output voltage, by the superposition theorem, is the sum of the responses to the individual step functions.

In order to describe the response of the network to a single rectangular pulse, it is generally adequate to consider three particular cases in relation to the pulse period T_p, namely

(a) $T_p = \tau$ (b) $T_p \gg \tau$ (c) $T_p \ll \tau$

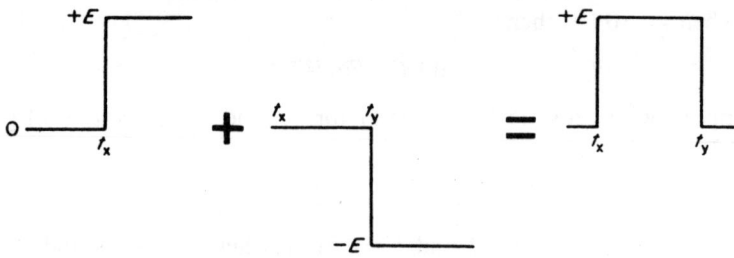

Fig. 5.3   Construction of a rectangular pulse.

$T_p = \tau$

The output voltage at time t after the application of input $+E$ is

$$V_2 = Ee^{-t/T_p}$$

and after a time interval $t = T_p$ the output voltage V_p is

$$V_p = Ee^{-T_p/T_p} = 0.368\,E \simeq 0.37\,E \qquad (5.10)$$

That is, the output voltage has fallen to about 37 per cent of its original value by the end of the input pulse period. If the input voltage had been maintained at $+E$ the output voltage would decay in the manner shown by curve A in Fig. 5.4. The response of the circuit to the application of a negative pulse at time T_p gives an output described by curve B. The net output response to the rectangular wave is the sum of curves A and B, and is shown by the curve drawn in full line in Fig. 5.4. From the figure

$$E = V_p - V_x$$

hence $$V_x = V_p - E = -0.63\,E \qquad (5.11)$$

Since only a single pulse is applied to the circuit, the output voltage decays from V_x to $0.01\,V_x$ in a time of approximately $4.6\,\tau$.

$T_p \gg \tau$

Here the time constant is short compared with the pulse period, so that the capacitor is fully charged well before the end of the pulse period. The circuit response in this case is illustrated in Fig. 5.5(a), the rise- and fall-times of the pulses after the trailing and leading edges, respectively, are $2.2\,\tau$. The effective pulse width for both positive- and negative-going output pulses is $4.6\,\tau$.

If the circuit could accurately differentiate the input waveform, the theoretical output signal would comprise a positive-going spike of infinite amplitude at the leading edge of the input wave and a negative-going spike at the trailing edge. The spikes would have zero width. From Fig. 5.5(a) we see that the smaller the circuit time constant, the closer the output

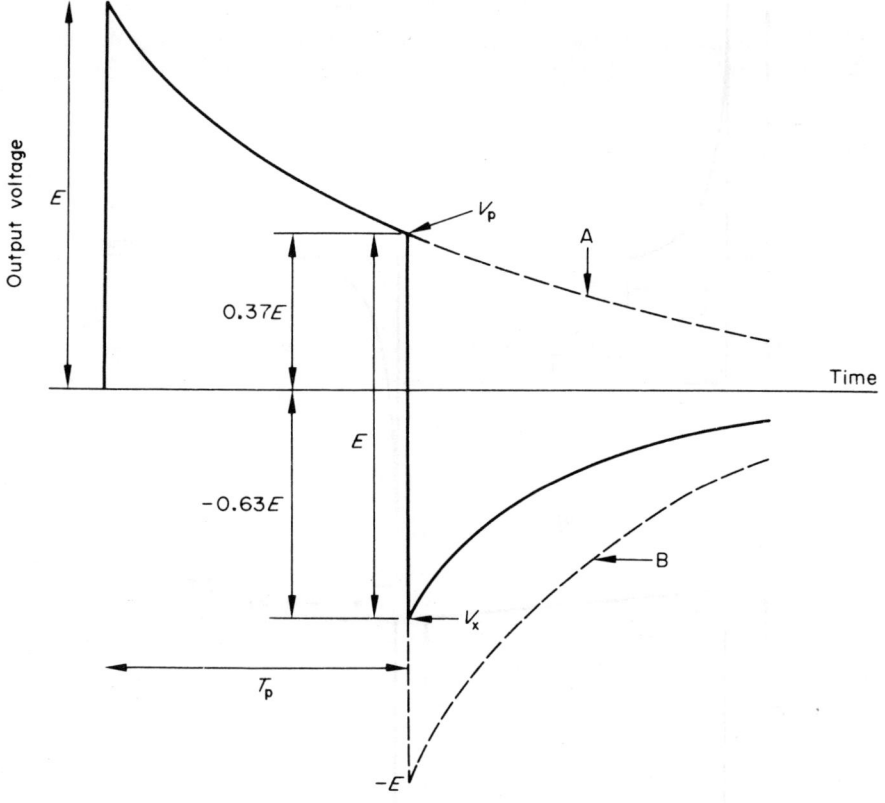

Fig. 5.4   Output for $T_p = \tau$.

waveform should approach the differential of the input signal. In general terms, the circuit acts as a differentiator if $\tau \ll T_p$.

The circuit has practical limitations as a differentiator in that if it is to approach the ideal then the time constant must be short. In this event, the circuit will impose a considerable transient load on the driving circuit while the capacitor is charging; in some instances this may not be acceptable. Also, the peak output voltage is restricted to E volts, which is further reduced in some cases (see section 5.3.3).

$T_p \ll \tau$

In this case, the circuit time constant is long compared with the pulse period, and very little charge is acquired by the capacitor during time T_p. The resulting output waveform is shown in Fig. 5.5(b), in which the slope of the top of the output waveform is $-E/\tau$ volts per second (see eq. (5.9)). The voltage *droop* or *sag* of the output wave during T_p is ET_p/τ volts. If, for

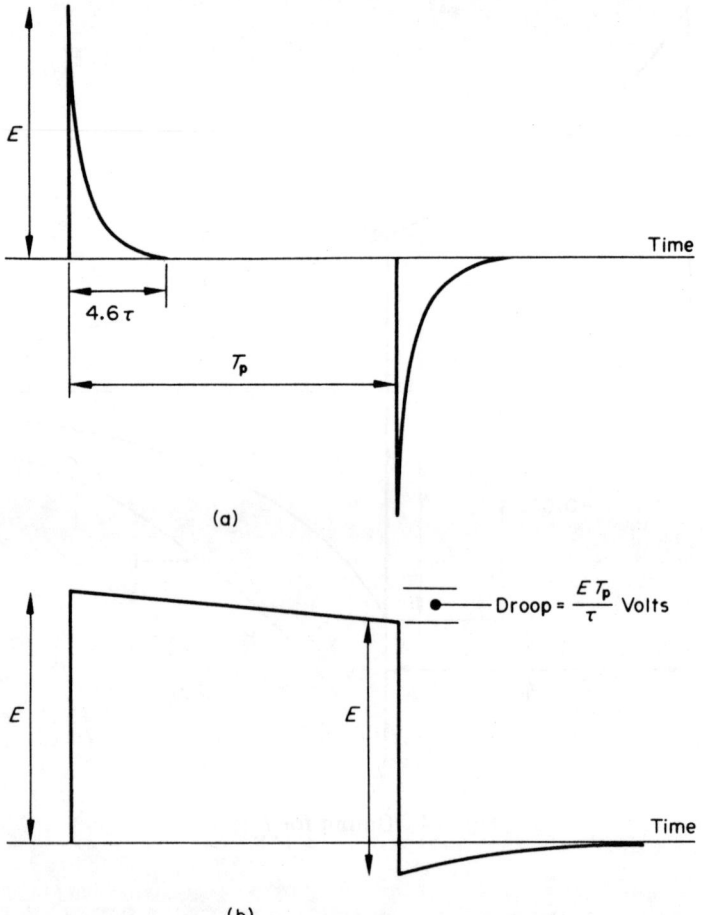

Fig. 5.5   Output for (a) $T_p \gg \tau$ and (b) $T_p \ll \tau$.

example, $E = 2$V, $T_p = 20\,\mu s$, and $\tau = 1000\,\mu s$, then the voltage droop is $2 \times 20/1000 = 0.04$ V.

At the trailing edge of the input pulse, the output voltage suddenly becomes $-ET_p/\tau$ (-0.04 V in the case considered). The output voltage then decays to 1 per cent of this value in a time equal to $4.6\,\tau$.

5.3.2   Response to an alternating rectangular wave train

The alternating component of any signal has zero average value, and the area under the positive half of the wave is equal to the area under the negative half of the wave, shown in Fig. 5.6(a).

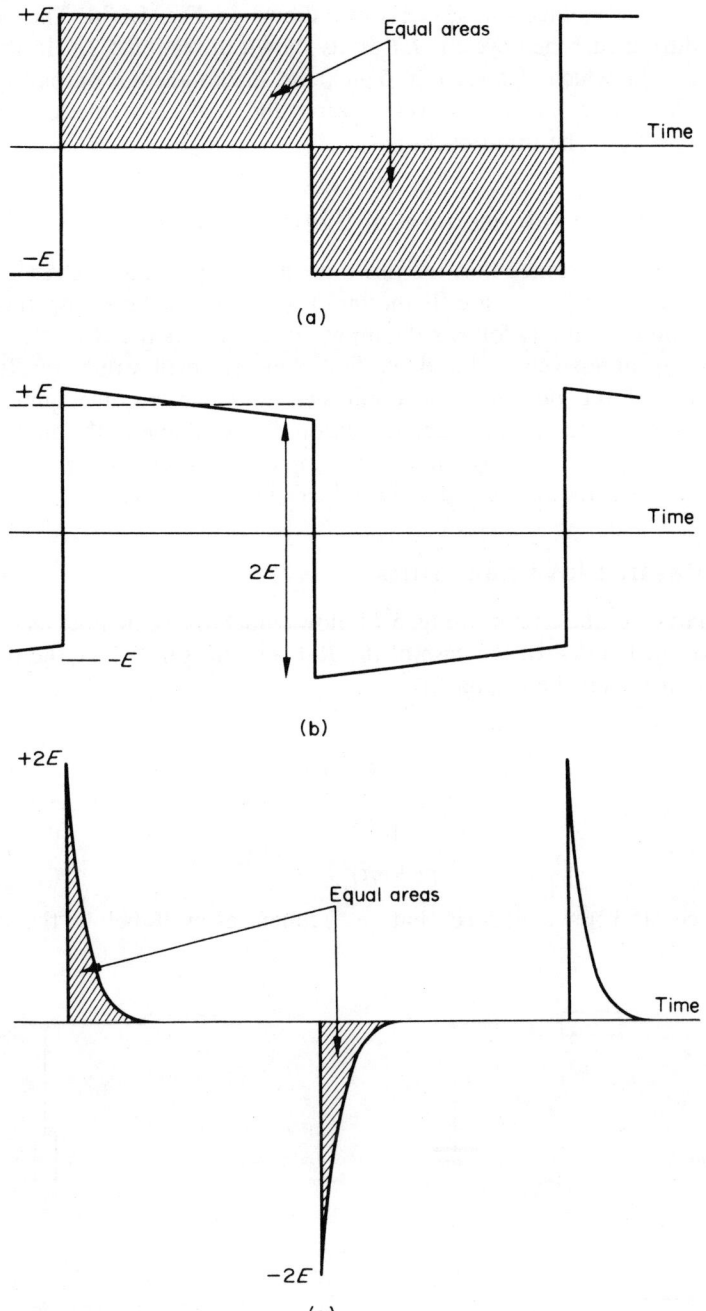

Fig. 5.6 (a) Input waveform to the circuit. Output waveforms for (b) $T_p \ll \tau$ and (c) $T_p \gg \tau$.

The output voltage for $T_p \ll \tau$ is as shown in Fig. 5.6(b). The voltage droop during each half cycle is ET_p/τ, as shown in section 5.3.1. In the case of circuits in which $T_p \gg \tau$, Fig. 5.6(c), the instantaneous output voltage during the major part of the cycle is zero, and is $\pm 2E$ at the leading and trailing edges of the input cycle.

5.3.3   Effect of the rise time of the input signal

So far we have assumed that the rise time of the input signal is zero, but in practice this is not the case. If the input signal has a finite rise time, the output voltage initially follows the input waveform. As the RC value of the circuit is progressively reduced we find that, at small values of RC, the amplitude of the output pulse diminishes and can become quite small compared with the input voltage transition. For example, if the circuit time constant is equal to the rise time of the input signal, then the peak output voltage is only about one-third of the peak input voltage.

5.4   Passive low-pass filter

The analysis of the circuit in Fig. 5.7 follows much the same lines as that for high-pass networks. In the case of the R–C circuit, Fig. 5.7(a), the transfer function for sinusoidal signals is

$$\frac{V_2}{V_1} = \frac{1}{1+j\omega CR} = \frac{1}{1+j\omega\tau}$$

$$= \frac{1}{\sqrt{[1+\omega\tau)^2]}} \angle -\tan^{-1}\omega\tau \tag{5.12}$$

From eq. (5.12) we deduce that at frequencies well below the cut-off

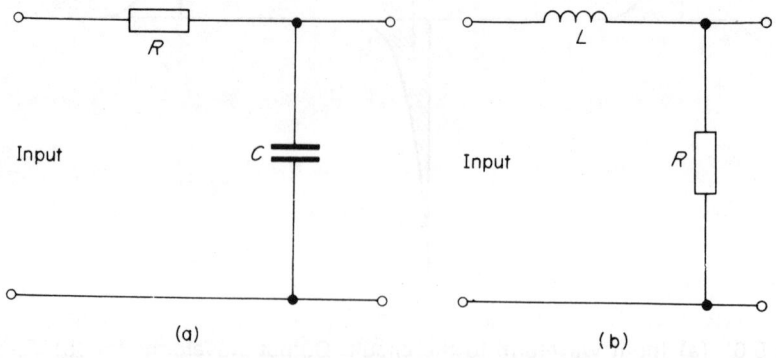

(a)   (b)

Fig. 5.7   Passive low-pass filter circuits.

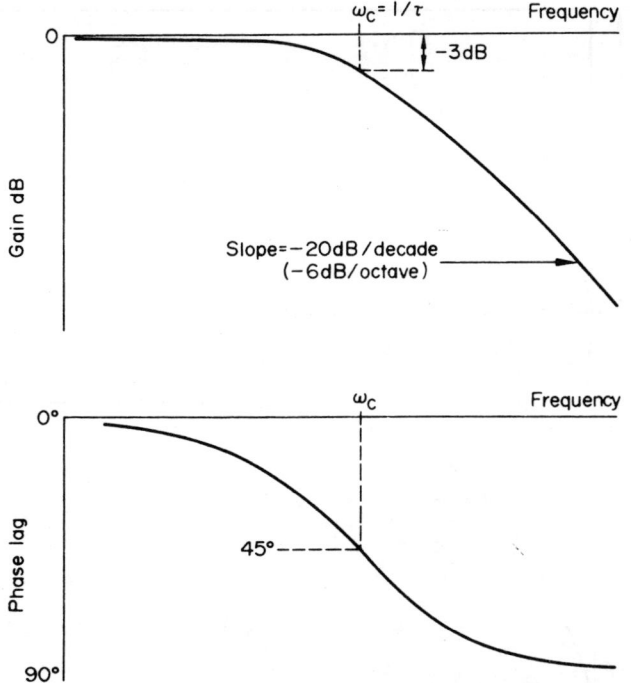

Fig. 5.8   Frequency response diagram for a low-pass filter.

frequency (when $\omega\tau \ll 1$) the voltage gain is approximately unity and that the phase shift is very small and is lagging. At frequencies well above the corner frequency ($\omega\tau \gg 1$) the gain is approximately $1/\omega\tau$ and reduces at the rate of 6 dB/octave or 20 dB/decade. The phase shift at high frequencies approaches 90 degrees lagging, shown in Fig. 5.8.

At the corner frequency, when $\omega_c = 1/\tau$, the gain is 0.707 (3 dB below the low-frequency value) and the phase shift is 45 degrees lagging.

The equation for the L–R circuit, Fig. 5.7(b), is given by eq. (5.12) with the exception that in this case $\tau = L/R$.

5.5 Response of a low-pass filter to a sudden change in input voltage

In the following, we will consider the R–C circuit and we will assume that the capacitor is initially uncharged. When the input voltage level is changed suddenly, the instantaneous output voltage is described by the equation

$$V_2 = E(1 - e^{-t/\tau}) \tag{5.13}$$

The input and output waveforms are shown in Fig. 5.9. The *rise time* t_r, as before, is $t_r = t_2 - t_1$, where t_1 is the time taken for the output voltage level

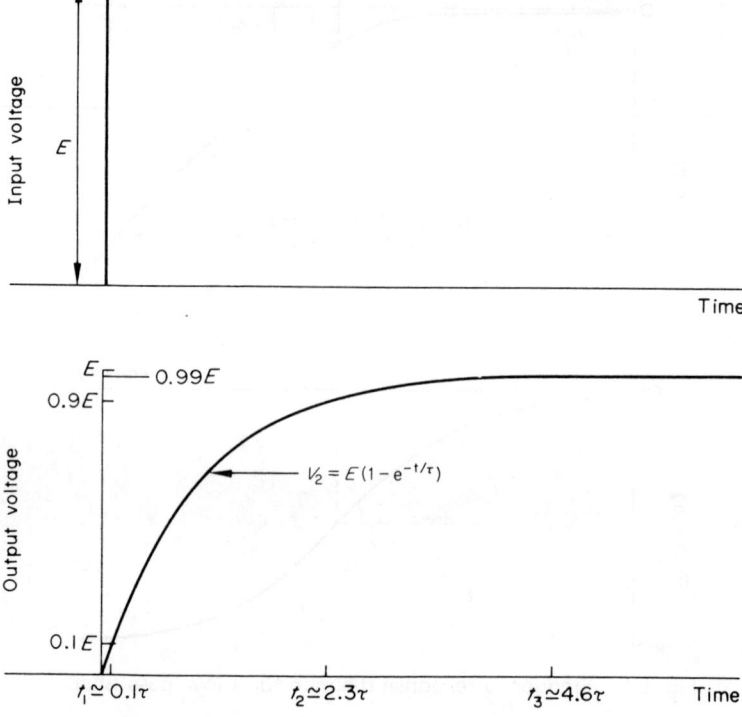

Fig. 5.9   Response to a sudden change in input voltage.

to reach $0.1\,E$ and t_2 is the time taken to reach $0.9\,E$. Calculations reveal that $t_1 \simeq 0.1\,\tau$ and $t_2 \simeq 2.3\,\tau$, so that

$$t_r = 2.2\,\tau \qquad\qquad (5.14)$$

It can also be shown that the time taken for the output voltage to change from zero to $0.99\,E$ is about $4.6\,\tau$ and that $t_r = 0.35/f_c$, where f_c is the cut-off frequency in hertz.

Also, for a sudden change in input voltage, the output voltage initially changes (an increase in the case considered here) at the rate of E/τ volts/second.

5.5.1 Impulsive response of a low-pass filter

$\mathbf{T_p = \tau}$

Substituting $T_p = \tau$ into eq. (5.13) shows that after a time equal to T_p the output voltage is $V_2 = 0.632\,E$, see Fig. 5.10(a). When the input signal is reduced to zero, the output voltage reduces in an exponential manner to zero (or nearly so) from $0.632\,E$ after a further $4.6\,\tau$ seconds.

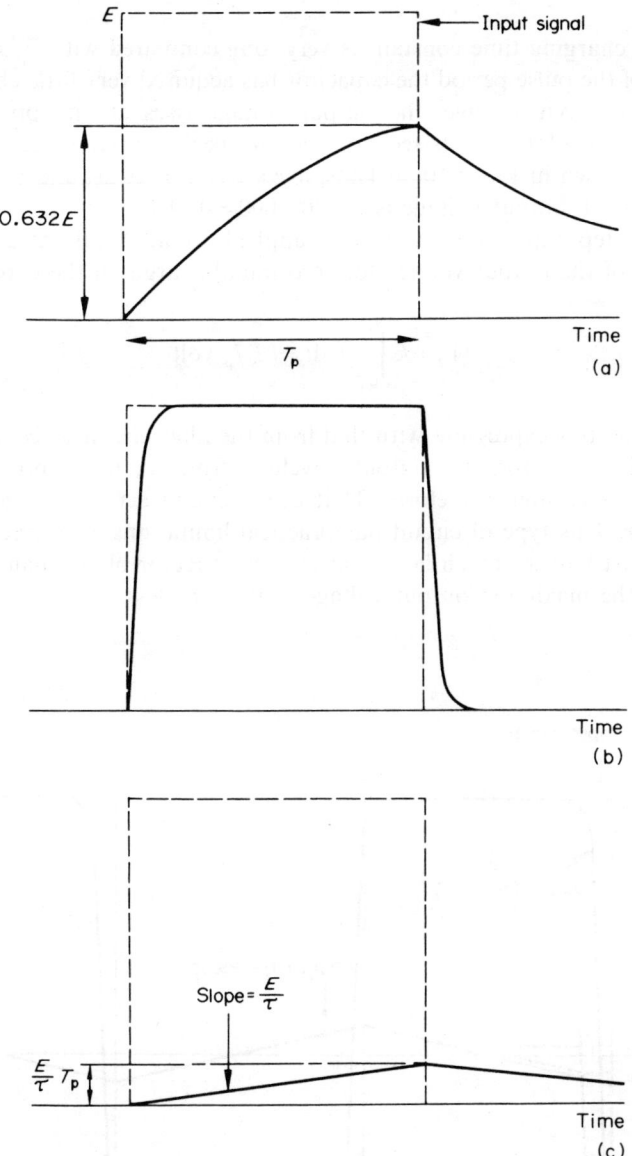

Fig. 5.10   Output waveforms for (a) $T_p=\tau$, (b) $T_p\gg\tau$, and (c) $T_p\ll\tau$.

$T_p\gg\tau$

In this case, the charging time constant is very short compared with the pulse period, so that the capacitor becomes fully charged shortly after the pulse is applied. As a result, the output waveform is a reasonable reproduction of the applied waveform, as in Fig. 5.10(b).

$T_p \ll \tau$

Here the charging time constant is very long compared with T_p so that by the end of the pulse period the capacitor has acquired very little charge. For a constant input voltage, the output voltage rises at an approximately constant rate of E/τ volts/second, and the peak output voltage is ET_p/τ volts, as shown in Fig. 5.10(c). Thus, if $E = 2V$, $T_p = 20\,\mu s$, and τ is $1000\,\mu s$, then the peak output voltage is $2 \times 20/1000 = 0.04$ V.

If the step input waveform was applied to an ideal integrator, the equation of the output voltage for zero initial charge on the capacitor is

$$V_2 = k \int_0^{T_p} E\, dt = kET_p \text{ volts}$$

Comparing this expression with that from the filter circuit when $T_p \ll \tau$, we see that if $k = 1/\tau$ then the output waveform from the filter approaches the integral of the input waveform. That is, the circuit acts as an *approximate integrator*. This type of circuit has practical limitations as an integrator in that τ must be very much larger (at least by a factor of 10) than T_p, which restricts the maximum output voltage to 0.1 E or less.

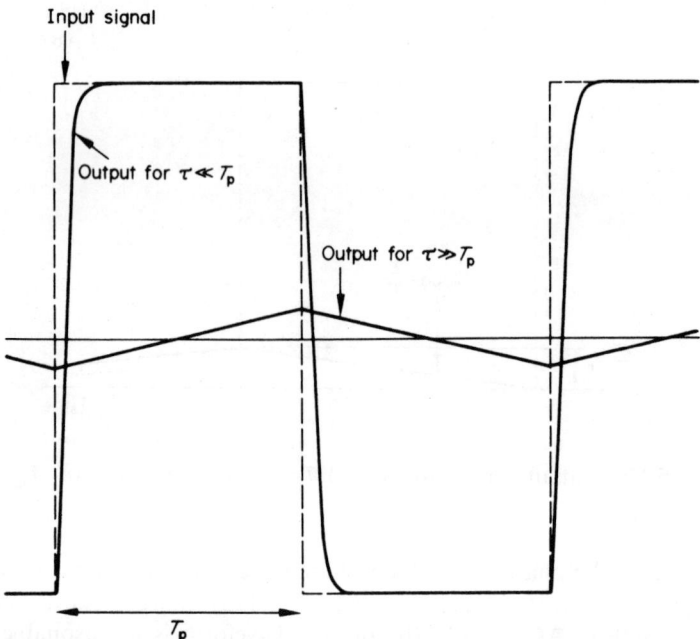

Fig. 5.11   Response of a low-pass filter to a rectangular wave train.

5.5.2   Response to a rectangular wave train

Typical output waveforms for two widely differing values of τ are shown in Fig. 5.11.

5.6   Clipping circuits or limiters

Diode clipping circuits are widely used in electronics and their function is to clip or limit the amplitude of an electrical signal to a specific range of values.

A basic form of *series clipping circuit* is shown in Fig. 5.12(a), and is simply a half-wave rectifier circuit which transmits only positive-going signals, as shown in Fig. 5.12(b). By including the battery E in series with R, shown in the inset to Fig. 5.12(a), the output waveform is as shown in (c).

If the diode (and battery) connections were reversed, the output waveform would comprise the negative-going parts of the input signal.

The effect of the forward p.d. of the diode is to reduce the output voltage

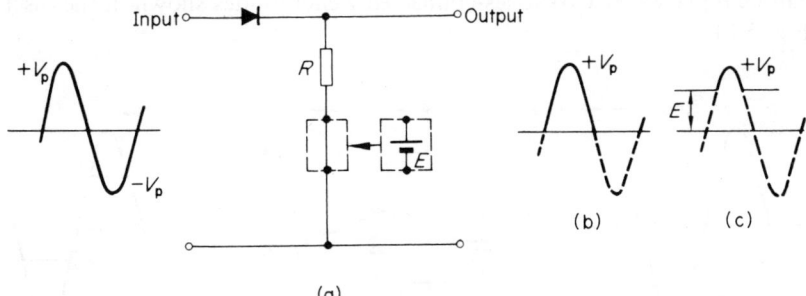

Fig. 5.12   A series clipping circuit.

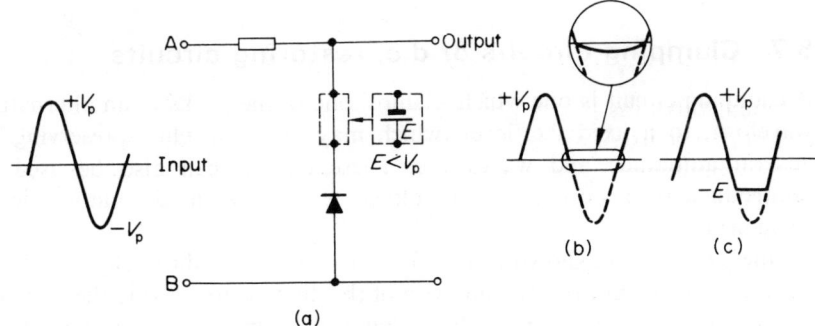

Fig. 5.13   A shunt clipping circuit.

during the period that the diode is conducting. If the slope resistance of the diode is r_a, the instantaneous output voltage for positive inputs is $Rv_1/(r_a+R)$, where v_1 is the instantaneous input voltage.

One version of *shunt clipping circuit* is shown in Fig. 5.13(a), in which the diode is shunted across the output terminals. When the diode conducts the output is ideally zero, and this occurs when terminal A is negative with respect to terminal B. When terminal A is positive with respect to B, the diode is reverse biased and the positive-going input signals are transmitted to the output. The effect of the forward p.d. of the diode on the output waveform is shown in the inset in Fig. 5.13(b).

By including a battery in series with the diode, shown in the inset in Fig. 5.13(a), the diode is reverse biased until terminal A is $-E$ volts with respect to terminal B. This permits a selected part of the negative half-cycle to be transmitted to the output, shown in Fig. 5.13(c).

An unsymmetrical shunt clipping circuit which clips the input waveform between the limits $+E_1$ and $-E_2$ is shown in Fig. 5.14. This type of circuit can be used to generate a pseudo square wave from a sinusoidal (or similar type) input signal of large amplitude. Alternatively, the diode-battery circuit can be replaced by two series-connected Zener diodes shown in the inset to Fig. 5.14.

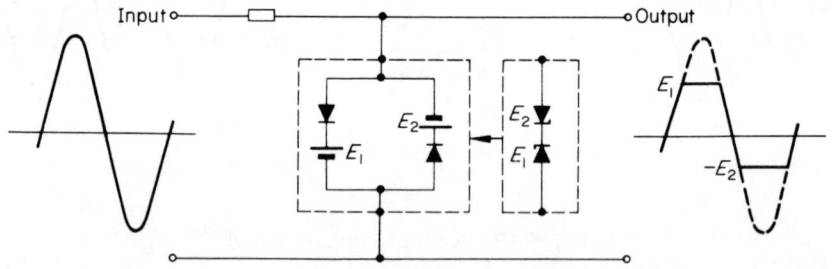

Fig. 5.14   An unsymmetrical shunt clipping circuit.

5.7   Clamping circuits or d.c. restoring circuits

A clamping circuit is one which 'clamps' one of the peaks of an alternating waveform to a fixed d.c. level (which may be zero) while preserving its general amplitude and waveshape. These circuits can also be used to reintroduce or restore a direct voltage level into an alternating signal waveform.

One basic circuit, shown in Fig. 5.15(a), combines a shunt clipping diode with an R–C integrator. The function of the diode is to provide the network with a small time constant with one polarity of input voltage and a large time constant with the opposite polarity input voltage. When terminal A is

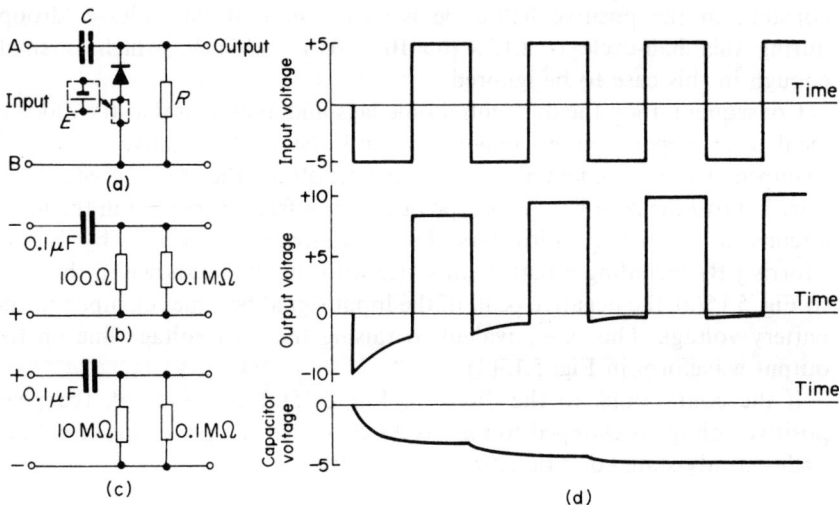

Fig. 5.15   (a) A clamping circuit. Equivalent circuit (b) applies when the diode is conducting and (c) applies when the diode is blocking. (d) Shows illustrative waveforms.

negative with respect to B, the diode conducts and the effective circuit time constant is $C \times r_a R/(r_a + R)$. The circuit elements are chosen so that $r_a \ll R$, thus in this part of the cycle the time constant is approximately Cr_a. Let us consider the case when $C = 0.1 \mu F$, $R = 0.1 M\Omega$, and $r_a = 100 \Omega$. Here the resistance of the parallel combination of r_a and R is approximately 100Ω, so that the time constant is $0.1 \times 10^{-6} \times 100 = 10 \mu s$. The equivalent circuit for this condition is shown in Fig. 5.15(b).

When terminal A is positive, as in Fig. 5.15(c), the diode is reverse biased and its resistance in this mode is assumed to be $10 M\Omega$. The circuit time constant is now $RC \simeq 0.1 \times 10^6 \times 0.1 \times 10^{-6} = 0.01$ s or $10^4 \mu s$.

For this circuit to operate satisfactorily, the time constant RC must be between ten and one hundred times greater than the pulse (or half-cycle) period of the input waveform. That is the half-cycle periodic time should be less than about 1000 to $100 \mu s$. For convenience, let us consider a square wave of periodic time $20 \mu s$ (half-cycle time of $10 \mu s$). Assuming that the capacitor is initially uncharged and that we switch on when the input voltage is passing through the zero point, the circuit waveforms are as shown in Fig. 5.15(d).

Initially, with $-5V$ applied, the capacitor charges through the diode so that at the end of the half-cycle (for which the charging time constant is equal to the pulse period) the output voltage has decayed to 37 per cent of the original value, i.e., to -1.85 V. At the beginning of the positive half-cycle, the output voltage rises to $10 - 1.85 = +8.15$ V. The circuit time constant in the positive half-cycle is 0.01 s, so that the voltage 'droop' during this half-cycle is $8.15 \times 10 \times 10^{-6}/0.01 = 8.15\,\text{mV}$, which is small enough in this case to be ignored.

Consequent upon the differential time lags and assuming that the diode is ideal, the output voltage builds up until the peak negative voltage is 'clamped' to zero potential, so that the whole of the cycle is effectively raised above the zero line. This type of circuit is frequently used in the input circuits of electronic voltmeters. [See *Industrial Electronics*, by N. M. Morris.] By including a battery in series with the diode, shown in the inset to Fig. 5.15(a), the negative peak of the input signal becomes clamped to the battery voltage. This is equivalent to raising the zero voltage line on the output waveform in Fig. 5.15(d).

If the connections to the diode in Fig. 5.15(a) are reversed, the peak positive voltage is clamped to zero voltage, so that the peak output voltage under steady-state conditions would be $-10\,\text{V}$.

5.8 Diode function generators

In this section, we will discuss function generators which incorporate diodes as voltage sensitive switching elements which either introduce resistance into or disconnect resistance from feedback amplifier circuits.

The basis forms of circuit are shown in Fig. 5.16 in which the variable resistance elements are formed using diode-resistance networks. From the figures

$$V_2 = -\frac{R_f}{R'_1}V_1 \quad \text{and} \quad V_4 = -\frac{R'_f}{R_1}V_3.$$

If identical diode networks are used in Figs. 5.16(a) and (b), then one will provide the inverse function of the other. For example, if the circuit in

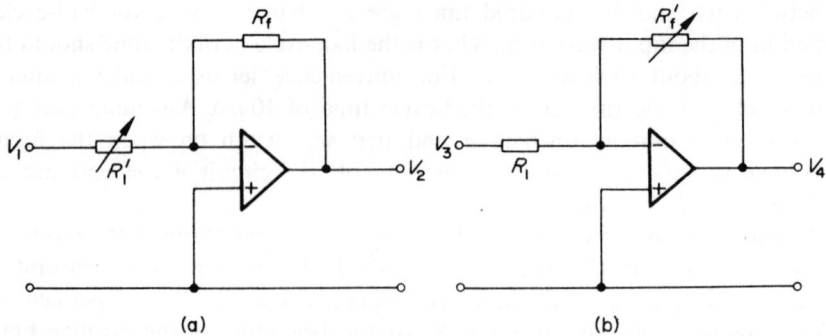

(a)   (b)

Fig. 5.16   Alternative forms of diode function generator.

Fig. 5.16(a) gives an output

$$V_2 = -\text{Constant} \times V_1{}^2$$

then the output from Fig. 5.16(b) using the diode network in the feedback

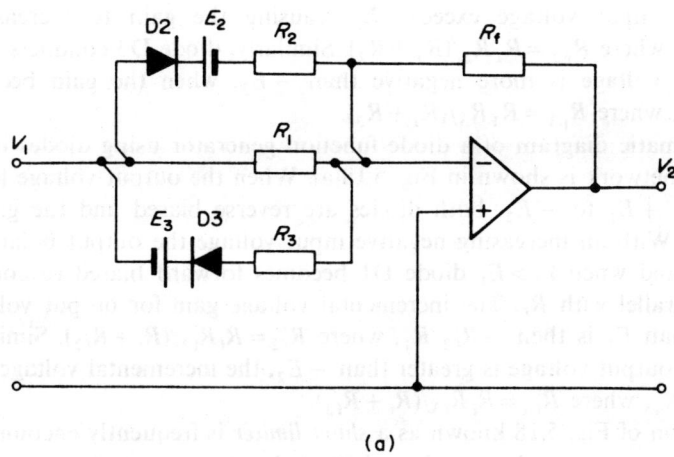

(a)

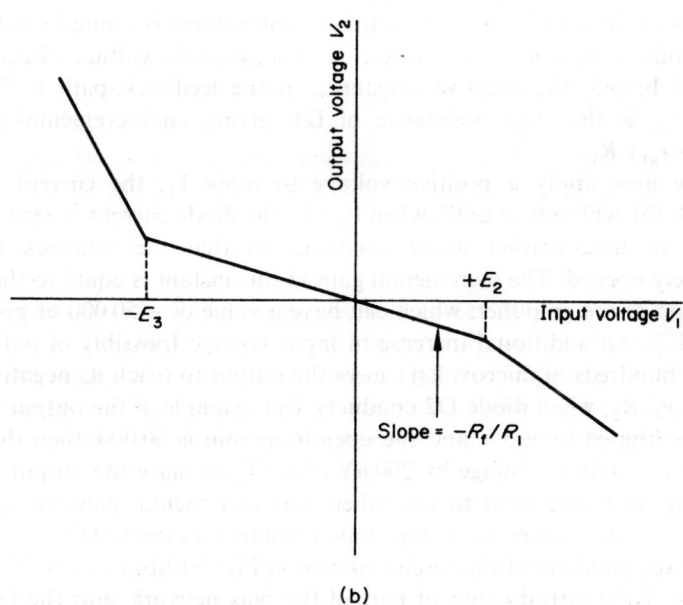

(b)

Fig. 5.17 Series-type diode function generator.

path is

$$V_4 = -\text{Constant} \times \sqrt{V_3}$$

A circuit of the type in Fig. 5.16(a) is shown in Fig. 5.17. In the case of input voltages in the range $+E_2$ to $-E_3$, diodes D2 and D3 are reversed biased so that the overall voltage gain is $-R_f/R_1$. Diode D2 is forward biased when the input voltage exceeds E_2, causing the gain to increase to $-R_f/R_{p2}$, where $R_{p2} = R_1 R_2/(R_1 + R_2)$. Similarly, diode D3 conducts when the input voltage is more negative than $-E_3$, when the gain becomes $-R_f/R_{p3}$, where $R_{p3} = R_1 R_3/(R_1 + R_3)$.

A schematic diagram of a diode function generator using diodes in the feedback network is shown in Fig. 5.18(a). When the output voltage lies in the range $+E_1$ to $-E_2$, both diodes are reverse biased and the gain is $-R_f/R_1$. With an increasing negative input voltage the output polarity is positive, and when $V_2 > E_1$ diode D1 becomes forward biased to connect R_{f2} in parallel with R_f. The incremental voltage gain for output voltages greater than E_1 is then $-R'_{f2}/R_1$, where $R'_{f2} = R_f R_{f2}/(R_f + R_{f2})$. Similarly, when the output voltage is greater than $-E_2$, the incremental voltage gain is $-R'_{f3}/R_1$, where $R'_{f3} = R_f R_{f3}/(R_f + R_{f3})$.

A version of Fig. 5.18 known as a *shunt limiter* is frequently encountered in comparator networks, see Fig. 5.19(a). In this circuit, V_r is a stable reference voltage with a negative polarity. When V_1 is zero, the reference voltage causes the output to be positive so that D1 is forward biased. An analysis of the circuit shows that the output voltage is limited under these conditions to $V_2 = R_4 V_L/R_5$, where V_L is a constant voltage. Since D1 is forward biased, the effective resistance in the feedback path is $R_4 + r_{a1}$, where r_{a1} is the slope resistance of D1, giving an incremental gain of $-(R_4 + r_{a1})/R_1$.

If we now apply a positive voltage to input V_1, the current flowing through D1 will reduce until, when $V_1 = V_r$, the diode current is zero. At this instant of time neither diode conducts, so that the feedback loop is effectively opened. The incremental gain at this instant is equal to the open-loop gain of the amplifier, which can have a value of $-50\,000$ or greater in linear ICs. An additional increase in input voltage (possibly of only a few tens or hundreds of microvolts) causes the output to reach its negative limit of $-R_3 V_L/R_2$, when diode D2 conducts. For example, if the output voltage swing is limited to ± 5 V and the open-loop gain is $50\,000$, then the input voltage has only to change by $200\,\mu$V about V_r to cause the output voltage to swing from one limit to the other. The incremental gain for $V_1 > V_r$ is $-(R_3 + r_{a2})/R_1$, where r_{a2} is the slope resistance of diode D2.

The characteristic of this circuit, shown in Fig. 5.19(b) has a 'soft' limit as a result of the introduction of part of the bias network into the feedback path. A 'hard' characteristic with practically zero gain in the limiting region

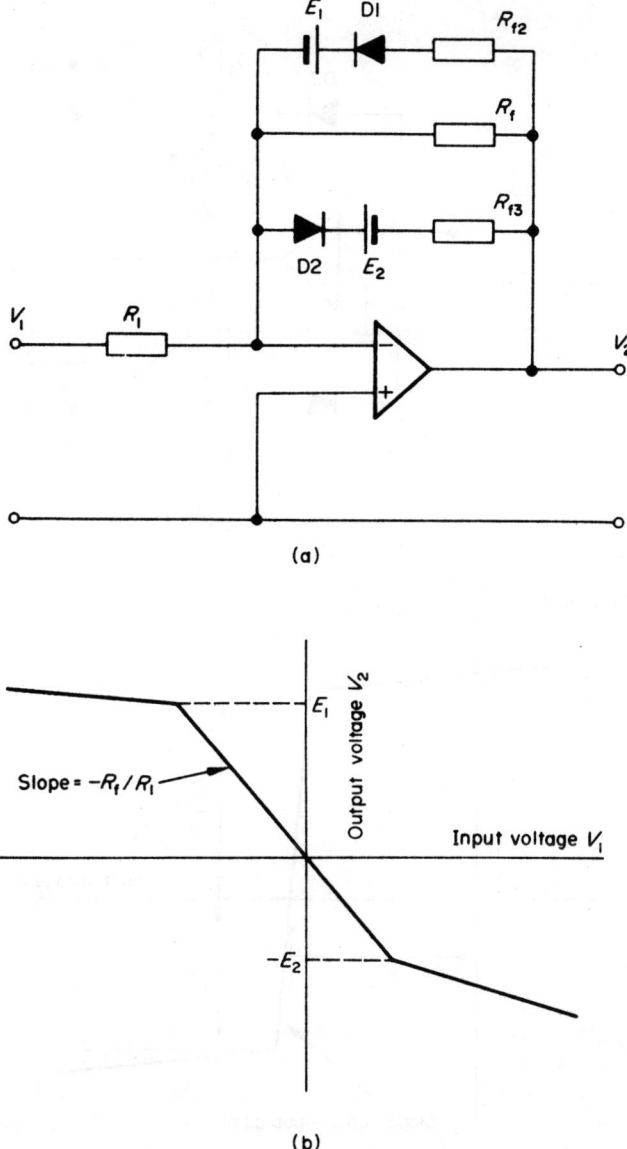

Fig. 5.18   Shunt-type diode function generator.

can be obtained if R_3 and R_4 are replaced by Zener diodes, shown in the inset to Fig. 5.19(a). In this case, the dynamic resistance of the Zener diodes is very low, so that the effective gain in the limiting region is very low.

If V_r has a positive polarity, the transition of the output voltage from one level to the other occurs at a negative input voltage.

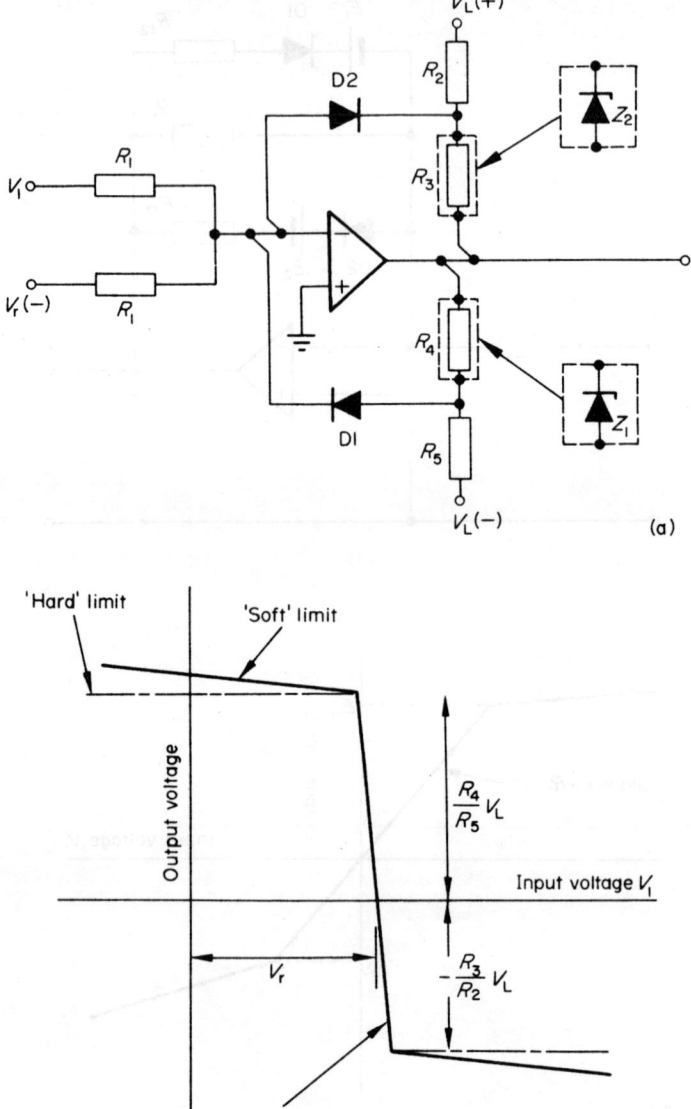

Fig. 5.19   (a) A voltage comparator circuit and (b) its transfer characteristic.

A disadvantage of this type of comparator is that it has little noise immunity, so that even a small noise voltage can cause the output to 'jitter'. This can be overcome by introducing a small amount of positive feedback into the amplifier via R_x and R_y, as shown in Fig. 5.20(a). This form of feedback also increases the voltage gain, so that at the point at which the output changes state the slope of the switching part of the characteristic is steeper. This gives the comparator a 'snap' action. Increasing the amount of positive feedback by adjusting potentiometer R_y increases the width of the 'hysteresis' loop on the transfer characteristic (Fig. 5.20(b)) at the expense of loss of accuracy of the point of comparison. This circuit is an operational amplifier version of the Schmitt trigger circuit described in section 5.14.

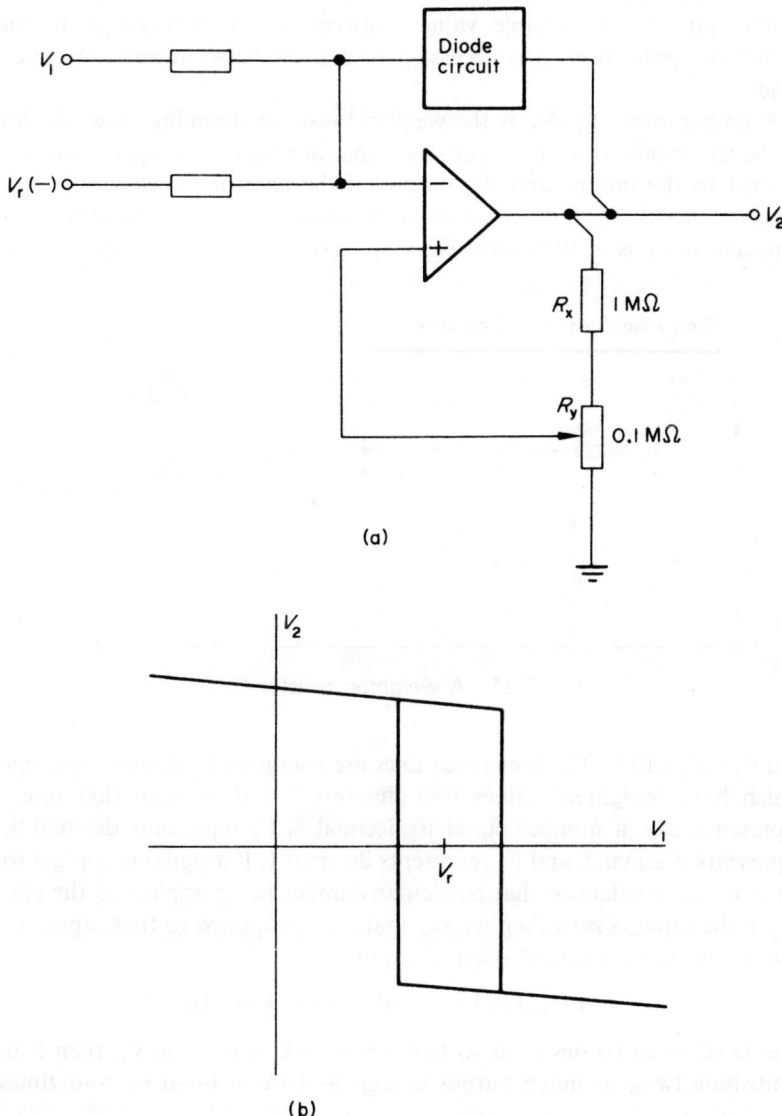

Fig. 5.20 A 'snap' action comparator with good noise immunity.

5.9   Digital-to-analogue convertors (DAC)*

In electronics we can easily identify two distinct methods of data representation. One is in the form of *digital data* in which a quantity is represented by a series of digits as, for example, the value 269. When a change occurs in the data it must do so in discrete steps, i.e., it must increase to 270 or decrease to 268. Information in *analogue* form appears as a steady voltage, and when the data changes it does so smoothly, having an infinite number of possible values between any two readings. In many industrial applications it is necessary to convert between one form and the other.

A simple form of DAC is the weighted resistor summing network shown in the left-hand half of Fig. 5.21. Here the four signals V_1, V_2, V_3, and V_4 are applied to the inputs, and the output of the resistor network is V'_{out}. The inputs either have zero value or a constant value. For example, if this constant value is $+10$ V, then we may have a condition where $V_1 = V_3 = 0$

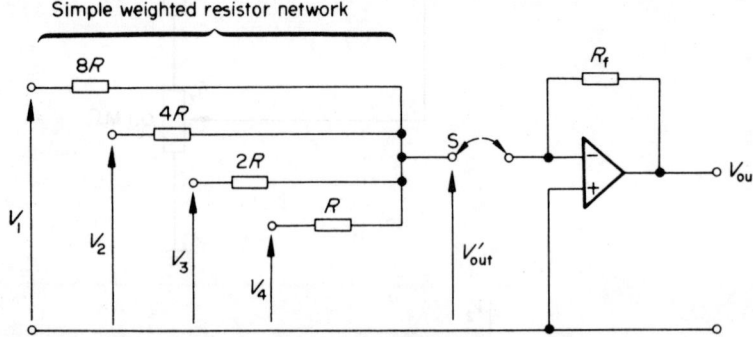

Fig. 5.21   A weighted resistor DAC.

and $V_2 = V_4 = 10$ V. The four input lines are energized by digital logic signals which have 'weighted' values (see chapters 7 and 8), such that input V_4 represents, say, a number equal to decimal 8, V_3 represents decimal 4, V_2 represents decimal 2, and V_1 represents decimal 1. If a signal is applied to an input it is equivalent to that particular number being applied to the circuit, and if the input is zero then we say that zero is applied to that input; in the above case the equivalent decimal input is

$$(0 \times 1) + (1 \times 2) + (0 \times 4) + (1 \times 8) = 10$$

The DAC must be designed so that when 10 V appears at V_4, then it must contribute twice as much output voltage as 10 V at input V_3, four times as

---

*See also chapter 9.

much as 10 V at V_2, and eight times as much as 10 V at V_1. This result is obtained by using resistors 'weighted' in the binary order shown in Fig. 5.21.

The nodal equation for summing junction S is

$$V'_{out}\left(\frac{1}{R}+\frac{1}{2R}+\frac{1}{4R}+\frac{1}{8R}\right)=\frac{V_1}{8R}+\frac{V_2}{4R}+\frac{V_3}{2R}+\frac{V_4}{R}$$

solving for V'_{out} yields

$$V'_{out}=(V_1+2V_2+4V_3+8V_4)/15$$

so that if $V_1=V_3=0$ and $V_2=V_4=10\,V$, then

$$V'_{out}=(20+80)/15=6.67\text{ V}$$

This circuit has some disadvantages which include the fact that the output resistance can be quite high and that the output signal is not a convenient multiple of the digital input value. For example, in the above case the equivalent decimal input is 10 and it would be convenient if the output was either a multiple or a sub-multiple of 10 V.

These disadvantages can be overcome by connecting a summing amplifier in the manner shown in Fig. 5.21. We determine the relative values of R and R_f as follows. Suppose we energize V_4 only, other inputs being connected to the zero line. Since V_4 has an equivalent decimal value of 8, let us compute the gain to give an output of -8 V when $V_4=10$ V.

$$\text{Gain}=-\frac{8}{10}=-\frac{R_f}{R}$$

hence

$$R_f=0.8\,R$$

Thus, if $R=10\,\text{k}\Omega$, then $R_f=8\,\text{k}\Omega$. The values of other input resistors are then 20 kΩ, 40 kΩ, and 80 kΩ.

This type of circuit is designed to deal with what is known as *parallel input* data, that is it accepts all the inputs simultaneously and provides an instantaneous output voltage which is related to the digital input value.

5.10 Analogue-to-digital convertors (ADC)*

Many forms of measuring and transducing equipment provide an output in analogue (continuous) form, and in some instances it is convenient to convert the information into digital form. The most popular types of ADCs include:

(a) Continuous balance convertors

*See also chapter 9.

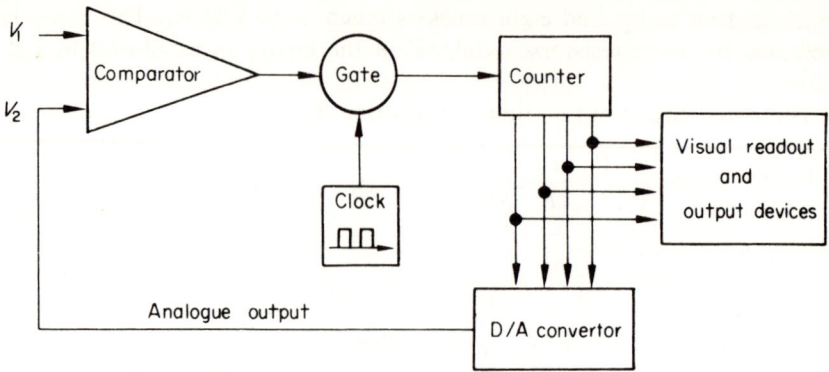

Fig. 5.22   A block diagram of a continuous balance ADC.

(b) Voltage-to-frequency convertors
(c) Dual ramp or integrating convertors.

The basis of one form of *continuous balance convertor* is shown in Fig. 5.22. The counter is initially set to zero so that signal V_2 from the DAC section of the circuit is zero. When the unknown input V_1 is greater than V_2, the comparator provides an output voltage which 'opens' the electronic 'gate' and allows pulses to be applied to the counter. So long as the gate remains open, clock pulses are fed to the counter and V_2 continues to increase. When $V_2 = V_1$, the comparator output falls to zero and closes the electronic gate. This 'freezes' the number stored in the counter, which can then be displayed on a digital readout device. In practical forms of the instrument, the counter is one which can count 'up' or 'down' to allow changes in V_1 to be followed.

The speed with which this type of convertor can convert the input voltage depends on the method adopted to reach final balance. In the case considered, the count begins at zero and increases by one step each time a pulse is applied to the counter. Thus, if 99.9 V is to be reached in 0.1 V steps then it is necessary to wait for 999 pulses to be applied before final balance is reached. A variant of the continuous balance convertor, known as a *successive approximation convertor*, lends itself to higher operating speeds than the simple system described above. In this type of convertor the first pulse from the clock generator sets the counter to, say, one-half of the maximum output, so that V_2 is one-half of its maximum value. If $V_1 > V_2$, the second pulse increases the number in the counter by half its value again, that is to three-quarters the maximum value. If now the new value of V_2 is greater than V_1, then the next pulse reduces the number in the counter to a value half-way between the two previous values, i.e., to 0.625 of its maximum value. In this way final balance is generally reached much more quickly than in simple systems, but only at the expense of an increase in system complexity.

Continuous balance convertors are capable of a measuring accuracy of 0.01 per cent. A disadvantage of this type is that they are sufficiently fast to record electrical noise voltages at their input and consequently may give a false output. The effects of noise voltages can be minimized by filtering the input signal, but only at the expense of reducing the speed of operation.

Voltage-to-frequency convertors incorporate a voltage controlled oscillator whose output frequency is proportional to the input voltage. The output from the oscillator is applied to a counter for a period of time controlled by a clock pulse generator. By a suitable choice of scaling factors the reading of the counter can be calibrated in terms of input voltage. For instance, with a gating period of 10 ms and a voltage-sensitive oscillator which generates 10 kHz per volt, an input of 2 V causes 200 pulses to be applied to the generator, allowing a calibration to three significant figures to be achieved.

Interference signals at mains supply frequency can be minimized in voltage-to-frequency convertors by using a gating period equal to the periodic time of the mains frequency (20 ms at 50 Hz and 16.67 ms at 60 Hz). However, it does not eliminate the effects of transient inputs nor does it account for the effects of small variations in gating period and in interference signal frequency.

The pin layout of a 9400 CT voltage-to-frequency convertor is shown in Fig. 5.23. The IC chip provides an output frequency in the range 10 Hz to 100 kHz, the frequency being linearly proportional to its input current I_{IN}. A schematic diagram illustrating its use is shown in Fig. 5.24. The input voltage V_{IN} (minimum value 10 mV, maximum value 10 V) is applied to terminal I_{IN} via a 1 MΩ resistor, corresponding to a full-scale input current

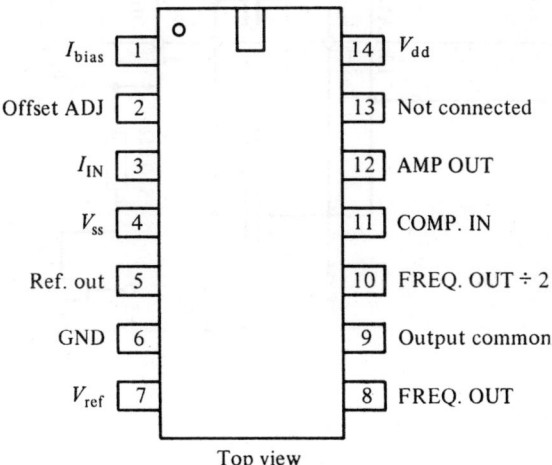

I_{bias}	1	14	V_{dd}
Offset ADJ	2	13	Not connected
I_{IN}	3	12	AMP OUT
V_{ss}	4	11	COMP. IN
Ref. out	5	10	FREQ. OUT ÷ 2
GND	6	9	Output common
V_{ref}	7	8	FREQ. OUT

Top view

Fig. 5.23   Pin connections of a typical voltage-to-frequency convertor.

(Courtesy of RS Components Ltd.)

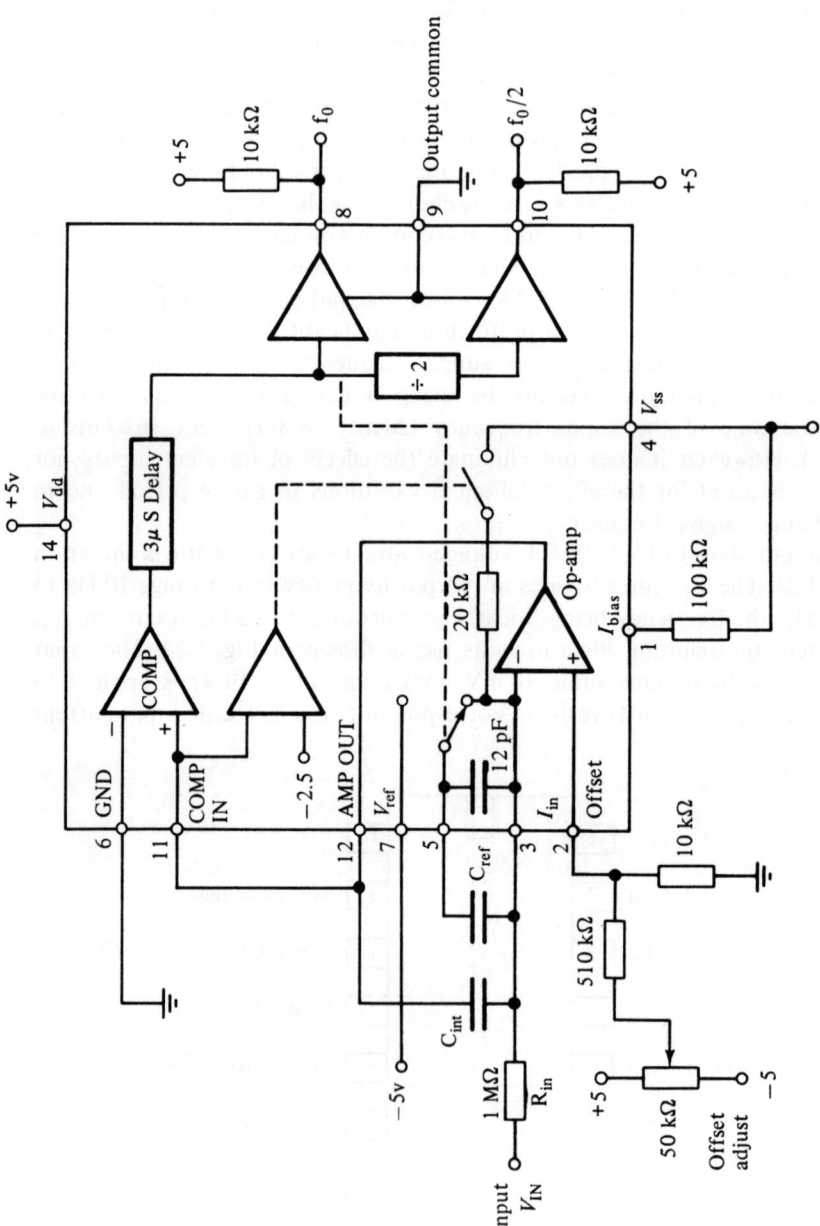

Fig. 5.24 Schematic diagram of a voltage-to-frequency convertor scheme. (Courtesy of RS Components Ltd.)

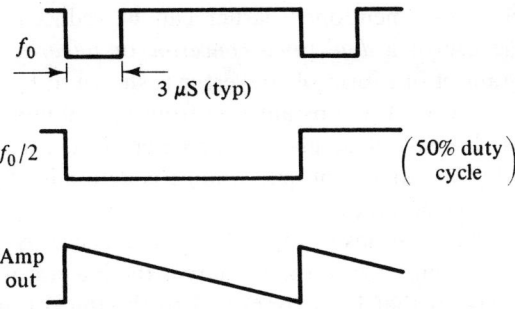

Fig. 5.25 Circuit waveforms from fig. 5.24.

(Courtesy of RS Components Ltd.)

of $10 \, \mu A$. The operational amplifier inside the chip is connected as an integrator, the chip having an internal switching circuit which causes the input voltage V_{IN} to be converted to a sawtooth waveform at the AMP OUT terminal (pin 12).

This signal is applied to the comparator input terminal, COMP IN (pin 11), which produces a rectangular output of frequency f_0 at the FREQ OUT terminal (pin 8) of the form shown in Fig. 5.25. This waveform is applied to a divide-by-two circuit to produce a 'square' waveform having a 50 per cent duty cycle at terminal FREQ. OUT $\div 2$ (pin 10).

The convertor described above can also be reconnected to function as a frequency-to-voltage convertor.

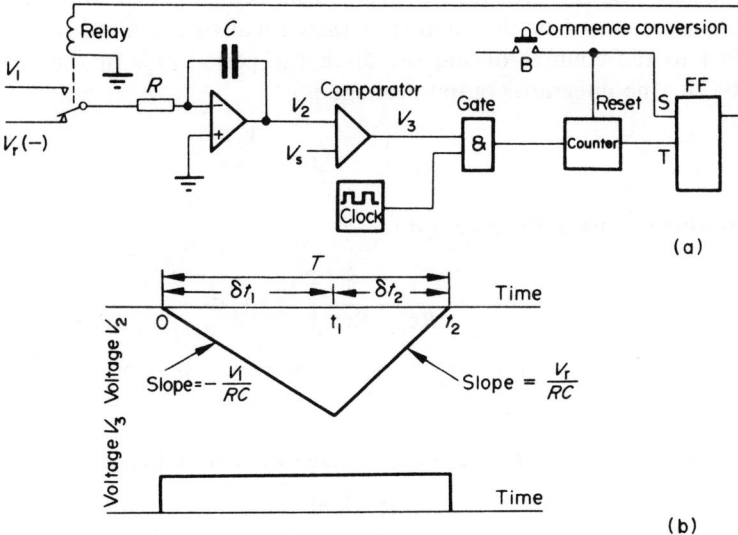

Fig. 5.26 One form of dual ramp ADC and (b) waveforms in the circuit.

Many of the errors mentioned earlier can be reduced to insignificant amounts by the use of a *dual ramp convertor* or *integrating convertor*. A schematic diagram of one form of convertor is shown in Fig. 5.26, in which capacitor C is charged at a constant rate from an unknown input voltage V_1, and is later discharged at another predetermined rate by a reference voltage source V_r. The digital output voltage is determined by the ratio of the discharge to charge times.

The sequence of operations in Fig. 5.26 is commenced by pressing button B which resets the counter to zero, discharges the integrator capacitor, and energizes the relay so that V_1 is connected to the integrator input. In this case, we assume that the input voltage has a positive polarity, so that the integrator output has a negative polarity. The comparator compares V_2 with V_s, which is a very small positive voltage in this case, and when $|V_2| > |V_s|$ then the comparator provides an output signal which opens the electronic gate to allow clock pulses to be applied to the counter. The time for which the capacitor charges is determined by the time taken for the counter to reach a predetermined count, often equal to the maximum storage capacity of the counter. For example, this would be 999 in a three digit counter. The next pulse causes the number stored in the counter to fall to zero and a '1' to be transmitted to the next higher decade stage of the counter, which is represented by flip-flop FF. The 'carry' pulse causes the output of FF to change state so that the relay is de-energized, causing the integrator to be connected to input $-V_r$.

This discharges the capacitor and causes the integrator output to change in the manner shown in Fig. 5.26(b). When $|V_2| < |V_s|$ the comparator output falls to zero and prevents further pulses from being applied to the counter. The number stored in the counter is then equal to the number of pulses applied to the counter during the discharge period δt_2. In the charging period δt_1 the integrator output voltage is

$$V_2 = -\frac{1}{RC}\int_0^{t_1} V_1 \, dt = -\frac{V_1 t_1}{RC}$$

and during the discharge period it is

$$V_2 = -\frac{V_1 t_1}{RC} - \frac{1}{RC}\int_{t_1}^{t_2}(-V_r)dt$$

$$= -\frac{V_1 t_1}{RC} + \frac{V_r(t_2 - t_1)}{RC} = \frac{-V_1 \delta t_1 + V_r \delta t_2}{RC}$$

After a length of time $T = \delta t_1 + \delta t_2$, voltage V_2 is zero, hence

$$V_1 \delta t_1 = V_r \delta t_2$$

or
$$V_1 = V_r \delta t_2 / \delta t$$

If the clock pulse generator produces n pulses/sec, then in time δt_1 the number of pulses applied to the counter is $N_1 = n\delta t_1$, and in time δt_2 the number of pulses is $N_2 = n\delta t_2$.

Hence
$$V_1 = V_r \frac{N_2/n}{N_1/n} = V_r \frac{N_2}{N_1}$$

Since N_1 is a constant equal to the storage capacity of the counter and V_r is a constant voltage, then

$$V_1 = kN_2$$

By suitable scaling, N_2 can represent V_1.

The errors from various causes in other types of convertor can be reduced to a minimum in the dual ramp convertor since the clock pulse controls both the charge and discharge periods. As a result, the effects of variations in clock frequency are practically eliminated. Also, since the circuit includes an integrator, the net effects of transient inputs are very slight.

5.11   Semiconductor diode switches

The maximum operating speeds of diodes and other semiconductor devices are restricted by the rate at which the diodes can be switched on and off.

In a reverse biased junction diode there are few free charge carriers in the region of the junction, and the flow of these constitutes the reverse leakage (saturation) current shown in Fig. 5.27. This current can be as low as a fraction of a nanoampere. The application of a forward voltage causes forward flow of current, and in this operating state a relatively large concentration of minority carriers occurs near to the junction. Further away from the junction the concentration is reduced by recombination effects.

When a reverse voltage is applied subsequent to forward conduction, the minority charge carriers in the region of the junction are attracted back across the junction by the reverse potential gradient, and a transient reverse current flow occurs. This is known as the *stored charge effect* since the reverse current is due to the removal of the charge held by the excess minority charge carriers in the regions near the junctions. The diode is said to be *turned off* when the stored charges have been swept out of the region of the junction.

The transient peak reverse current is restricted to $-E/R$, where R is the circuit resistance. The time required to remove the stored charge is known as the *storage time* t_s, which may have a value of a fraction of a nanosecond in high-speed switching diodes up to about a millisecond in a high current diode. When the stored charge is removed, the reverse current then decreases to its leakage value.

Schottky barrier diodes (see section 1.23) do not exhibit the charge storage effect, and their switching speed is very fast.

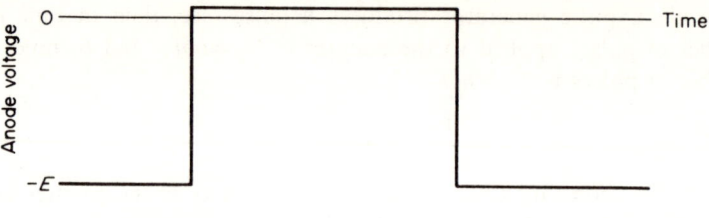

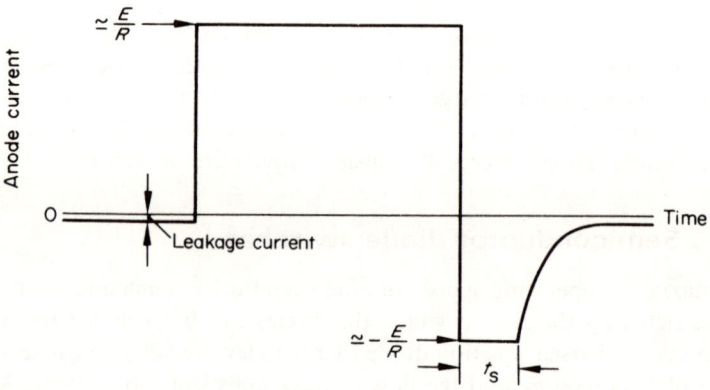

Fig. 5.27   Waveforms in diode circuits.

5.12   BJT switches

Transistors are not ideal switching devices, and in this section of the book we discuss some of their limitations.

When BJTs are used as switches they can be operated in different modes. The most popular is known as *saturated mode operation* in which the two stable operating states are shown at A and B in Fig. 5.28. At point A the base current is reduced to zero (or it may be reversed in some cases) so that the collector voltage approaches V_{CC}. At the other extreme of the load line, the base current is sufficiently great for the transistor to be saturated (i.e., both collector and emitter junctions are forward biased), when the operating point moves to point B on the characteristic. At point A, the transistor is equivalent to a switch which is *open* or OFF, since it passes only a small collector current (typically a few nanoamperes in silicon devices) and supports the supply voltage across it. At point B on the characteristic, the BJT is said to be ON since it passes a large current and supports only a small voltage at its collector. This voltage is the collector-emitter saturation voltage $V_{CE(sat)}$, and may have a value between about 0.15 V and 0.6 V. Clearly the BJT is not an ideal switch because of the limitations of leakage

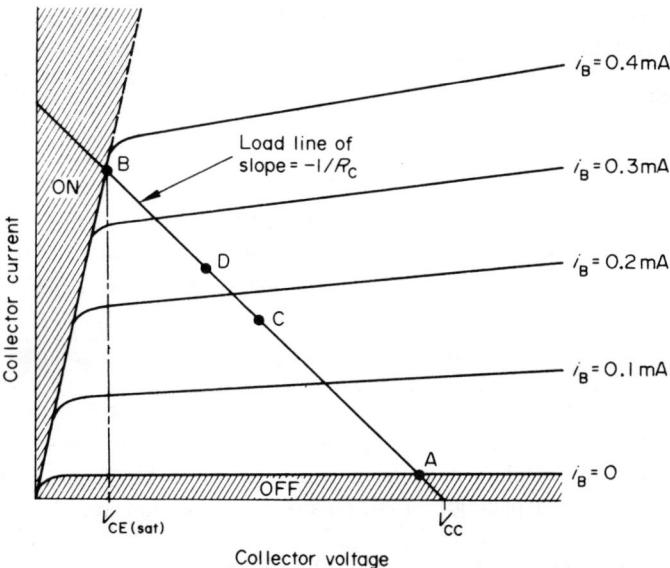

Fig. 5.28 Operating states of a BJT switch.

current and $V_{CE(sat)}$, but it is possible to design switching circuits which account for departures from the ideal.

The calculation of component values for switching circuits can be complex if factors such as switching speed, drive capability, and component and supply tolerances are taken into account. In this chapter we shall consider basic design principles. Assuming that the base-emitter saturation voltage $V_{BE(sat)}$ and $V_{CE(sat)}$ are small compared with V_{CC}, and that the input voltage V_1 in Fig. 5.29 is equal to V_{CC} when the transistor is driven into saturation, then

$$I_B \simeq V_{CC}/R_B$$

and
$$I_C \simeq V_{CC}/R_C$$

Also
$$I_C = h_{FE(sat)}I_B$$

where $h_{FE(sat)}$ is the current gain of the transistor in the saturated mode. Solving between the above equations for R_B yields

$$R_B = h_{FE(sat)}R_C$$

To allow for possible variations in the factors listed in the paragraph above, the value of R_B selected should be about one-half to one-quarter of the theoretical value. This ensures that the transistor remains in saturation over a range of input voltages when component and supply tolerances are at their limiting values.

Example 5.1: Estimate suitable component values for a switching circuit similar to that in Fig. 5.29 given that $V_{CC} = 5$ V, the maximum collector current is 5 mA and $h_{FE(sat)}$ is 40.

Solution:

$$R_C \simeq V_{CC}/I_{C(max)} = 5 \text{ volts}/5 \text{ mA} = 1 \text{ k}\Omega$$
$$R_B = 40 \times 1 = 40 \text{ k}\Omega$$

This value of R_B allows the transistor to operate at the edge of saturation if the component parameters and supply voltage are as stated. To allow for variations from these values, a value between 10 kΩ and 18 kΩ should be selected for R_B.

5.12.1   Dynamic operation of the BJT switch

The dynamic operation of the BJT switch is illustrated in Fig. 5.29. Ideally, semiconductor switches have zero transition time between the ON and OFF states but, in practice, small delays do occur and are defined in Fig. 5.29.

After the input voltage is first applied there is an initial *delay time* t_d, which is the time taken for the collector current to reach 10 per cent of its final value. This delay is due to the time taken for charge carriers to spread across the base region and is dependent on the collector and emitter junction depletion capacitances. The *rise time* t_r is the time taken for the collector current to rise from 10 to 90 per cent of its final value (that is, the time taken for the collector voltage to fall from 90 to 10 per cent of its initial value with a resistive load), and is dependent both on the magnitude of the base current and on the high frequency parameters of the transistor. When the input voltage is reduced to zero again, the collector current continues for some time at a more or less constant value due to the minority charge carrier storage delay, which is generally similar to the effect described in the case of diodes in section 5.11. This is accounted for by the *storage time delay* t_s. The *fall time* or *turn-off transition* t_f is the time taken for the collector current to fall from 90 to 10 per cent of its maximum value. The *turn-on time* t_{on} and *turn-off time* t_{off} of the circuit are defined in Fig. 5.29, typical values for a 2N3832 n-p-n silicon transistor being $t_{on} = 4.5$ ns, $t_{off} = 8$ ns.

In specifying the switching properties of logic circuits, the delays defined above can be simplified by stating the *propagation delay* t_{pd} of the gate. This is the *mean* time taken for a signal to propagate through a gate, and a convenient method of measuring t_{pd} is illustrated in Fig. 5.30 in which two identical circuits are cascaded. The propagation delay is measured at the 50 per cent voltage levels, and since the waveforms contain two t_{on} times and two t_{off} times, then

$$t_{pd} = (t_A + t_B)/4$$

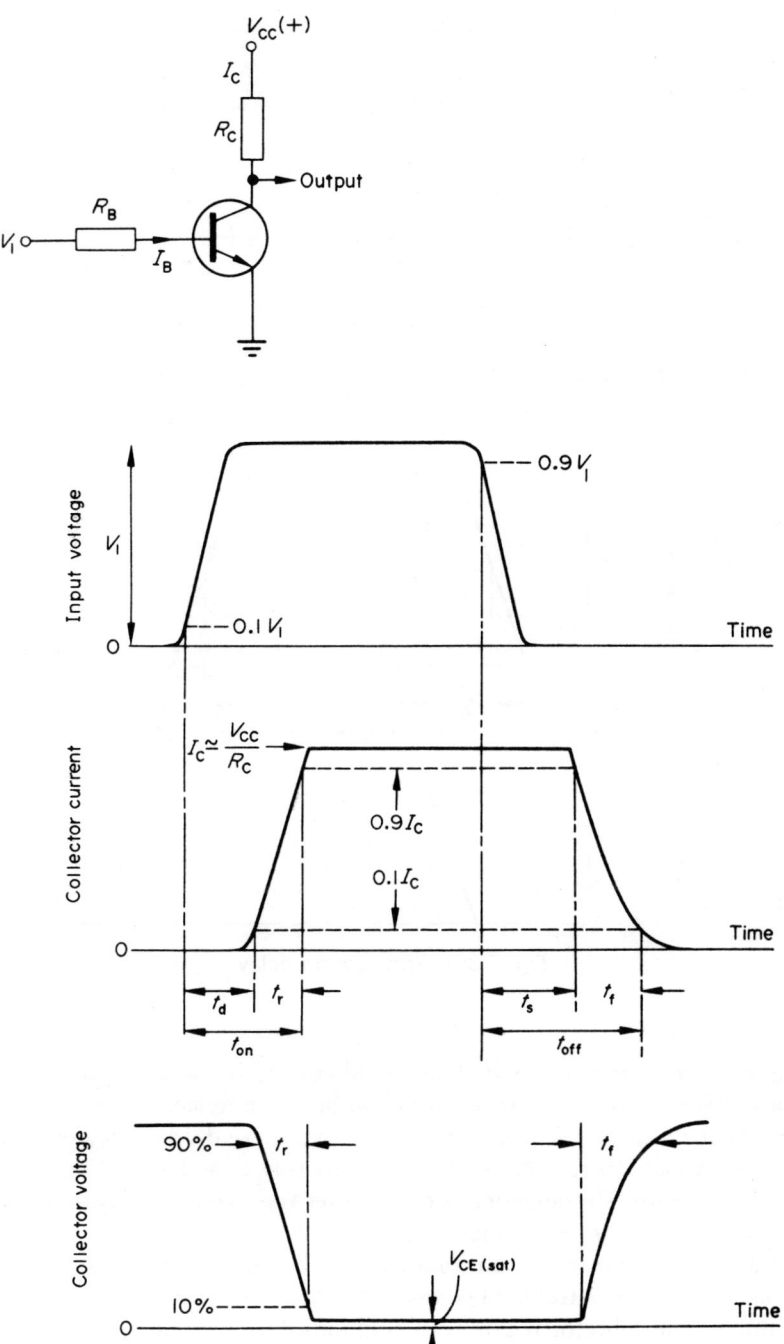

Fig. 5.29 Definition of switching time.

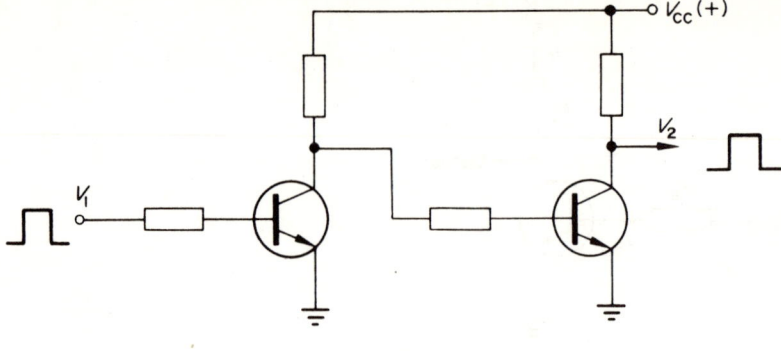

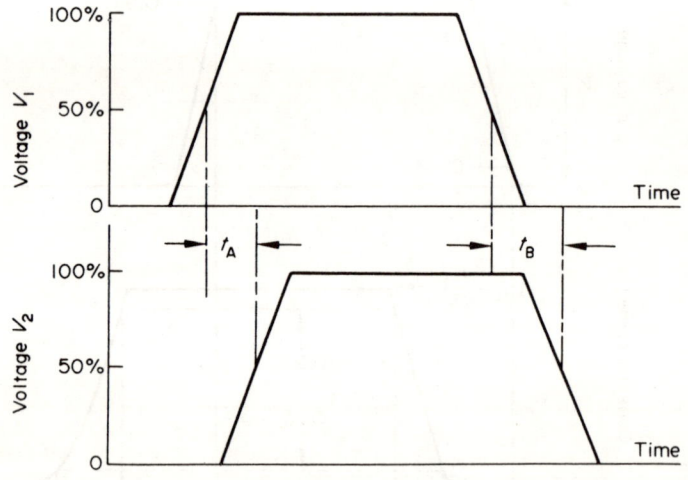

Fig. 5.30   Propagation delay.

Some improvement in switching speed can be brought about by using *non-saturating circuits*. Modifications can be made to saturating circuits to prevent them from going into saturation or, alternatively, circuits can be designed which operate between two points (say C and D in Fig. 5.28) in the active region. Certain circuits of the latter type have propagation delays as low as a fraction of a nanosecond.

The turn-off time of a given transistor can be reduced from microseconds to nanoseconds by a technique known as *gold doping*. This is brought about by doping the collector region with gold ions during manufacture (see also section 1.11), and has the effect of reducing the lifetime of the charge carriers.

5.13 FET switches

In principle, FETs can carry out the same switching operations as BJTs, the drain current waveform for a square wave input being generally similar to the collector current waveform of the BJT in Fig. 5.29.

Since FETs are majority carrier devices, they do not exhibit the display carrier storage effects observed in BJTs but they do display effects due to interelectrode capacitances. As a result, FETs display a *turn-on delay* $t_{d(on)}$ which is equivalent to t_d in Fig. 5.29, and a *turn-off delay* $t_{d(off)}$ which is equivalent to t_s in BJTs. Typical values of turn-on time t_{on} and turn-off time t_{off} (defined in terms of Fig. 5.29) for a 2N4029 n-channel JUGFET are 6.5 ns and 18 ns, respectively.

5.14 Schmitt trigger circuit

The Schmitt trigger circuit, which is a voltage level discriminator, is a version of the emitter coupled amplifier described in chapter 2. A popular BJT version is illustrated in Fig. 5.31. In this circuit, the voltage level at the collector of Q_2 is low (not zero) when the input voltage is low. When the input voltage exceeds a predetermined value, the output voltage abruptly changes to a high level even though the input voltage may only change slowly. The circuit characteristic contains a 'hysteresis' effect so that the turn-OFF voltage level is below that of the turn-ON level. This feature gives the trigger circuit a certain amount of noise immunity.

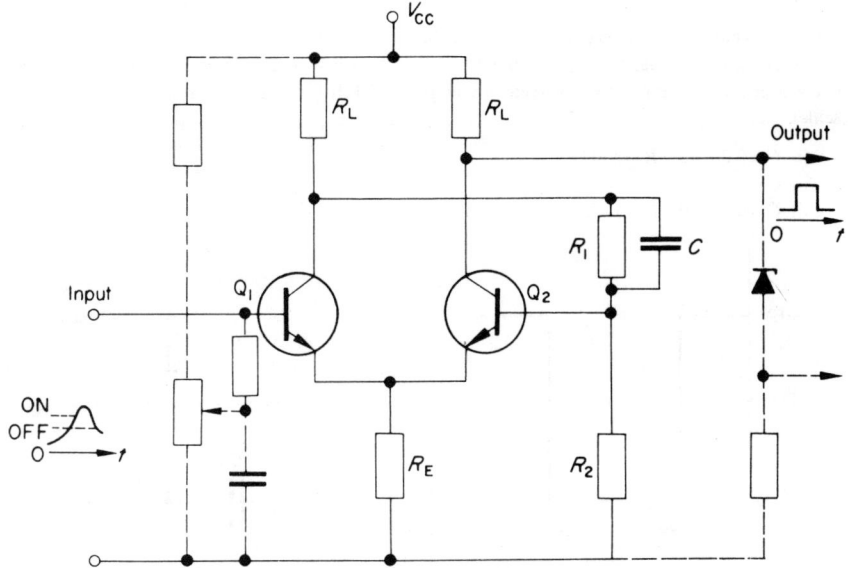

Fig. 5.31 A BJT Schmitt trigger circuit.

The amplifier has positive feedback applied between the collector of Q_1 and the base of Q_2 via R_1 and R_2. In operation only one of the transistors conducts at any time, so that the emitter potential is always positive with respect to the zero line. When the input voltage is lower than the emitter voltage, Q_1 is cut off and, by regenerative feedback, Q_2 is saturated. As the input voltage rises above the emitter voltage, Q_1 begins to turn ON and this has two immediate effects. Firstly, the emitter potential begins to rise by emitter follower action and, secondly, the collector potential of Q_1 begins to fall. The cumulative effect causes Q_2 to be rapidly cut off so that its collector potential rises to its highest value. The change in the collector potential of Q_1 is not nearly so rapid, so that this signal is not normally used as an output. The switching speed of the circuit can be increased by the use of speed-up capacitor C, which has a small value, typically 20–200 pF.

When the input voltage falls again, Q_1 is turned off and, by regenerative feedback, Q_2 is rapidly turned ON.

A feature of the Schmitt trigger circuit is that both output levels are non-zero. A simple method of reducing the output voltage level to zero when the input is zero is to take the output from the anode of the Zener diode shown. Alternatively, voltage levels compatible with logic circuits can be obtained by adding an additional switching circuit at the output of the Schmitt trigger.

Problems

5.1 (a) What is (i) a differentiating circuit, (ii) an integrating circuit?

(b) The pulse in Fig. 5.32(a) is applied to input A of Fig. 5.32(b). For each of the following, draw a graph of the resulting waveforms at point B (all three graphs should be to the same scale).

   (i) $C = 0.001 \ \mu F$, $R = 250 \ \Omega$
   (ii) $C = 0.001 \ \mu F$, $R = 500 \ \Omega$
   (iii) $C = 0.001 \ \mu F$, $R = 1 \ k\Omega$.

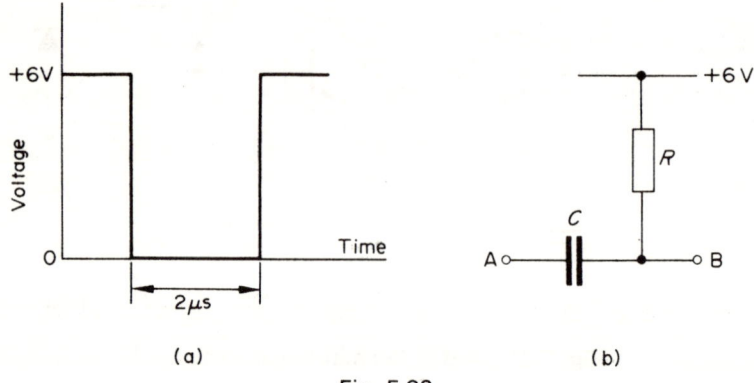

(a)   (b)

Fig. 5.32.

5.2 A pulse of magnitude -5 V and duration 1 μs is fed through a 0.001 μF capacitor to a load resistor R. Draw a graph of the resultant voltage waveform across R for the following values of R; (i) 0.5 kΩ, (ii) 1 kΩ, (iii) 1 MΩ. Show all calculations.

5.3 A signal source of 1 kΩ output resistance is connected to a 1 kΩ load through a 1 μF capacitor. The signal generator applies a single pulse of 2 V amplitude and 1 ms duration to the circuit. Sketch the waveform of the p.d. across the load, indicating the magnitude of the voltage at the beginning and end of the pulse.

5.4 A square wave pulse train of 20 V peak-to-peak amplitude and frequency 10 kHz is applied to a low-pass R–C network. If $R = 0.1$ MΩ, $C = 0.1$ μF, determine the peak-to-peak output voltage change. Sketch the input and output voltage waveforms.

5.5 Explain why 'frequency compensated' attenuators are required at the input of high quality electronic instruments. In the frequency compensated attenuator in Fig. 5.33, $C_2 = 45$ pF, $R_1 = 900$ kΩ, and $R_2 = 100$ kΩ. Determine the value of C_1 required to give correct frequency compensation.

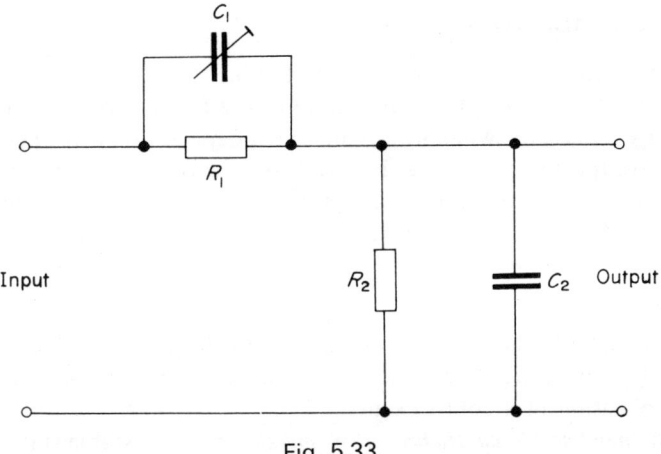

Fig. 5.33.

5.6 In an attenuator with the values given in problem 5.5, it is found that the transient overshoot on the output voltage is 5 per cent when a rectangular wave is applied to the input terminals. Calculate the value of C_1 employed in the attenuator.

5.7 Draw a block diagram of a digital-to-analogue convertor and (i) explain its operation, (ii) state how the conversion accuracy is determined.

5.8 (a) With the aid of diagrams, explain the principle of ONE method of digital-to-analogue conversion.

(b) Briefly suggest how a digital-to-analogue convertor could be used to produce an analogue-to-digital convertor.

5.9 Explain fully, with the aid of suitable diagrams, the basic principle of operation of *one* form of digital voltmeter.

Give *three* reasons why the use of digital instruments may be preferred to analogue instruments in measurement systems.

6. Multivibrators, non-sinusoidal oscillators, and sweep generators

6.1 Multivibrators*

The name multivibrator is derived from the fact that the output waveform contains multiple vibrations (harmonics) of the fundamental frequency.

They are a class of switching circuit which depend for their operation on regenerative (positive) feedback. In operation, the transistors are driven into saturation for a part of the operating time and are cut off for the remainder of the time. Oscillators employing this mode of operation are also known as *relaxation oscillators*. There are three principal types of multivibrator as follows:

(a) The *bistable multivibrator* or *set-reset (S–R) flip-flop*† which has two stable operating states, and is triggered from one to the other by the application of a control signal.

(b) The *monostable multivibrator* or *one-shot* has one stable state and one quasi-stable state. The circuit is switched by an external signal into its quasi-stable state, in which it remains for a period of time determined by the circuit time constants, after which it returns to its stable state.

(c) The *astable multivibrator* or *free-running multivibrator* which has two quasi-stable states, one following the other in succession.

6.2 The bistable multivibrator or *S–R* flip-flop

A popular form of circuit is shown in Fig. 6.1, and comprises two cross-connected BJT switches. To improve the switching performance, each half of the circuit incorporates a speed-up capacitor C and a reverse bias supply V_{BB}.

*See also Sections 6.11 to 6.14 inclusive.

† This name is derived from the fact that the output flips from one stable state to the other upon demand.

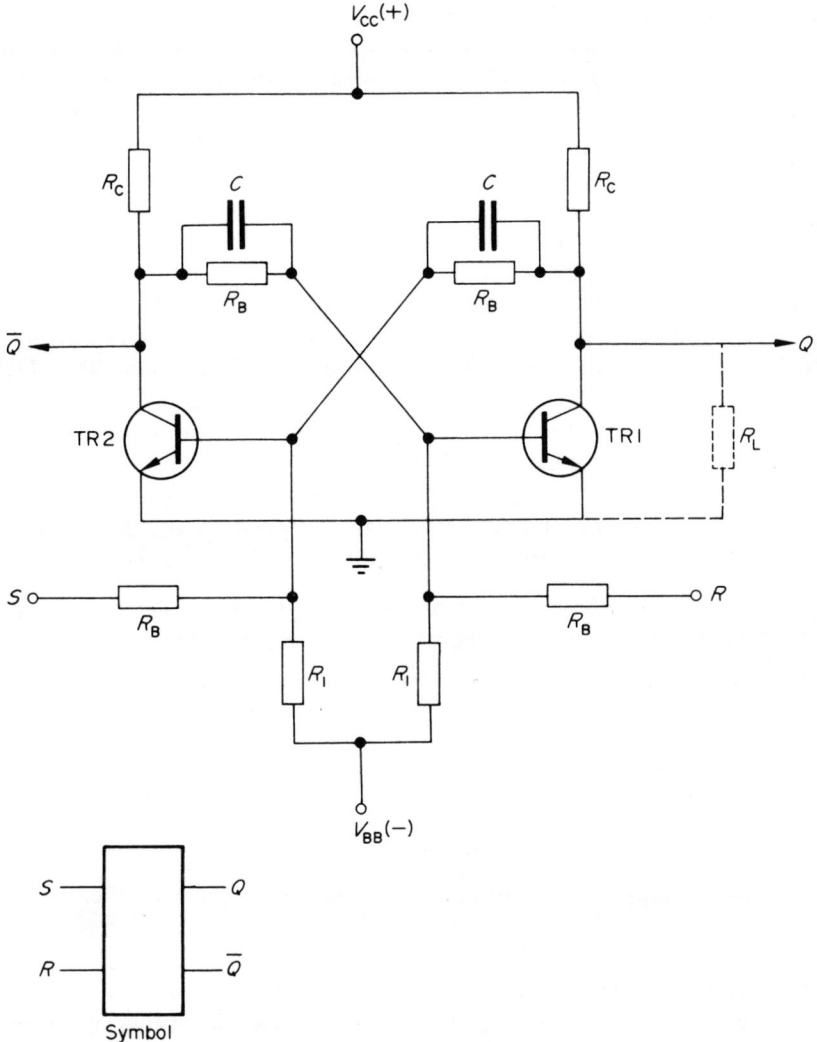

Fig. 6.1   An S–R flip-flop.

At the instant of initial switch-on, both transistors begin to carry current and, due to differences in circuit parameters, one transistor will pass more current than the other. Let us assume that TR1 begins to turn on more rapidly than TR2, so that its collector potential is lower than that of TR2. After a time equal to the turn-on time of TR1 the collector potential of TR2 is at its highest value, and it provides sufficient base drive to TR1 to ensure that it is driven well into saturation. This maintains the collector voltage of TR1 at practically zero. Consequent upon the feedback between TR1 and TR2, the base of TR2 is held at a negative potential by the bias supply so that TR2 remains cut off. This, then, is one of the stable operating states with TR1 saturated and TR2 cut off. Output Q is practically zero (equal to $V_{CE(sat)}$) and output $\bar{Q}$ is high* (usually between about $0.9\,V_{CC}$ and $0.5\,V_{CC}$, depending on the loading).

The function of the S input line, or *set* line is explained in the following. The application of a positive potential of adequate amplitude to the S-line (with zero voltage applied to the R-line) causes TR2 to be driven into saturation, when its collector voltage falls to zero. This, in turn, reduces the base drive to TR1 to zero, cutting it off. Now that TR1 is cut off, its collector voltage rises and acts as a source of current to the base of TR2. When the signal applied to the S-line is reduced to zero, the regenerative feedback from the collector of TR1 to the base of TR2 ensures that TR2 remains saturated. This is the second stable state in which TR1 is cut off and TR2 is saturated. In this way, either a steady voltage or an impulsive signal of a few microseconds duration applied to the S-line can *set* output $\bar{Q}$ to the high voltage level (and also sets $\bar{Q}$ to the low voltage level).

A similar reasoning shows that a signal applied to the R-line *resets* output Q to zero (or, alternatively, it sets $\bar{Q}$ to a high voltage level).

Design procedure: Since the circuit comprises two cross-connected satu-rated BJT switches, we can use the results of section 5.12. Here it was shown that $R_B < h_{FE(sat)}R_C$, and an empiric equation is

$$R_B = h_{FE(sat)}R_C/4$$

The value of R_C can be selected on the basis of the maximum collector current I_{CM} as follows

$$R_C = V_{CC}/I_{CM}$$

For example, suppose $V_{CC} = 10$ V, $I_{CM} = 5$ mA, and $h_{FE(sat)} = 50$, then

$$R_C = 10\ \text{V}/5\ \text{mA} = 2\ \text{k}\Omega$$

Since the *maximum* collector current is to be 5 mA, we select a collector resistance of preferred value 2.2 kΩ to ensure that this value of current cannot be exceeded. Now,

$$R_B = 50 \times 2.2/4 = 27.5\ \text{k}\Omega$$

and a preferred value of 27 kΩ is selected. The value of R_1 is computed on the basis of the reverse bias required to turn off the transistor. As a general rule of thumb, R_1 has a value in the range 10 to 20 R_B.

The voltage at the collector of the saturated transistor is equal to $V_{CE(sat)}$ (i.e., between about 0.15 and 0.6 V), and the potential at the collector of the

---

*Q is the 'normal' output and $\bar{Q}$ is the 'logical complement' of Q (see also Chapter 7); that is, when the potential of Q is high then the potential of $\bar{Q}$ is low and vice versa.

cut-off transistor is $R_L V_{CC}/(R_C + R_L)$, where R_L is the connected load. If each transistor has an external load of 4.7 kΩ, then the maximum output voltage for $V_{CC} = 10$ V is

$$4.7 \times 10/(2.2 + 4.7) = 6.8 \text{ V}$$

6.3 Fan-in, fan-out, and noise immunity

Important parameters of switching circuits are the fan-in, fan-out, and noise immunity figures, which are related to the allowable range of 'low' and 'high' voltage levels. The meaning of both 'low' and 'high' voltages in connection with logic circuits is discussed in detail in chapter 7, and it will suffice for our purposes here to define a 'low' output as one which is within about one-half of a volt of zero potential, and a 'high' output as one above about $0.5 V_{CC}$. With a 6 V collector supply this range may be, for example, 3 to 5.7 V.

The *fan-out* of a switching circuit is the maximum number of similar circuits which can be driven by the circuit without jeopardizing the maximum 'low' or the minimum 'high' output levels.

We can calculate the fan-out of the *S–R* flip-flop designed in section 6.2 as follows. Suppose that, with $V_{CC} = 10$ V, the minimum 'high' output voltage that is acceptable is to be 5 V, then from the previous section

$$5 = R_L V_{CC}/(R_C + R_L)$$

or $$R_L = 2.2 \text{ k}\Omega$$

Since this circuit is driving M other circuits, where M is the fan-out, each having an input resistance approximately equal to R_B, then

$$2.2 = R_B/M$$

or $$M = 27/2.2 = 12.3$$

The value of M is the integral part of the calculated value, giving an effective fan-out of 12.

The *fan-in* of a switching circuit is the maximum number of inputs that may be connected before the 'high' or 'low' output voltages are jeopardized.

One definition of *noise immunity* is the maximum electrical noise signal which, when superimposed on the input signal, causes the output voltage level to change from 'high' to 'low' or vice versa.

6.4 The monostable multivibrator or one-shot

A popular monostable multivibrator circuit is shown in Fig. 6.2(a). In its stable operating mode, TR2 is saturated by base drive through R, so that its collector potential is low. The negative bias applied to TR1 causes it to be

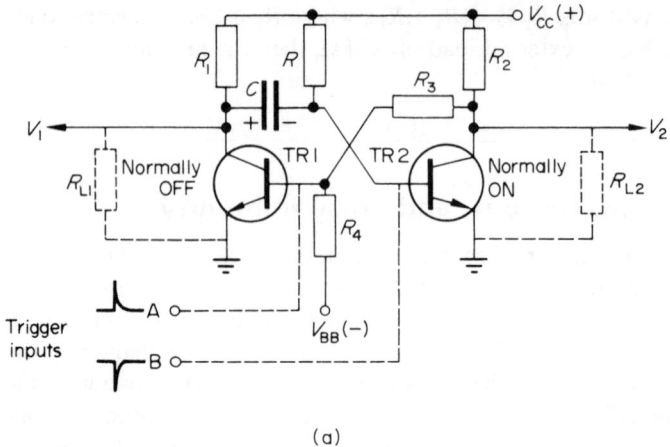

(a)

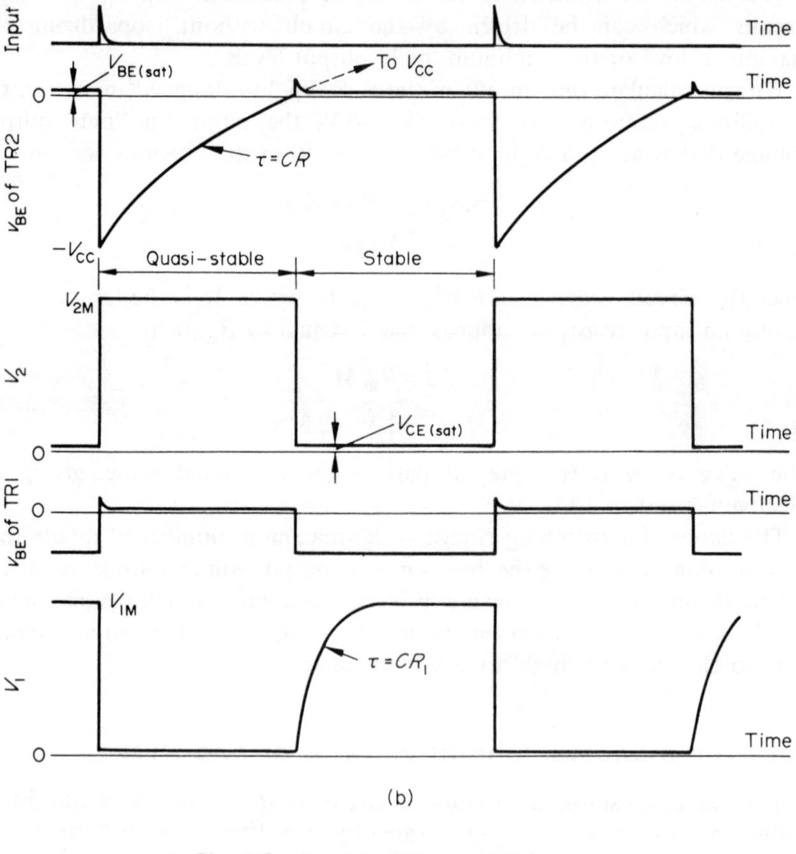

(b)

Fig. 6.2   A monostable multivibrator.

cut off so that its maximum collector voltage is

$$V_{1M} = R_{L1} V_{CC}/(R_{L1} + R_1)$$

If the load is small, i.e., $R_{L1} \gg R_1$, then $V_{1M} \rightarrow V_{CC}$.

The circuit is triggered into its quasi-stable state either by turning TR1 ON or by turning TR2 OFF. This can be accomplished in a number of ways, two of which are illustrated in the figure. Transistor TR1 can be turned ON by applying a positive pulse to its base via input A, and TR2 can be cut off by applying a negative pulse to input B. Triggering by injecting pulses into the emitter or collector lines can be employed. In the following we assume that a positive pulse at input A is used to trigger the circuit. (**Note:** The trigger pulse duration should be short compared with the multivibrator pulse duration, but large compared with the turn-on time of the transistors.)

Prior to the application of the trigger pulse, output V_1 approaches the value of V_{CC} and the base voltage of TR2 is practically zero, so that capacitor C charges to approximately V_{CC} with the polarity shown. The trigger pulse drives TR1 into saturation, connecting the left-hand plate of C to earth. This, in turn, causes the base of TR2 to be driven to a voltage equal to the voltage across C, i.e., approximately $-V_{CC}$ instantly cutting off TR2. The consequent sudden rise in V_2 is fed back to the base of TR1, thereby maintaining TR1 in saturation so long as TR2 is cut off. The circuit is now in its quasi-stable state with TR2 cut off and TR1 saturated.

During this period of operation the potential of the right-hand plate of capacitor C changes from $-V_{CC}$ (or $-V_{1M}$ if TR1 is loaded to any extent) towards $+V_{CC}$ in an exponential manner with a time constant of CR. Eventually, the base voltage of TR2 will reach a threshold voltage at which the emitter junction becomes forward biased and TR2 begins to turn on. Its collector voltage then falls and, by regenerative feedback, TR1 is rapidly cut off and the circuit quickly resumes its stable state.

In the case of an unloaded circuit, the pulse width is calculated as follows. In the quasi-stable state the capacitor discharges from $-V_{CC}$ towards $+V_{CC}$ with a time constant of CR, so that the instantaneous base voltage of TR2 is

$$v_{B2} = -V_{CC} + 2V_{CC}(1 - e^{-t/RC})$$

Let us assume that TR2 begins to conduct when v_{B2} reaches zero. This occurs after time T_p. Substituting these values into the above equation yields

$$T_p = CR \ln 2 \simeq 0.7 \, CR \qquad (6.1)$$

If $C = 0.01 \, \mu F$ and $R = 15 \, k\Omega$, the pulse width is

$$T_p = 0.7 \times 0.01 \times 10^{-6} \times 15 \times 10^3 = 105 \, \mu s$$

After the quasi-stable state the rate of rise of output voltage V_1 is restricted by the rate at which capacitor C can be charged from V_{CC} through R_1 (during this period of operation the right-hand plate of C is effectively earthed through the forward biased emitter junction of TR2). The circuit should not again be triggered into operation until V_1 has fully recovered to its final value. From the work in chapter 5 we see that V_1 reaches 99 per cent of its final value after a time of $4.6\,R_1C$. It is reasonable to assume therefore that the collector voltage has fully recovered after about $5R_1C$, so that the minimum time t_m that has to be allowed between trigger pulses is

$$t_m = T_p + 5R_1C = 0.7RC + 5R_1C$$

In some circuits it is desirable to employ a large collector voltage and, as this is applied to the base of TR2 at the instant of triggering, there is a risk of reverse breakdown of the emitter junction. This problem can be overcome by including a diode in series with the base of TR2.

A frequent problem in industrial situations is the generation of a pulse which is time delayed with respect to an input signal. This might occur in a machine carrying out a series of operations in time sequence on a component. Such a pulse can be generated by cascading two monostable multivibrators, the first providing the required time delay and the second controlling the width of the output pulse.

6.5   An astable multivibrator

A popular form of astable multivibrator is shown in Fig. 6.3, comprising two saturated inverting switches with regenerative feedback applied via capacitors. The operation is as follows. At the instant TR2 saturates, the drop in its collector voltage is transferred to the base of TR1, so cutting it off. So long as TR1 is cut off, TR2 is saturated and the circuit rests in one of its quasi-stable states. During this period of time, capacitor C_2 charges from $-V_{CC}$ towards $+V_{CC}$ exponentially with a time constant of R_4C_2. The time T_1 taken for the voltage across the capacitor to reach zero is, as for the monostable circuit,

$$T_1 \simeq 0.7\,R_4C_2$$

As soon as this happens, TR1 begins to conduct and, by regenerative action TR2 is cut off. Transistor TR2 begins to conduct again after a period of time T_2, where

$$T_2 \simeq 0.7\,R_3C_1$$

This process is repeated indefinitely with a periodic time of

$$T = 0.7\,(R_4C_2 + R_3C_1)$$

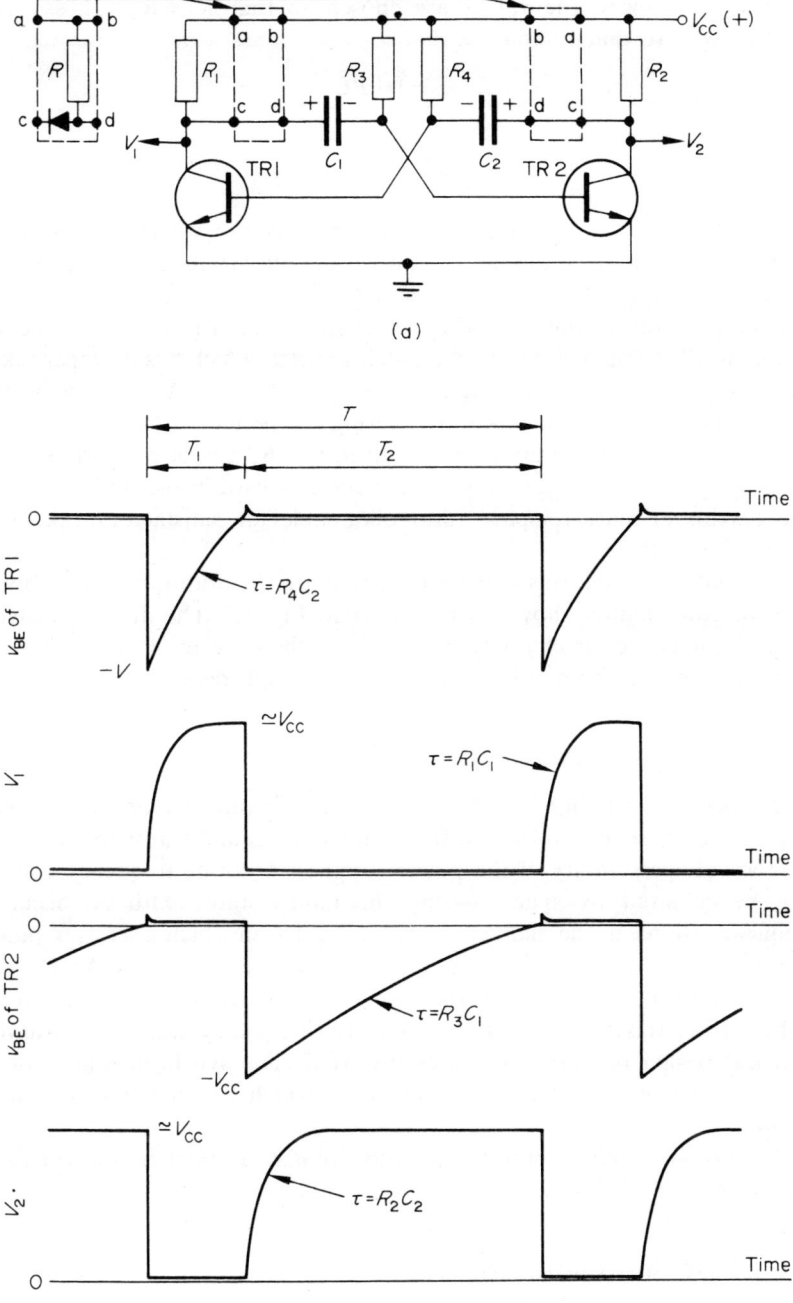

(a)

Fig. 6.3   One form of astable multivibrator.

In many instances, the values are chosen so that $R_3 = R_4 = R$, say and $C_1 = C_2 = C$, so that

$$T = 1.4\,RC$$

giving a frequency of

$$f = 1/T = 1/1.4\,RC \qquad \text{Hz}$$

The ratio T_1/T_2 is the mark-to-space ratio of the waveform. Where a precise 1:1 ratio is required it is necessary either to employ precision resistors and capacitors in the timing circuit or to provide some means of adjusting the ratio. One method of achieving the latter is to connect the top of R_3 and R_4 to opposite ends of a potentiometer, which has its wiper taken to the supply voltage. Adjusting the wiper position alters the relative values of T_1 and T_2 without significantly altering the periodic time T.

The mark-to-space ratio of the circuit in Fig. 6.3 can also be adjusted by altering the values of the individual components used in the timing circuits, the maximum mark-to-space ratio obtainable for satisfactory operation being about 10:1.

The output waveforms can be improved by the addition of the diode-resistor combination shown in the insert to Fig. 6.3. The diode effectively isolates the capacitor charging current from the collector load so that the capacitor charges from $-V_{CC}$ towards V_{CC} through resistor R.

6.5.1   Astable multivibrator synchronization

The frequency stability of the basic astable multivibrator is not only dependent on the parameters of the timing circuits and transistors but also on the voltage stability of the power supplies. Accurate frequency control can be achieved by synchronizing the multivibrator with an accurate frequency source in the manner shown in Fig. 6.4, in which a series of pulses are superimposed on the base voltage of one of the transistors. As a result, the pulse duration is T'_1, compared with T_1 in the free-running circuit. When using this technique to improve the frequency stability, an asymmetrical design of multivibrator is preferred. The synchronizing pulse is then applied to the base of the transistor which has the longest timing period.

Synchronized multivibrators are also popular as frequency dividers in many forms of electronic musical instruments.

6.6   Blocking oscillators

A blocking oscillator is a pulse generator using a single transistor with transformer feedback. The circuit can be designed to operate either in an astable or monostable mode.

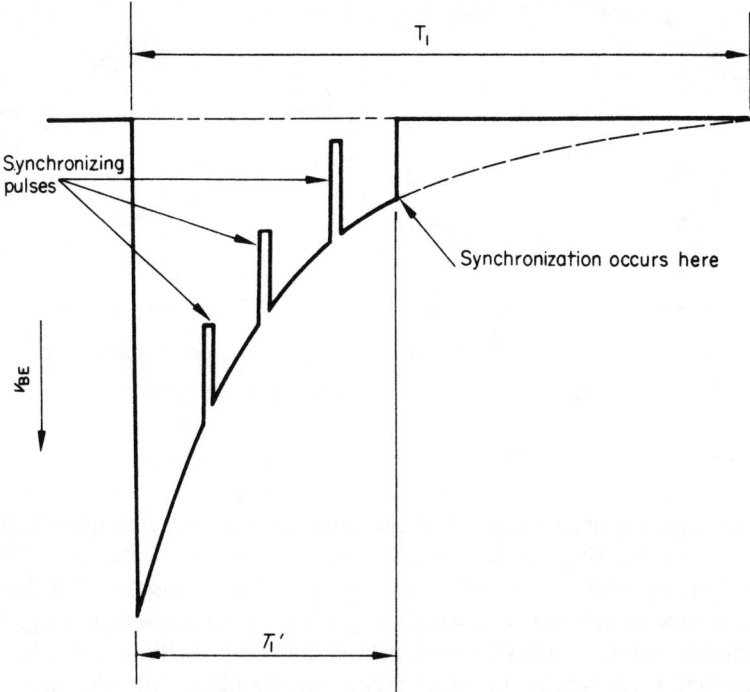

Fig. 6.4   One method of synchronizing an astable multivibrator.

The advantages of the blocking oscillator over many other forms of pulse generator include the fact that it can generate pulses with rise times that are difficult to attain in other circuits with identical transistors, and that a mark-to-space ratio of up to 1000:1 is attainable. In the blocking oscillator, the transistor conducts for only a very short period of time so that even though the peak output power may be large, the mean power dissipated in the transistor is small.

Applications of these circuits include their use as high energy pulse generators, as frequency dividers, as thyristor trigger pulse generators, and as capacitor dischargers in time base circuits.

6.6.1   Basic circuits

Transistor blocking oscillator circuits are developments of the earlier valve versions first patented in 1923 by Appleton, Herd, and Watson-Watt. Basic common-emitter, common-base, and common-collector versions are shown in Figs. 6.5(a), (b) and (c), respectively. The circuits can be operated either in saturated or non-saturated modes, the former being the more popular since circuit design is simplified and also the output impedance is low during the period that the pulse is generated.

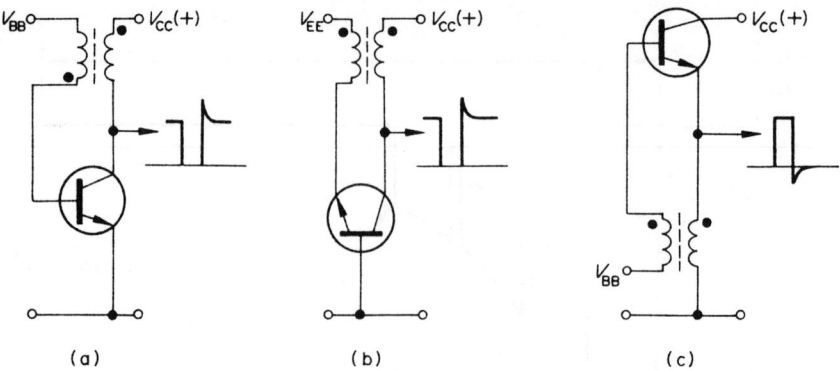

Fig. 6.5   Basic forms of blocking oscillator.

In *monostable circuits* the transistor is arranged to be cut off in its stable state, so that in the monostable versions of Figs. 6.5(a) and (c) the bias supply V_{BB} has a negative polarity; the circuit can be triggered into its quasi-stable state by the application of a positive pulse to the base of the transistor. In *astable circuits* the transistor is initially biased on, so that V_{BB} in the above astable versions would be positive to ensure self-starting. As in the case of astable multivibrators, astable blocking oscillators can be used as frequency dividers by injecting synchronizing pulses into the circuit.

In blocking oscillators, regenerative feedback ensures that once the transistor begins to conduct it is rapidly driven into saturation, as follows: when the transistor begins to conduct, current flows into the collector and the polarity of the voltage induced in the base winding causes the base current to increase. The turns ratio of the transformer is about 4:1 so that the base current rapidly drives the transistor hard into saturation. At this instant the voltage across the transistor falls to its saturation value, causing the collector winding of the transformer to support practically the whole of the supply voltage. From basic transformer theory, this gives

$$V_{CC} \simeq L \, di_C/dt$$

where L is the collector winding inductance and di_C/dt is the rate of change of collector current. Now

$$di_C/dt = V_{CC}/L$$

which has a constant value since V_{CC} and L are constants. That is, the collector current begins to increase at a constant rate. During this period of its operation, the transistor is saturated so that $h_{FE}i_B > i_C$. As the collector current rises a state will be reached when $i_C = h_{FE}i_B$, at which point the transistor begins to come out of saturation and the collector current can no longer continue to increase. This causes the e.m.f. induced in the base winding to be reduced, so that the collector current also begins to fall.

When this happens the collector voltage begins to rise, and this voltage change is transferred into the base circuit by transformer action and the transistor is cut off more rapidly. The cumulative effect is to rapidly force the collector current to zero.

The rapid reduction in the current in the collector winding causes a voltage spike to be developed at the collector, as shown in Fig. 6.5. The magnitude of the spike can be reduced in a number of ways including shunting the collector winding by a resistor or diode-resistor combination. This protects the transistor against overvoltage by causing the core flux to decay more slowly but, for the same reason, causes the recovery time of the circuit to increase. In some cases, the self-capacitance of the collector winding causes the output voltage to *ring* or oscillate for a short time after the end of the pulse.

The output from the circuit can either be taken from the points shown in Fig. 6.5, or from a tertiary winding on the pulse transformer.

The pulse is terminated when the transformer magnetizing current cannot increase further and, in the above case, was assumed to be due to the finite value of h_{FE}. Other factors which can affect the pulse width are core saturation and the bias voltage in R–C networks. None of the above factors provide particularly accurate methods of controlling the pulse width. Where accurate control is required other techniques are used including (a) the application of a turn-off trigger pulse, (b) the feedback of a turn-off pulse from a delay line fed by the blocking oscillator output, and (c) the feedback of a turn-off pulse from an oscillatory circuit energized by the blocking oscillator.

6.6.2   Blocking oscillator circuits

An example of a monostable circuit is shown in Fig. 6.6(a), and one of an astable circuit is shown in Fig. 6.6(b).In the stable operating state of the common-base monostable circuit, the potential at the junction of R_1 and R_2 cuts the transistor off ($R_2 \simeq R_1/100$). The application of a negative-going pulse to the trigger input causes the transistor to conduct, and the output pulse is generated in the manner described in section 6.6.1. Resistor R_3 (typically 1 kΩ) is included to provide a minimum value of input resistance. The collector overshoot voltage is limited by diode D and damping resistor R_4 (which has a value between about 1 and 10 kΩ).

The common-emitter astable circuit, Fig. 6.6(b), uses a series R–C timing circuit which controls the length of time for which the transistor is cut off. At the instant of switch-on, base current flows via resistor R to cause the transistor to conduct, and regeneration occurs. At the instant of cut-off a large negative voltage appears at the base, and this slowly changes towards V_{CC} with a time constant of RC. When the base voltage becomes positive, base current again flows and causes the cycle to be repeated.

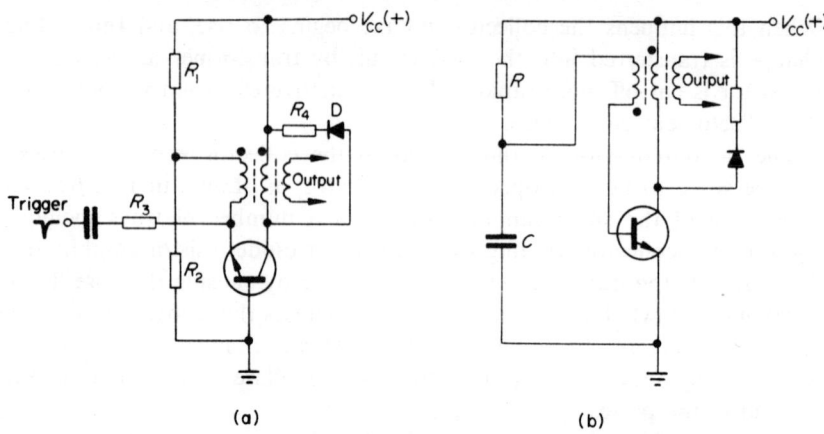

(a)   (b)

Fig. 6.6   (a) Monostable and (b) astable blocking oscillator circuits.

6.7   A unijunction transistor (UJT) relaxation oscillator

A simple pulse generator employing an n-bar UJT is shown in Fig. 6.7(a). At the instant of switch-on the capacitor is uncharged, so that the p.d. between its terminals rises exponentially towards V_{BB} with a time constant of RC. The equation for the voltage at X is

$$V_X = V_{BB}(1 - e^{-t/RC})$$

When $V_X < V_P$, where V_P is the peak-point voltage of the UJT, the emitter-to-B1 path of the UJT has a very high resistance. However, when V_X is equal to V_P resistance falls to a very low value and capacitor C is rapidly discharged. If we take V_P to be approximately equal to ηV_{BB}, then emitter conduction commences when

$$\eta V_{BB} = V_{BB}(1 - e^{-T/RC})$$

where T is the periodic time of the pulses, and

$$T = RC \ln 1/(1-\eta) = 2.3\, RC \lg 1/(1-\eta)$$

A typical value of η is 0.55 so that $T \simeq 0.8\, RC$. Once the UJT has turned on, the capacitor is rapidly discharged, providing a pulse at Y and a non-linear sawtooth waveform at X, shown in Fig. 6.7(b).

This type of circuit is frequently used as a thyristor trigger circuit, the thyristor gate being connected to B1. If it is necessary to isolate the thyristor (or other load) from the UJT, it can be done by replacing R_{B1} with the pulse transformer shown in the inset to Fig. 6.7(a).

The pulse repetition rate of the circuit can be controlled by including a variable resistor as part of R. When the resistance of the variable resistor is reduced to zero, the pulse repetition rate is at its maximum value. Resistor R_{B2} is included in the circuit for reasons of thermal stability.

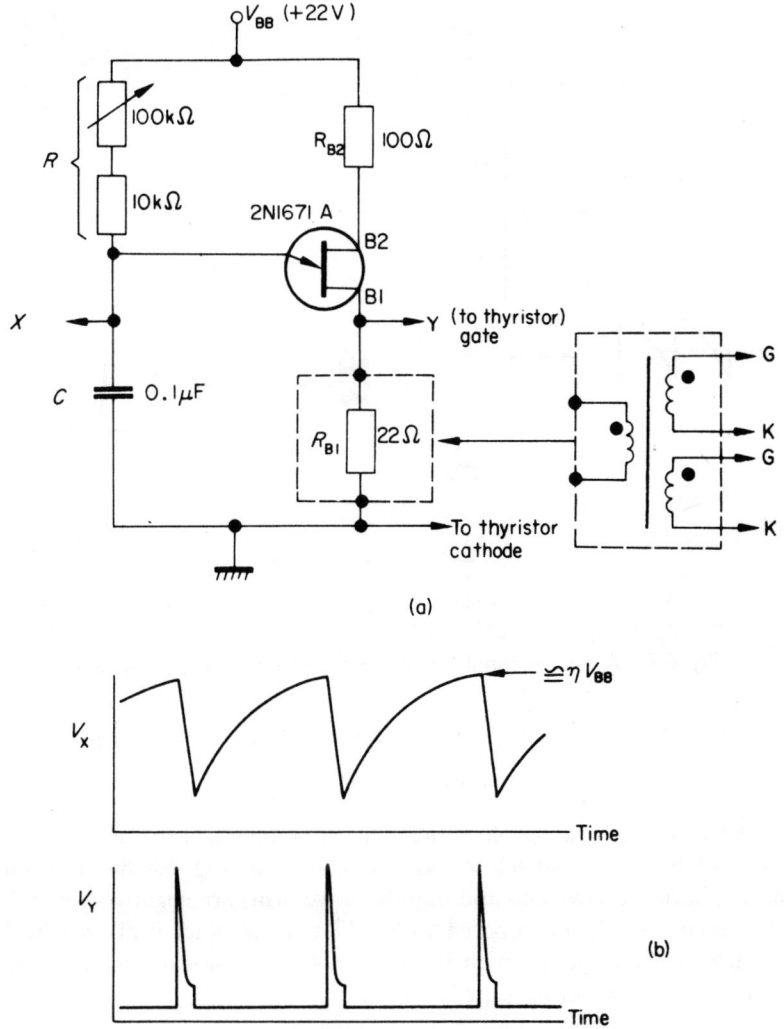

Fig. 6.7 A UJT relaxation oscillator.

6.8 A bi-directional breakdown diode (diac) relaxation oscillator

One form of circuit is shown in Fig. 6.8. The operation is generally similar to the UJT circuit insomuch that the *diac* is used as a voltage-sensitive switch to discharge capacitor C when the voltage at X reaches the break-down voltage of the device. When this occurs

$$V_{BR} = V_S(1 - e^{-T/RC})$$

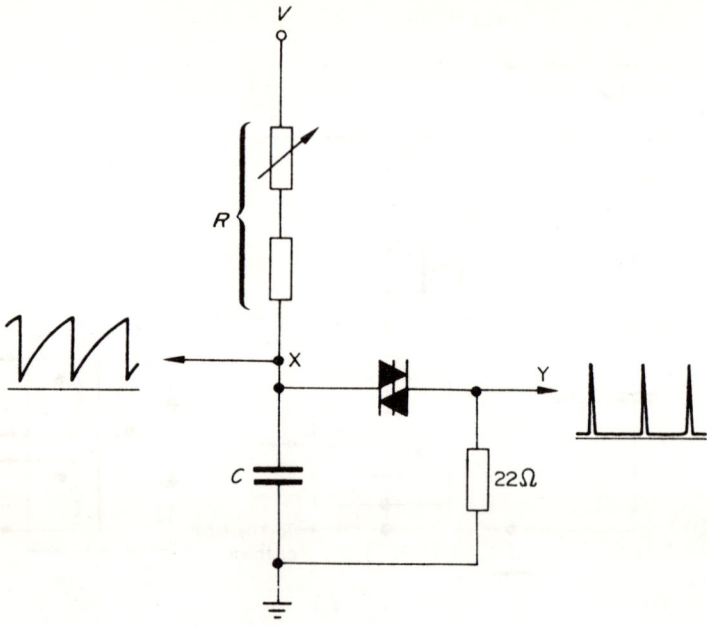

Fig. 6.8   A bidirectional breakdown diode relaxation oscillator.

where T is the periodic time of the output waveform. Solving for T yields

$$T = RC \ln V_{\mathrm{S}}/(V_{\mathrm{S}} - V_{\mathrm{BR}})$$

In otherwise identical applications, the components used in the timing circuits of Figs. 6.7 and 6.8 are identical. If V_{S} in Fig. 6.8 has a negative polarity, both the sawtooth and impulse waveforms are negative-going. This is an advantage when compared with UJT circuits, since it allows the diac to be used in conjunction with *bi-directional breakdown thyristor* (*triacs*) in the control of alternating supplies.

6.9   A Miller integrator sweep circuit

It was shown in chapter 3 that an RC circuit in combination with a high-gain amplifier could simulate an ideal integrator. When the input to such a circuit is a constant voltage, the output is in the form of a linear *ramp* or *sweep* waveform. A block diagram of a basic Miller integrator circuit is illustrated in Fig. 6.9, and provides a *run down* ramp with a slope of $-V_{\mathrm{CC}}/RC$ volts/sec from a voltage of about V_{CC}.

The circuit is designed so that when the electronic switch is open the active device in the amplifier is forward biased and has a low input resistance (this includes BJT and FET circuits). Consequently, V_{x} is a small

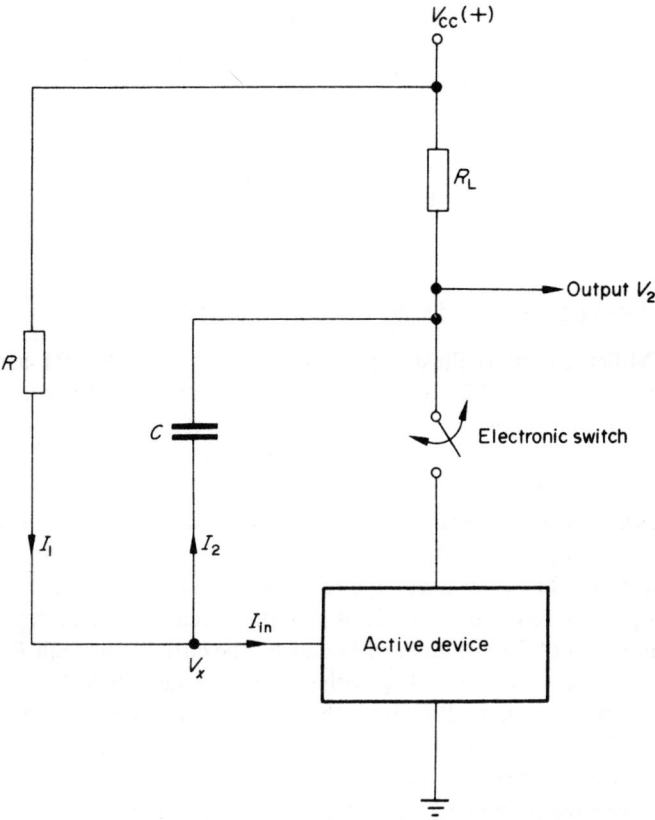

Fig. 6.9   The basis of the Miller integrator ramp generator.

positive voltage. Thus, if the input resistance of the amplifier in this state can be ignored, capacitor C charges through R_L to a voltage of approximately V_{CC}, the charging time constant being $R_L C$.

When the electronic switch closes, R_L is connected into the amplifier and the circuit forms a conventional resistance loaded amplifier. Since V_x is initially a small positive voltage, the output voltage drops abruptly at first due to the inverting action of the circuit. The initial voltage drop depends on the type of active device used, and ranges from a few tenths of a volt in BJT's to a few volts in JUGFETs. The reduction in output voltage is transferred directly to the input via capacitor C, driving the active device towards cut off. Equilibrium is finally reached when voltage V_x is just sufficient to maintain amplifying action. In this state, the amplifier input current has dwindled to an insignificant value so that $I_1 \simeq -I_2$.

Now
$$I_1 \simeq V_{CC}/R$$

and
$$I_2 \simeq C \, dV_2/dt$$

therefore
$$\frac{V_{CC}}{R} = -C \frac{dV_2}{dt}$$

or
$$\frac{dV_2}{dt} = -\frac{V_{CC}}{RC} \quad \text{volts/sec}$$

6.9.1   Semiconductor Miller sweep generators

A BJT Miller circuit is illustrated in Fig. 6.10, in which TR1 acts as an electronic switch and TR2 as the active circuit element. When the trigger signal is sufficiently positive, TR1 saturates and TR2 is switched into circuit. At this instant of time, the output voltage falls by an amount δV as TR2 is driven towards cut-off. After equilibrium conditions have been reached, the output voltage runs down in the manner described above at a linear rate of approximately V_{CC}/RC and, with resistance R_{min} in circuit, the transistor finally bottoms. The output voltage remains constant at this value until the trigger signal is reduced to zero. When this occurs, TR1 is cut off. The emitter junction of TR2 is forward biased by the current through R, and the voltage across C recovers to V_{CC} with a time constant of $R_L C$.

If a very large value of R is used, the effect on the output waveform is as shown for R_{max} in Fig. 6.10.

Using a FET to replace TR2, shown in the inset to Fig. 6.10, the circuit provides waveforms which are not greatly different to those of the BJT circuit with the exception that the initial drop δV is of the order of several volts.

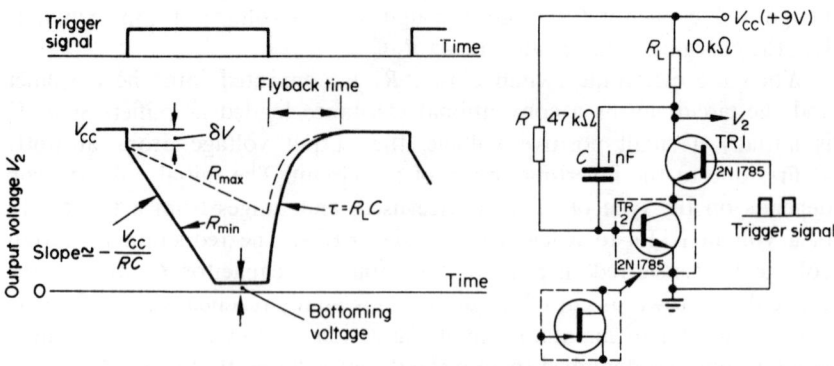

Fig. 6.10   A Miller sweep generator.

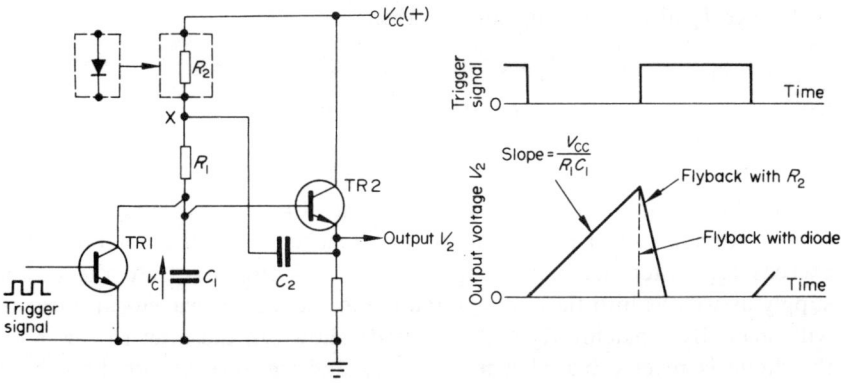

Fig. 6.11   A BJT bootstrap sweep generator.

6.10   A bootstrap sweep generator

Bootstrap sweep generators utilize the basic principle of bootstrap feedback outlined in chapter 3, one form of circuit being shown in Fig. 6.11. The timing components for the voltage sweep, which is of the *run up* type, are R_1 and C_1. Since the output stage is an emitter follower, the output voltage is approximately equal to the voltage across C_1.

Without bootstrap feedback (supplied via C_2), the voltage across C_1 would increase in a non-linear fashion, following an exponential curve. It is possible to force v_C to rise in a linear fashion by driving a constant current through resistor R_1, which means that a constant voltage must be maintained across R_1 during the sweep period. This is brought about in Fig. 6.11 by the use of bootstrap feedback between the output terminal and the upper connection of R_1. The circuit operation is as follows.

When a positive voltage is applied to the base of TR1, the transistor is saturated and C_1 is discharged, so that V_2 is zero. In this mode of operation, TR2 is cut off and its base current is minimal. As a consequence, the voltage across C_2 is practically equal to V_{CC}. Reducing the trigger voltage to zero cuts TR1 off and C_2 begins to charge towards V_{CC}. By emitter follower action voltage V_2 also begins to increase and, by bootstrap feedback, causes the potential of junction X to begin to rise. Ideally, for any given value of v_C the voltage at X is $(V_{CC}+v_C)$, so that the voltage across R_1 is equal to V_{CC}. In effect, capacitor C_2 is equivalent to a battery of e.m.f. V_{CC}; for this equivalence to be maintained throughout the sweep period, capacitor C_2 must not discharge appreciably during this period of the operation. This latter condition is brought about by making the value of C_2 much greater than the value of C_1. If the current flowing in R_1 is I_1, then

$$I_1 \simeq V_{CC}/R_1$$

and, since I_1 also flows through C_1, then

$$I_1 = C_1 \, dv_C/dt \simeq C_1 \, dV_2/dt$$

hence $\qquad\qquad\qquad C_1 \, dV_2/dt = V_{CC}/R_1$

or $\qquad\qquad\qquad\quad dV_2/dt = V_{CC}/R_1 C_1$

that is, the output voltage runs up at the rate of $V_{CC}/R_1 C_1$ volts/sec. As the output voltage rises a point will soon be reached when the potential at X exceeds V_{CC}. After this point, capacitor C_2 will discharge into the power supply as well as into the timing circuit, and the output waveform linearity will suffer. By replacing R_2 with the diode shown in the inset to Fig. 6.11, the diode is reverse biased when $V_X > V_{CC}$ and the linearity of the output waveform is improved when compared with the original circuit.

In the original circuit in Fig. 6.11, when the trigger signal again becomes positive, C_1 is discharged and the flyback time depends on the time taken for capacitor C_2 to charge to its final value via R_2. When R_2 is replaced by a diode, the flyback time is small since the effective charging resistance is r_a of the diode.

6.11   IC timers

A number of monolithic timers are available which operate in any one of several modes including monostable and astable modes. The best known of these are the 555, the 556 and the ZN1034.

The 555 timer is a multi-function timing element which is described in section 6.12 and is housed in an 8-pin DIP assembly. The 556 is a chip containing two identical 555 timers in a 14-pin DIP assembly. The timing accuracy of a 555 circuit is typically ± 0.5 per cent for a time period of a few microseconds and about ± 2 per cent for a time period of about 3 min. The ZN1034 timer is a high accuracy timer housed in a 14-pin DIP and can provide precise timing intervals in the range 50 ms to several days with an accuracy of about ± 0.01 per cent; at a lower accuracy, it is capable of providing a timing interval of several months.

6.12   The 555 timer

The 555 timer (see Fig. 6.12) can operate from a supply voltage in the range $+5\,V$ to $+18\,V$ and can either 'sink' or 'source' a load current of up to $0.2\,A$. The IC contains two comparators CP1 and CP2 whose function is described below.

A resistor chain comprising three equal resistors R connected in series, applies a voltage of $\frac{2}{3}V_{CC}$ (where V_{CC} is the supply voltage) to the '$-$' input (the 'inverting' input) of comparator CP1. Whenever the input signal to

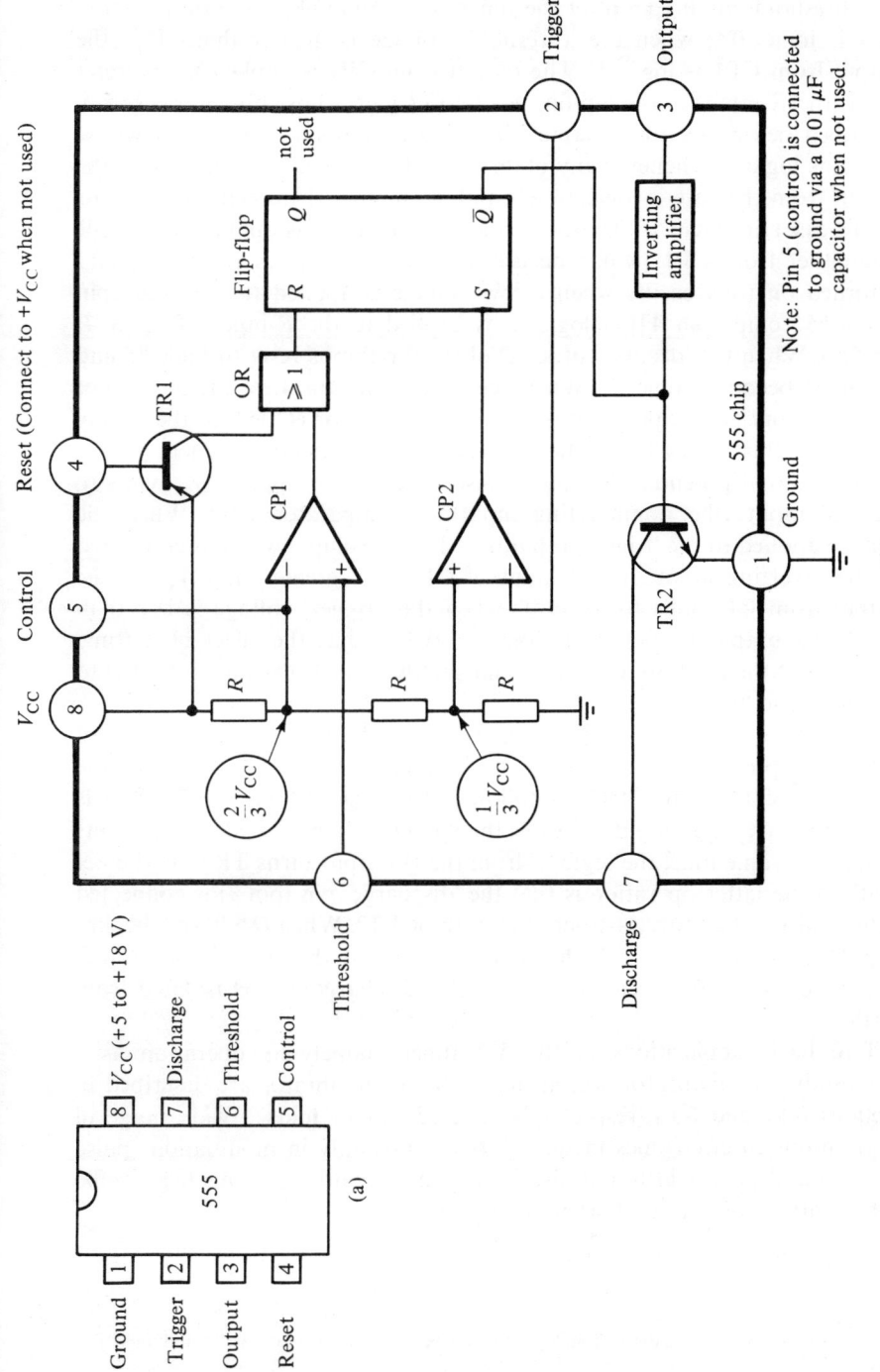

Fig. 6.12 555 timer: (a) pin connections, (b) internal block diagram.

the 'threshold' input (pin 6) of the timer is less than $\frac{2}{3}V_{CC}$, the output from CP1 is logic '0'*; when the 'threshold' voltage is greater than $\frac{2}{3}V_{CC}$, the output from CP1 is logic '1'. The output from CP1 is applied to the reset input (the R-input) of the flip-flop via an OR gate whose function is briefly described below (see also chapter 7). An OR gate is a logic circuit whose output is logic '1' whenever any of its inputs has a logic '1' applied to it; the output from the gate is logic '0' when all its inputs (there being two inputs in this case) are logic '0'. Thus, when a logic '1' signal is applied to the OR gate either from CP1 (which occurs when $V_{threshold} > \frac{2}{3}V_{CC}$) OR when TR1 is turned on (this occurs when a low voltage is applied to the 'reset' pin of the 555 chip [pin 4]), a logic '1' is applied to the R-input of the $S-R$ flip-flop. When this occurs, output Q of the flip-flop is reset to logic '0' and output $\bar{Q}$ becomes logic '1'. When the 'reset' pin (pin 4) of the 555 is not needed in any given application, it is connected to or is 'tied' to the supply line $+V_{CC}$; this causes TR1 to be turned off and it does not conduct.

The internal potential dropping resistor chain also applies $\frac{1}{3}V_{CC}$ volts to the '+' input (the noninverting input) of comparator CP2. When the voltage applied to the 'trigger' input (pin 2) of the chip – which is connected to the inverting input or '−' input of CP2 – is greater than $V_{CC}/3$, the output from CP2 falls to logic '0'; when the 'trigger' voltage is less than $V_{CC}/3$, the output from CP2 is logic '1' (which has the effect of 'setting' output Q of the flip-flop to logic '1' and simultaneously causing output $\bar{Q}$ to become logic '0').

The output $\bar{Q}$ from the flip-flop is used to control not only the 'output' voltage at pin 3 via an inverting buffer amplifier, but also controls the operation of transistor TR2 as follows. When $\bar{Q} = 1$, i.e., the flip-flop is 'reset', the inverting amplifier causes the voltage at pin 3 to fall to zero volts and, at the same time, the logic '1' from the $\bar{Q}$ output turns TR2 on; the net result of the latter operation is that the 'discharge' pin (pin 7) is connected to ground via the collector-to-emitter path of TR2. When $\bar{Q} = 0$, i.e., the flip-flop is in the 'set' state, a high voltage appears at the 'output' pin 3 and, simultaneously, TR2 is cut off so that the 'discharge' pin is isolated from earth.

Two basic applications of the 555 timer, namely its operation as a monostable multivibrator and as an astable multivibrator are described in sections 6.13 and 6.14, respectively. The 555 timer finds a wide range of applications in electronics including its use not only in modulation (pulse position and pulse width) but also in sequential control systems (e.g., traffic light control), bleeper and siren circuits, etc.

---

*For the purpose of this discussion, logic 'O' may be assumed to be zero volts, and logic '1' is $+V_{CC}$.

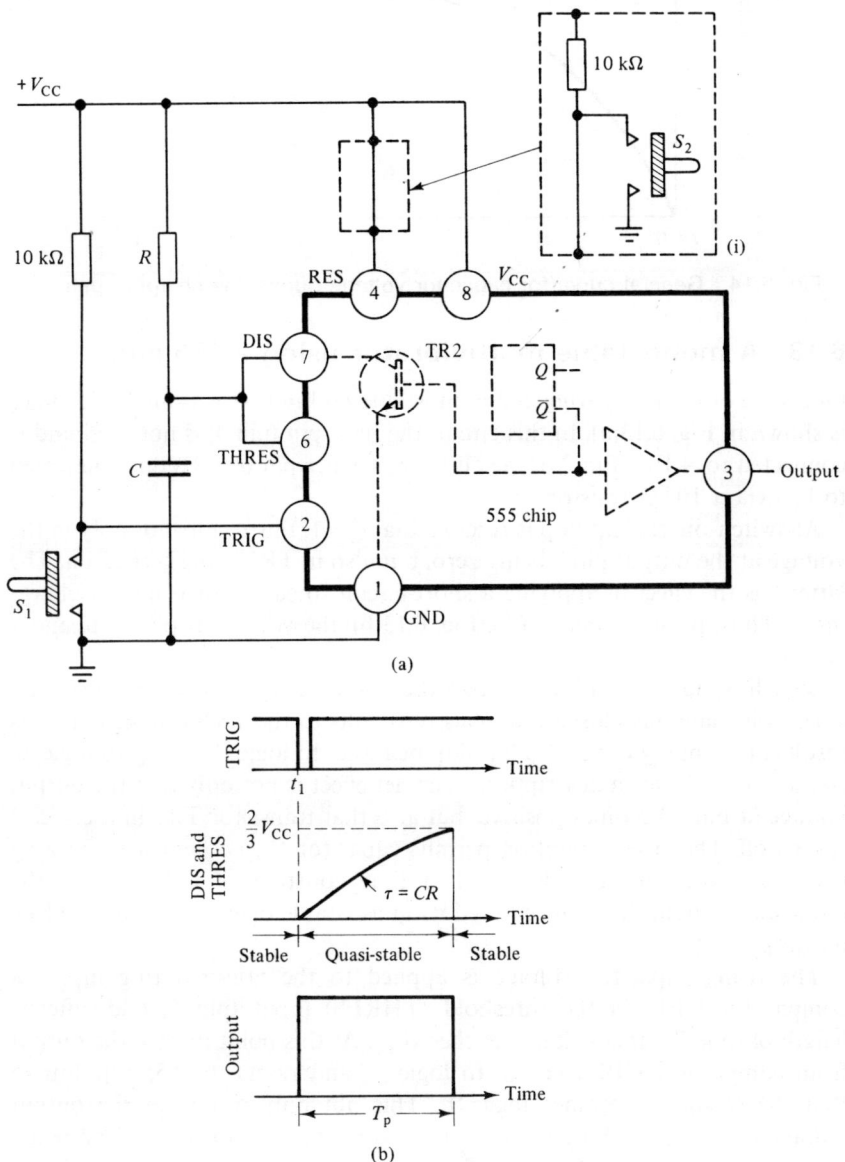

Fig. 6.13(a)   555 timer connected as a monostable multivibrator and (b) typical waveforms.

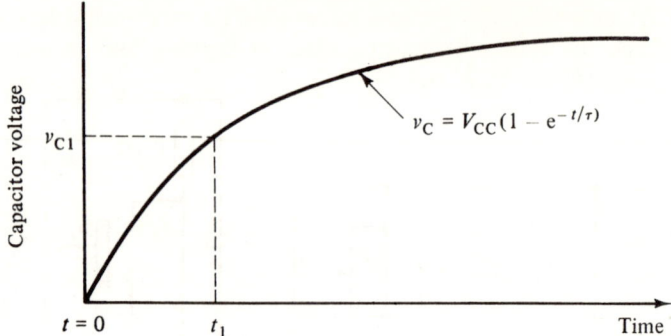

Fig. 6.14   General curve for capacitor voltage during the charging period.

6.13   A monostable multivibrator using a 555 chip

One form of manually triggered monostable multivibrator using a 555 timer is shown in Fig. 6.13(a). In this circuit the reset pin (pin 4) is not used and is connected to $+V_{CC}$. Pin 2 of the chip (the trigger pin of TRIG) is connected to V_{CC} via a $10\,k\Omega$ resistor.

At switch-on, the flip-flop is reset so that $\bar{Q}=1$; this results not only in the voltage at the output pin 3 being zero, but also in TR2 being turned on. The latter has the effect of applying a short-circuit to capacitor C and discharging it. Thus, prior to time t_1 (see Fig. 6.13(b)) the voltage at pins 3, 6, and 7 is zero.

Switch S_1 in Fig. 6.13(a) controls the initiation of the output pulse, and when it is manually closed it applies zero volts to the TRIG input pin. This results in output Q of the 555 flip-flop being set to logic '1' and $\bar{Q}$ to logic '0' (see section 6.12 for a description). The net effect is not only that the output voltage at pin 3 becomes positive, but also that transistor TR2 in the 555 is turned off. The latter condition permits capacitor C to commence charging from zero volts towards V_{CC} with a time constant of CR seconds; the monostable circuit has now entered its quasi-stable operating state (see Fig. 6.13(b)).

The rising capacitor voltage is applied to the non-inverting input of comparator CP1 via the threshold (THRES) input (pin 6) and, after a length of time T_p, this voltage reaches $\frac{2}{3}V_{CC}$. At this point in time the output from comparator CP1 changes to logic '1' and resets the 555 flip-flop so that the $\bar{Q}$ signal becomes logic '1'. This not only results in the output voltage from pin 3 falling to zero but also in the 555 transistor TR2 being turned on. The latter has the effect of short-circuiting capacitor C and discharging it. Thus after a period of time T_p, the monostable returns to its stable state once more.

The operation of the circuit can be interrupted at any time in the interval T_p if the circuit in inset (i) is connected in the 'reset' (RES) input line in Fig.

6.13(a). When switch S_2 is pressed, transistor TR1 (see Fig. 6.12(b)) is turned on; this has the effect of resetting the 555 flip-flop and terminating the output pulse as described above.

6.13.1 Analysis of the 555 monostable multivibrator

General solution
Since an RC circuit is used for timing purposes, the curve of the voltage across the capacitor plotted to a base of time is as shown in Fig. 6.14. The capacitor voltage applied to pins 6 and 7 of the 555 timer in Fig. 6.13 is given by

$$v_C = V_{CC}(1 - e^{-t/\tau})$$

where

v_C = capacitor voltage at time t after the start
V_{CC} = supply voltage to the RC circuit
e = base of Naperian logarithms = 2.71828
τ = time constant of the RC network = RC (R in Ω, C in farads)

Thus the voltage v_{C1} across the capacitor at time t_1 is given by

$$v_{C1} = V_{CC}(1 - e^{-t_1/\tau})$$

and the time t_1 taken to reach voltage v_{C1} is

$$t_1 = -\tau \log_e \left(\frac{V_{CC} - v_{C1}}{V_{CC}} \right) \tag{6.2}$$

Particular solution
In the case of the 555 monostable multivibrator circuit in Fig. 6.13, the following conditions apply.

$$v_{C1} = \tfrac{2}{3} V_{CC}$$

$$t_1 = T_p$$

$$\tau = RC$$

Substituting the above values in eq. (6.2) gives

$$T_p = -RC \log_e \left(\frac{V_{CC} - \tfrac{2}{3} V_{CC}}{V_{CC}} \right) = -RC \log_e \tfrac{1}{3}$$

$$= 1.0986\tau \simeq 1.1\tau$$

Thus if $R = 1$ MΩ, $C = 1$ μF, then $T_p = 1.1$ s.

6.14   An astable multivibrator using a 555 chip

One form of 555 astable multivibrator circuit is shown in Fig. 6.15(a). At switch-on, the 555 internal flip-flop is reset so that $\bar{Q} = 1$; this results in the capacitor being discharged via resistor R_B. Arising from this, the voltage applied to the TRIG input at this time is zero, and results in comparator CP2 in the 555 (see Fig. 6.12) applying a logic '1' to the S-input of the S–R flip-flop. This causes transistor TR2 in the 555 to be turned off so that the capacitor begins to charge towards V_{CC} with a time constant of

$$\tau_1 = (R_A + R_B)C$$

The THREShold pin on the 555 senses when the capacitor voltage reaches $\frac{2}{3}V_{CC}$ and, at this time, the 555 flip-flop is reset from the logic '1' signal from the internal comparator CP1. This has two effects; firstly, transistor TR2 is turned on and, secondly, the output voltage falls to zero. The former operation causes the capacitor to be discharged via resistor R_B and TR2. The capacitor discharge time constant is therefore

$$\tau_2 = R_B C$$

The output voltage of zero volts is maintained until the TRIGger pin on the 555 senses that the capacitor voltage has fallen to $\frac{1}{3}V_{CC}$. At this instant of time, the 555 flip-flop is 'set' by the signal from the internal comparator CP2 (see Fig. 6.12). This has the effect of turning TR2 off, allowing the capacitor to begin to charge once more with a time constant of $(R_A + R_B)C$ – see above.

Consequently, the voltage across the capacitor changes exponentially between $\frac{1}{3}V_{CC}$ and $\frac{2}{3}V_{CC}$, and the output voltage at pin 3 in a rectangular waveform having an 'on' period or 'mark' period of

$$t_m = t_c - t_b$$

and an 'off' period or 'spare' period of

$$t_s = t_d - t_c$$

6.14.1   Analysis of the 555 astable multivibrator

Capacitor charging period (the time interval between t_b and t_c)
It was shown in the general analysis of the monostable multivibrator that the time taken for the capacitor voltage to rise from zero to some voltage v_{C1} is given by

$$t_1 = -\tau \log_e \left(\frac{V_{CC} - v_{C1}}{V_{CC}} \right)$$

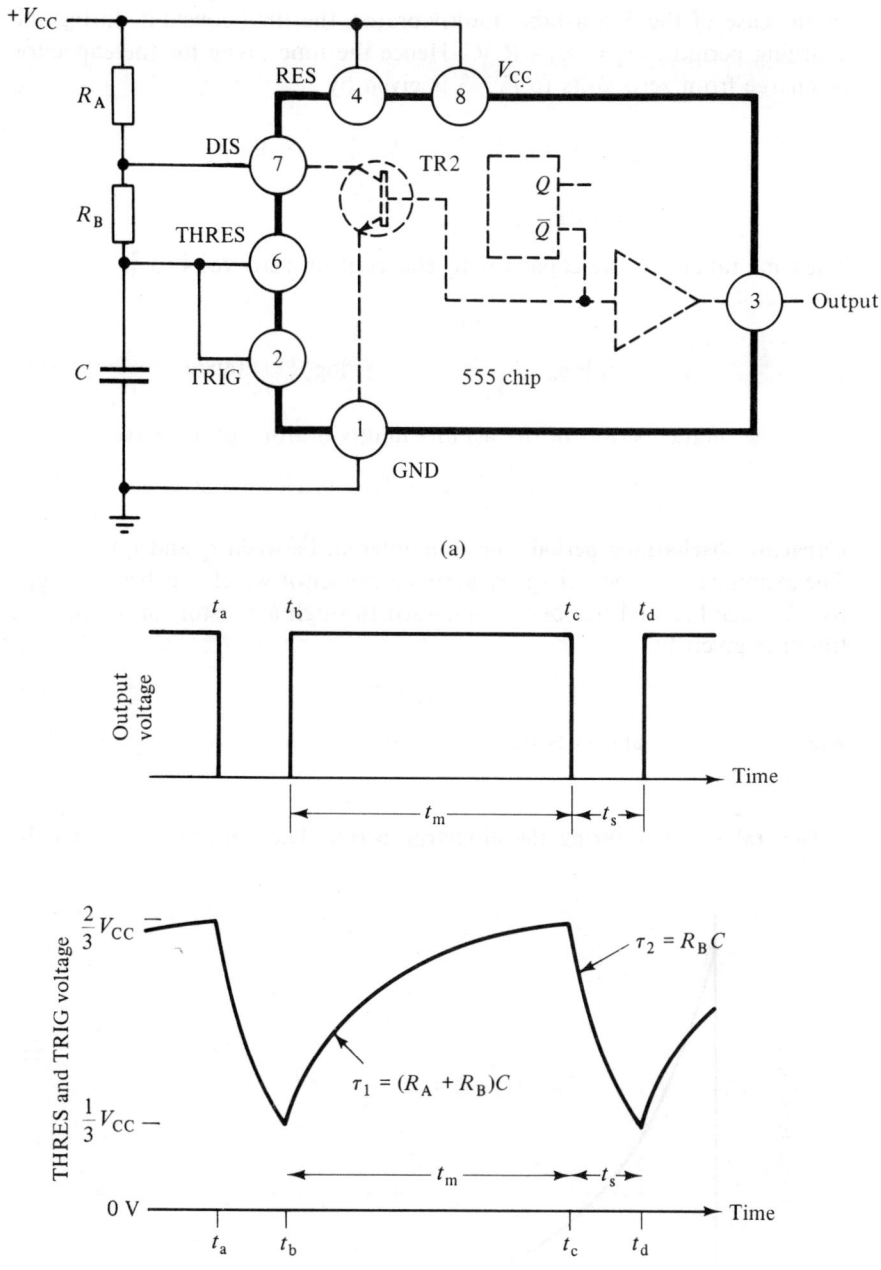

Fig. 6.15(a) 555 timer connected as an astable multivibrator and (b) typical waveforms. Pin 5 (control) may either be left disconnected or may be connected via a 0.01 $F capacitor to ground.

In the case of the 555 astable multivibrator, the time constant during the charging period is $\tau_1 = (R_A + R_B)C$. Hence the time taken for the capacitor to charge from zero volts to $V_{CC}/3$ is given by

$$t_b = -\tau_1 \log_e \left(\frac{V_{CC} - V_{CC}/3}{V_{CC}} \right) = -\tau_1 \log_e \tfrac{2}{3}$$

$$= 0.4054\tau_1$$

The time taken for the capacitor to charge from zero volts to $\tfrac{2}{3}V_{CC}$ is

$$t_c = -\tau_1 \log_e \frac{V_{CC} - \dfrac{2}{3}V_{CC}}{V_{CC}} = -\tau_1 \log_e \tfrac{1}{3} = 1.0986\tau_1$$

hence the 'mark' period of the astable multivibrator output wave is

$$t_m = t_c - t_b = 0.6932\tau_1 \simeq 0.7\tau_1 = 0.7(R_A + R_B)C$$

Capacitor discharging period (the time interval between t_c and t_d)
The expression for the voltage v_C across a capacitor which has been charged to a voltage V_{CC} and has been discharged through a resistor for a length of time t is given by

$$v_C = V_{CC}e^{-t\tau_2} \tag{6.3}$$

where $e = 2.71828$ and τ_2 is the time constant of the discharge circuit $= R_B C$ (see Fig. 6.16).

General solution during the discharge period The voltage v_{C1} across the

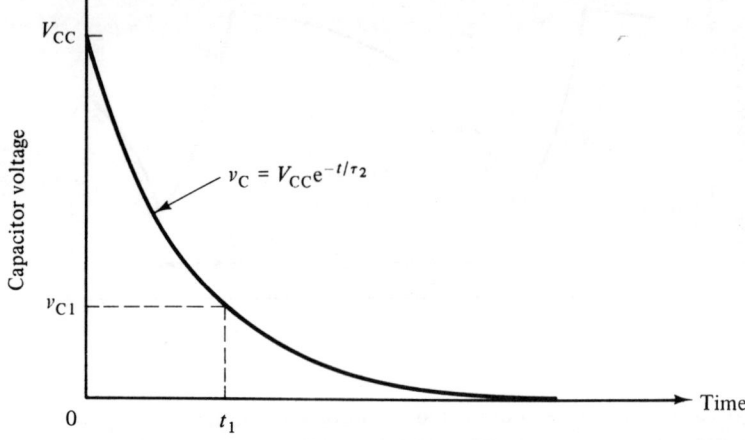

Fig. 6.16   Graph of the voltage across a capacitor during the discharge period.

capacitor at time t_1 after commencement of the discharge is given by

$$v_{C1} = V_{CC} e^{-t_1/\tau_2}$$

The time t_1 taken for the voltage to fall from V_{CC} to v_{C1} is

$$t_1 = -\tau_2 \log_e \frac{v_{C1}}{V_{CC}}$$

Particular solution during the discharge period (see also Fig. 6.15(b)) The time taken for the capacitor voltage to fall from V_{CC} to $\frac{2}{3}V_{CC}$ is given by

$$t_c = -\tau_2 \log_e \frac{\frac{2}{3}V_{CC}}{V_{CC}} = -\tau_2 \log_e \tfrac{2}{3} = 0.4054\tau_2$$

The time taken for the capacitor voltage to fall from V_{CC} to $V_{CC}/3$ is

$$t_d = -\tau_2 \log_e \frac{V_{CC}/3}{V_{CC}} = -\tau_2 \log_e \tfrac{1}{3} = 1.0986\tau_2$$

hence the 'space' period of the astable multivibrator output waveform is

$$t_s = t_d - t_c = 0.6932\tau_2 \simeq 0.7\tau_2 = 0.7R_BC$$

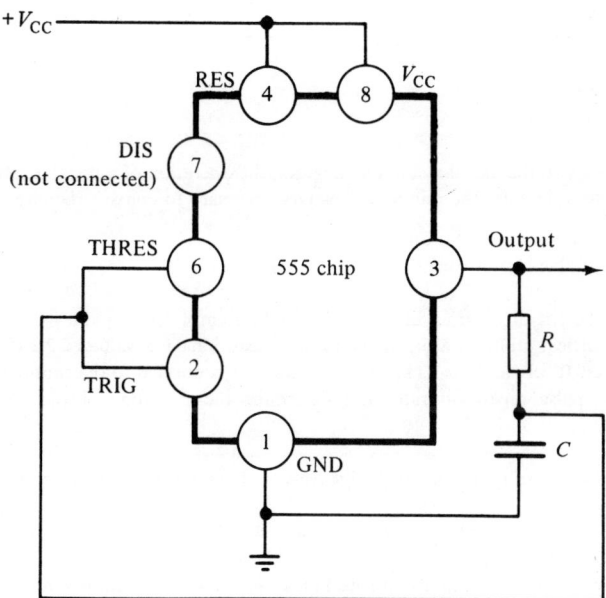

Fig. 6.17   A 555 astable multivibrator with a 50 per cent duty cycle.

Periodic time and frequency of the 555 astable multivibrator circuit

The periodic time, T, of the astable multivibrator is given by

$$T = t_m + t_s = 0.7\tau_1 + 0.7\tau_2 = 0.7(\tau_1 + \tau_2)$$
$$= 0.7[(R_A + R_B)C + R_B C] = 0.7(R_A + 2R_B)C \quad \text{s} \qquad (6.4)$$

and the frequency is given by

$$f = 1/T = 1.43/[(R_A + 2R_B)C] \quad \text{Hz} \qquad (6.5)$$

The reader should note that the basic astable circuit in Fig. 6.15(a) always provides a 'mark' period which is greater than the 'space' period (see the equations above). Modifications to the basic circuit allow a duty cycle in the range 1.0 per cent to 99 per cent to be obtained. A circuit providing a 50 per cent duty cycle is illustrated in Fig. 6.17 having a periodic time of $T = 1.4CR$.

Problems

6.1   Explain, with the aid of a sketch, the principle of operation of a p-n-p transistor. Explain the meaning of the term *hole storage* and how it is caused.

Indicate briefly why the hole storage time may be a disadvantage in digital systems, and show how the hole storage time can be reduced.

6.2   Explain the following terms:

(a) thermal runaway
(b) clamping diode
(c) fan out
(d) parity bit
(e) speed-up capacitors.

6.3   Describe, with the aid of a suitable diagram, the operation of a bistable circuit using two p-n-p transistors. Discuss the minimum changes necessary to convert the circuit into

(a) a free-running multivibrator
(b) a monostable circuit (one-shot multivibrator).

6.4   Sketch the circuit diagram of an astable multivibrator and explain its operation.

In a symmetrical multivibrator, the collector resistors have a value of 2.2 kΩ, the coupling capacitors are 0.01 μF, and the 43 kΩ base resistors are connected to the supply rail at one end. Estimate the pulse repetition rate, and determine the rise time of the collector voltage waveforms.

6.5   Explain, with the aid of time-related waveform diagrams, the operation of the oscillator in Fig. 6.18.

Determine *either* (a) graphically, *or* (b) from first principles, the frequency of oscillation.

6.6   Consider an astable multivibrator using transistors in the common-emitter configuration.

(a) With the aid of a circuit diagram and sketches of the voltage waveforms, describe the action of the multivibrator.

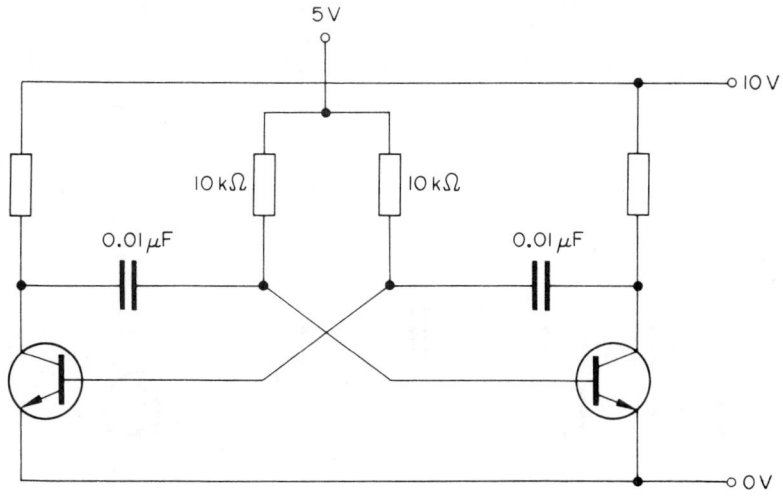

Fig. 6.18.

(b) Develop a simple expression for one of the two quasi-stable periods, stating any assumptions made.

(c) Calculate the values of the circuit components in such a device if the required pulse rate is 1 MHz and the mark-space ratio is 1:2. Assume that the collector current $I_{C(sat)}$ for each transistor is 5 mA, that the minimum current gain of each transistor is 20, and that the d.c. supply voltage is $+10$ V.

6.7 Components with the following values are used in the monostable circuit in Fig. 6.2; $R_1 = R_2 = 4.7$ kΩ, $R_3 = R_4 = R = 100$ kΩ, $C = 0.01$ μF. If $V_{CC} = +10$ V and $V_{BB} = -5$ V, determine (i) the duration of the output pulse, and (ii) its rise time.

6.8 In the circuit in problem 6.7, evaluate the collector and base voltages, together with the currents, all in the quiescent state. Assume that $V_{C(sat)} = 0$ V, and $V_{BE(sat)} = 0.6$ V.

6.9 Explain the action of the circuit in Fig. 6.19 when a trigger pulse is applied to the input. Illustrate your answer with scaled time and voltage waveform diagrams at the base and collector of T2. It may be assumed that the collector and base saturation voltages are zero.

6.10 (a) With the aid of suitable diagrams, explain in detail the operation of a transistor one-shot blocking oscillator.

(b) Draw the waveforms to be expected at the input and output, showing the relationship between them.

(c) How can the output pulse duration (width) be varied?

6.11 A unijunction transistor is used in a circuit of the type shown in Fig. 6.7(a). Determine the pulse repetition rate if $\eta = 0.5$, $R = 10$ kΩ, and $C = 0.1$ μF. If may be assumed that the capacitor is completely discharged at the end of each cycle.

6.12 Determine the pulse repetition rate of the bi-directional breakdown diode relaxation oscillator in Fig. 6.8, given that $R = 10$ kΩ, $C = 0.1$ μF, $V_S = 50$ V, and $V_{BR} = 25$ V. Assume that the capacitor is completely discharged at the end of each cycle.

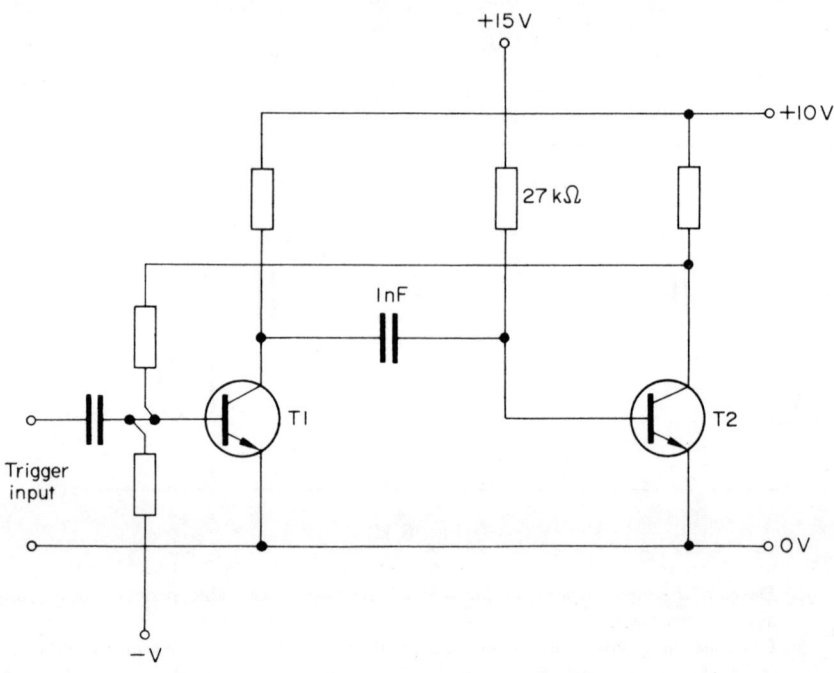

Fig. 6.19.

6.13 Given a Miller integrator of the type in Fig. 6.10, evaluate the rate of rundown of the output voltage if $V_{CC} = 10$ V, $R = 100$ kΩ, and $C = 0.001$ μF.

6.14 In the case of problem 6.13, estimate the rundown time if $\delta V = 0.3$ V and the bottoming voltage is 0.2 V.

6.15 A bootstrap sawtooth generator similar to that in Fig. 6.11 has $V_{CC} = 10$ V, $R_1 = 100$ kΩ, $C_1 = 0.001$ μF. If the trigger frequency is 20 kHz, determine the approximate maximum value of the output voltage waveform.

7. Logic circuits

7.1 Engineering logic

The word 'logic' in its engineering sense has come to mean the representation of switching circuits by equations or equivalent algebraic expressions.

Switching circuits can be divided into two broad categories, namely *combinational switching circuits* and *sequential switching circuits*. A combinational switching circuit is designed to provide an output at the instant that a predetermined combination of input conditions occurs. Examples of this type are burglar alarms and smoke detection systems. In sequential systems, the output is dependent on the sequence of events which has already occurred in the system. Examples are counters and automatic machine tool control systems. Sequential systems are dealt with in chapter 8, and here we discuss aspects of combinational systems.

7.2 Basic logic functions

The three basic logic connectives with which engineers are primarily concerned are the AND function, the OR function, and the NOT (INVERT) function. In practice, combinations of the basic functions are used.

7.3 The AND gate

The simplest form of electrical AND circuit comprises a number of series-connected relays, shown in Fig. 7.1(a). The circuit is also known as a *gate* since, dependent on the input conditions, it is either open to or closed to the flow of *information*. In the relay circuit, a voltage appears at the output only

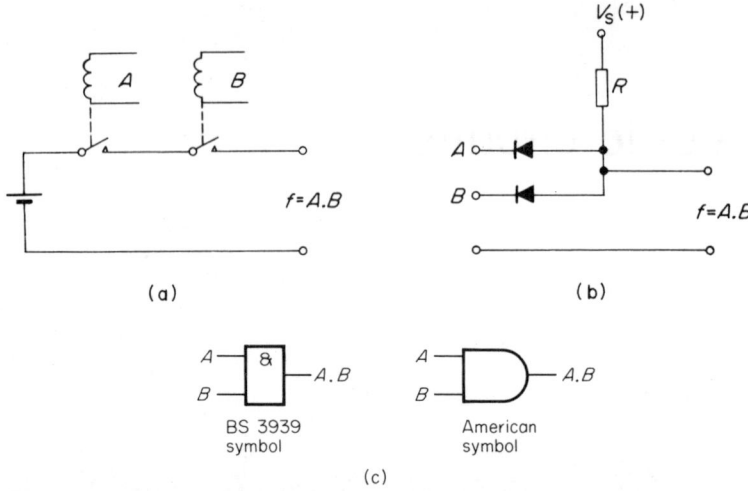

Fg. 7.1 Logic AND gates: (a) relay circuit, (b) diode-resistor logic (DRL) circuit, and (c) some circuit symbols.

when both relay coils are energized, i.e., both input A AND input B have signals applied to them. If the voltage applied to either coil is zero, then the output is zero. We can represent the operation of the circuit by what is known as a *truth table* in which all possible combinations of input and output states are shown. The truth table for a two-input AND gate is given in Table 7.1. In the notation used here, when a voltage is applied to one of the inputs we say that the input has a logic '1' signal applied to it, subject only to the limitation that the voltage exceeds some predetermined level. When the input voltage is zero we say that the input has logic '0' applied to it. These logic levels also apply to the output. In the circuit considered we

Table 7.1

The truth table for a two-input AND gate

Inputs A B		Output f=A.B
0	0	0
0	1	0
1	0	0
1	1	1

are concerned with two *input variables* each of which can have the value '1' or '0', so that four ($2^2 = 4$) combinations of the input variables exist, and are shown in Table 7.1. Only in the case when $A = 1$ AND $B = 1$ is there an output; in all other cases the output is '0'. The output from the circuit is described by the expression

$$f = A.B$$

where the 'dot' (.) is known as the logic AND connective.

A diode version of the AND gate is shown in Fig. 7.1(b), and in this case the signals A and B can be connected separately either to zero voltage (logic '0') or to a positive potential (logic '1'). When either input terminal is earthed, the diode in that line is forward biased and its anode voltage is practically zero. Since the output terminal is connected to the diode anodes, the output voltage is zero when either of the inputs signals are zero. This condition satisfies the first three rows of the truth table. Only when input A AND input B are both at a positive potential (logic '1') are the diodes reverse biased. In this event, the output voltage rises to the logic '1' level, which satisfies the final row of the truth table.

Gates with multiple inputs are often used in industrial applications and, in the case of an N-input AND gate, the logical expression describing its operation is

$$f = A.B.C. \dots L.M.N.$$

7.4 The OR gate

A relay OR gate is shown in Fig. 7.2(a), and it can be seen that a voltage appears at the output if either input A OR input B is energized by logic '1' signals. The truth table for the circuit is given in Table 7.2. This circuit is also known as an INCLUSIVE-OR gate since it includes in its truth table

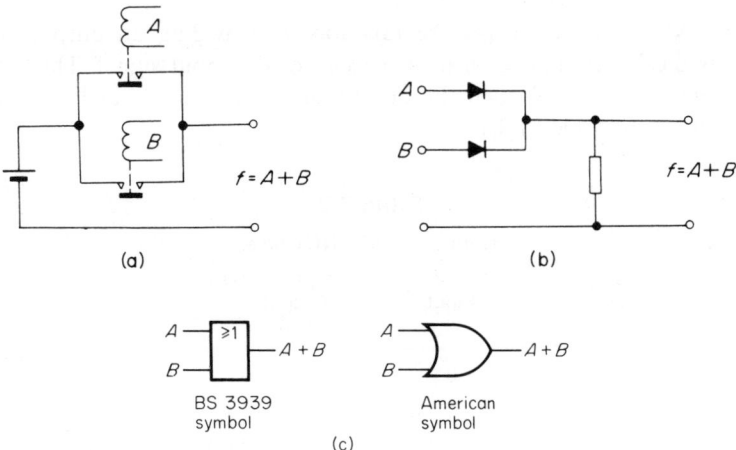

Fig. 7.2 Logic OR gates: (a) relay circuit, (b) diode-resistor logic circuit, and (c) some circuit symbols.

Table 7.2

Truth table for a two-input OR gate

| Inputs | | Output |
A	B	$f=A+B$
0	0	0
0	1	1
1	0	1
1	1	1

the condition that the output terminal is energized when both inputs are energized. From the circuit we also see that when both inputs are zero, then the output is zero (see the first row of the truth table).

In the diode OR gate, circuit 7.2(b), the diodes act as voltage operated switches which close whenever the input (anode) voltage is positive. Clearly, when both inputs are at zero potential then the output voltage is zero (corresponding to the first row of the truth table). Whenever any input line is energized, the diode in that line is forward biased and a signal is transmitted to the output terminal (as shown in the last three rows of the truth table).

The symbol used by engineers to describe the logic OR function is the 'plus' (+) sign. This symbol should not be confused with the mathematical 'plus' sign which has quite a different meaning, as witness the final line of the OR gate truth table. To avoid confusion a 'vee' ('v') symbol is sometimes used.

7.5 The NOT (INVERT) gate

A logic NOT gate performs the function of providing an output signal which is the logical inversion (*complement*) of the input signal. Thus, if the input signal is logic '0', then the output signal is logic '1', and vice versa, giving the truth table in Table 7.3.

Table 7.3

Truth table for a NOT gate

| Input | Output |
A	$f=\bar{A}$
0	1
1	0

In the relay NOT gate, Fig. 7.3(a), when the signal applied to input A is zero the relay contacts remain open and a high voltage (logic '1') appears at

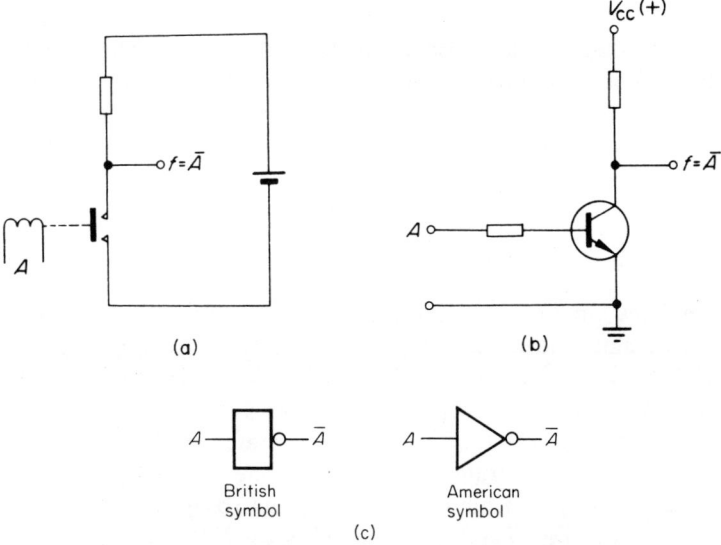

Fig. 7.3 Logic NOT gate: (a) relay circuit, (b) resistor-transistor logic (RTL) gate, and (c) some circuit symbols.

the output. When a signal (logic '1') is applied to input A, the relay contacts close so that the output voltage is zero ($f=0$). Hence

$$f = \text{NOT } A = \bar{A}$$

where the bar over the A represents the NOT function.

From the work in section 5.12, we already know that the circuit in Fig. 7.3(b) can be used as an inverting switching circuit, in much the same way as the relay circuit in Fig. 7.3(a) operates. Consequently, the transistor circuit also operates as a *resistor-transistor logic* (RTL) NOT gate, with the truth table given in Table 7.3.

7.6 Logic circuit block diagrams

The design of logic systems is dealt with using a block diagram technique, and in this section we shall consider basic design methods. Let us consider the design of a circuit which satisfies the truth table below:

Inputs A	B	Output f	Notes
0	0	0	$\bar{A}.\bar{B}=0$
0	1	1	$\bar{A}.B=1$
1	0	1	$A.\bar{B}=1$
1	1	0	$A.B=0$

From the truth table, the circuit output must be zero when $A=0$ (i.e., $\bar{A}=1$) AND $B=0$ (i.e., $\bar{B}=1$) and also when $A=1$ AND $B=1$. The circuit output must be '1' when $A=0$ (i.e., $\bar{A}=1$) AND $B=1$, OR when $A=1$ AND $B=0$ (i.e., $\bar{B}=1$). These states are shown in the notes to the right-hand side of the truth table. Hence, for an output signal to be generated ($f=1$) the circuit must satisfy the function

$$f=(\bar{A} \text{ AND } B) \text{ OR } (A \text{ AND } \bar{B})$$
$$=\bar{A}.B+A.\bar{B}$$

The completed circuit is developed in three steps, shown in Fig. 7.4. We will assume that signals A and B are developed by *transducers* or *sensors* such as photoelectric devices, ultrasonic devices, proximity detectors, etc. To generate signals $\bar{A}$ and $\bar{B}$ we employ NOT gates G1 and G3, see Figs. 7.4(a) and (b), respectively. The function $\bar{A}.B$ is generated by AND gate G2, while the function $A.\bar{B}$ is generated by G4. The complete system is realized by connecting the outputs from circuits 7.4(a) and (b) to the inputs of an OR gate, as shown in Fig. 7.4(c) in which G5 is the OR gate.

The logic network in Fig. 7.4 is known as an EXCLUSIVE-OR gate, and is a widely used electronic circuit. The EXCLUSIVE-OR function described by the truth table at the beginning of this section is such a useful function that all microprocessors include the EXCLUSIVE-OR instruction in their instruction set (see also chapter 9).

The function generated by the circuit in Fig. 7.4 is also known as a NOT-

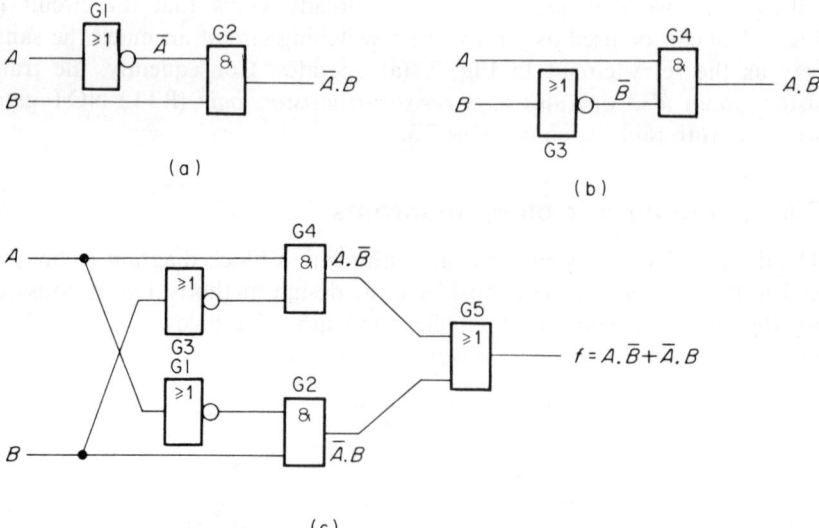

Fig. 7.4 Development of the logic block diagram for the function $A.\bar{B}+\bar{A}.B$.

EQUIVALENT network for the reason that the output is logic '1' when input A is NOT EQUIVALENT to input B. That is, $f=1$ when $A=1$ AND $B=0$ or when $A=0$ AND $B=1$. The symbol '∀' is often used to describe the EXCLUSIVE-OR function as follows.

$$f=A.\bar{B}+\bar{A}.B=A \vee B$$

7.7 Karnaugh maps

The Karnaugh map, first proposed by M. Karnaugh in 1953, is a means of plotting the truth table in the form of a map of rectangular shape. Two advantages of the Karnaugh map, when compared with the truth table representation, are that it is possible to 'see' the general form of the problem and it also allows the problem to be simplified quickly. A disadvantage of the map method is that it is difficult to handle problems which involve more than four or five variables.

A Karnaugh map for two variables is shown in Fig. 7.5, in which two methods of identifying the locations or *cells* on the map are shown. Since there are two variables and each can have two possible values, viz, '1' or '0', there are four combinations, that is $A.\bar{B}$, $A.B$, $\bar{A}.\bar{B}$, and $\bar{A}.B$. These combinations are, in fact, listed by the side of the truth table in section 7.6 (page 231). From the above, we see that function A is associated with half of the combinations, so that on the Karnaugh map the function A acts as a cross-reference to one-half of the cells. The map itself is divided into four cells, and information about variable A is contained in the pair of cells on the right-hand side. Clearly, information about $\bar{A}$ (NOT A) is contained in the pair of cells on the left-hand side. Also, since variable B appears in the terms $A.B$ and $\bar{A}.B$, it too must define one-half of the total number of cells. This is achieved by allowing variable B to define the two cells in the lower row of the Karnaugh map. Again, $\bar{B}$ is everything that is NOT B, i.e., the two cells in the top row of Fig. 7.5(a).

The map in Fig. 7.5(b) differs from that in 7.5(a) only in the respect that variable A is replaced by a '1' above the right-hand column, and $\bar{A}$ is replaced by a '0' above the left-hand column. Similarly, B and $\bar{B}$ are replaced by '1' and '0', respectively, to define the two rows of the map.

Each cell on the maps in Fig. 7.5 is defined by the intersection of two variables. Thus, the top right-hand cell is defined by the intersection of variables A AND $\bar{B}$, and is defined as $A.\bar{B}$. The lower right-hand cell is the intersection of the variables A AND B and is, therefore, the cell $A.B$. Using this method of identifying cells let us now map the function

$$f=\bar{A}.B+A.\bar{B}$$

Clearly, this function has the value '1' either if $\bar{A}.B=1$ OR if $A.\bar{B}=1$, and is

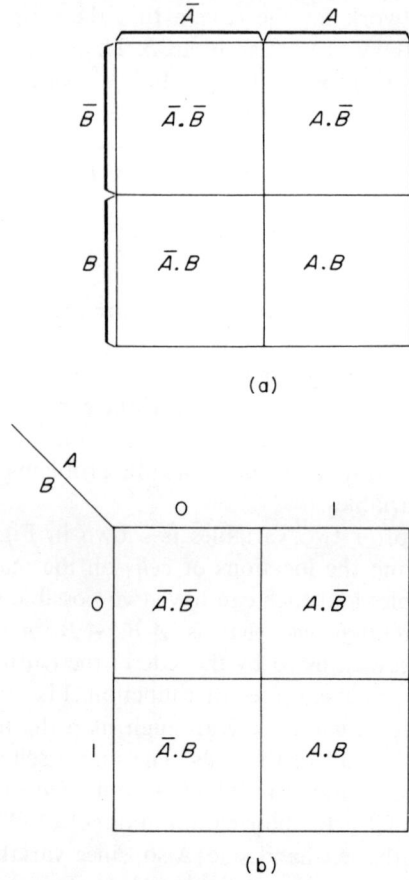

Fig. 7.5 The Karnaugh map for two variables.

mapped in Fig. 7.6. The remaining cells in which the function has zero value (cells $\bar{A}.\bar{B}$ and $A.B$) contain zeros.

The sequence in which the variables are positioned on the map is relatively unimportant, i.e., columns A and $\bar{A}$ can be interchanged as can B and $\bar{B}$. What is important is that each variable defines one-half of the total available cells and that it links with one-half of the cells associated with *each* of the other variables.

Maps for three and four variables are illustrated in Figs. 7.7(a) and (b), respectively. In both cases, the requirement of the number of cells covered by each variable defined above is satisfied, so that each cell is defined uniquely in terms of all the variables. Let us now develop the Karnaugh map for the function

$$f = \bar{A}.\bar{B}.C + A$$

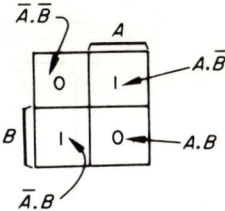

Fig. 7.6 Karnaugh map of the function $\bar{A}.B+A.\bar{B}$.

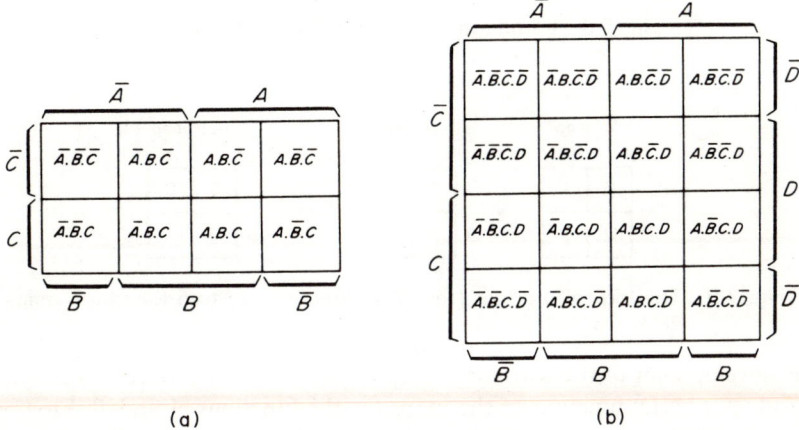

(a)   (b)

Fig. 7.7 Karnaugh map for (a) three variables and (b) four variables.

Since this is a three-variable problem, it is best mapped using the configuration of Fig. 7.7(a) (it can also be mapped on a diagram of the type in Fig. 7.7(b)). The term $\bar{A}.\bar{B}.C$ defines one cell only and a '1' is placed in that cell, see Fig. 7.8. The equation also tells us that if $A=1$ then the function has the value '1' whatever the values of B and C. As a result, we mark a '1' in the four cells defined by the variable A on the Karnaugh map, i.e., cells $A.B.\bar{C}$, $A.B.C$, $A.\bar{B}.C$, and $A.\bar{B}.\bar{C}$.

7.8 Operations with Karnaugh maps

In some instances it may be necessary to determine the result of AND gating two or more functions, each of which may be a more-or-less complex logic function. The result can be quickly obtained by the use of Karnaugh maps. This technique is also valuable in the verification of the laws and theorems of logic (see also section 7.9). As a simple example, let us consider

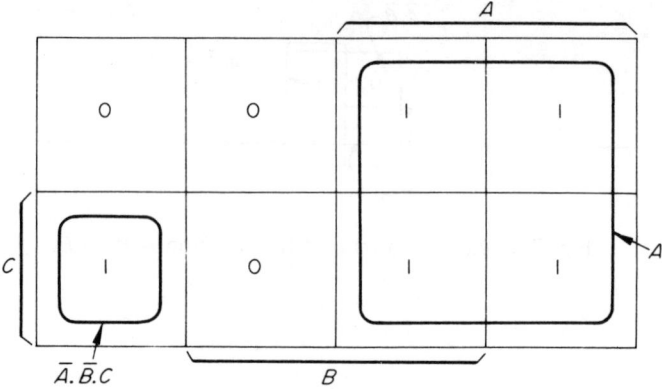

Fig. 7.8   Karnaugh map of the function $\bar{A}.\bar{B}.C+A$.

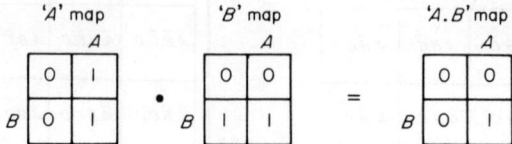

Fig. 7.9   The logical product (the AND function) of two Karnaugh maps.

the development of the Karnaugh map of the function $A.B$ from the *logical* 'product' (AND) function of the maps for the functions A and B. First, we map the 'A' function on a single map, see Fig. 7.9, and the 'B' function on a second map. The logical values placed in corresponding cells on each map are then combined according to the AND function truth table, Table 7.1, to give $A.B$ map on the right-hand side of Fig. 7.9.

In other cases, it may be necessary to generate the OR function of two more-or-less complex functions. Once more, this can be done by the use of Karnaugh maps. An example is given in Fig. 7.10, in which the Karnaugh map of the function $\bar{A}+\bar{B}$ is developed from maps of the functions $\bar{A}$ and $\bar{B}$. Here, the values placed in corresponding cells of the two left-hand maps are combined according to the OR function truth table, Table 7.2, to give the values in the cells in the right-hand map.

Additionally, the Karnaugh map can be used to determine the logical *complement* (i.e., the INVERTED function) of an expression. For example, the function mapped in Fig. 7.10 is

$$f=\bar{A}+\bar{B}$$

If it is necessary to evaluate the logical expression representing $\bar{f}$ (i.e., NOT f), then we can do so by determining which cells are defined by zeros since

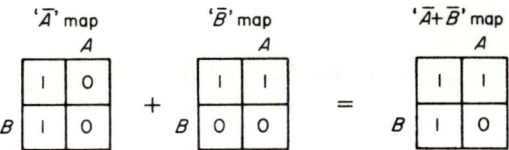

Fig. 7.10 The logical sum (the OR function) of two Karnaugh maps.

these are NOT the cells defined by 1's. In this case, the only zero occurs in the single cell $A.B$, hence for Fig. 7.10

$$\bar{f} = A.B$$

Further examples of this technique are given in sections 7.11 and 7.14.

7.9 Some laws of logic

In the following, a number of basic theorems and laws are listed. Most can be verified using the Karnaugh map techniques outlined in section 7.8, and formal proofs are not given here. [A more extensive treatment of laws of logic is given in *Logic Circuits* by N. M. Morris (McGraw-Hill).]

Theorem 1 $A+0=A$
Theorem 2 $A.0=0$
Theorem 3 $A+1=1$
Theorem 4 $A.1=A$
Theorem 5 $A+A=A$
Theorem 6 $A.A=A$
Theorem 7 $A+\bar{A}=1$
Theorem 8 $A.\bar{A}=0$
Theorem 9 $\bar{\bar{A}}=A$

Commutative law

$$A+B+C=C+A+B=B+C+A$$
$$A.B.C=C.A.B=B.C.A$$

Associative law

$$A+B+C=(A+B)+C=A+(B+C)=B+(A+C)$$
$$A.B.C=(A.B).C=A.(B.C)=B.(A.C)$$

Distributive law

$$A+(B.C.D....)=(A+B).(A+C).(A+D)....$$
$$A.(B+C+D+...)=A.B+A.C+A.D+...$$

De Morgan's Theorem

$$\overline{A+B+C\ldots} = \bar{A}.\bar{B}.\bar{C}.\ \ldots$$

$$\overline{A.B.C.\ \ldots} = \bar{A}+\bar{B}+\bar{C}\ \ldots$$

De Morgan's theorem is a very powerful tool in logic circuit design, and it leads to a general theorem for the inversion of logic functions as follows:

The logical complement of a function is obtained if, in the original function, the 'dots' are changed for 'plusses' and 'plusses' for 'dots', and each term in the expression is complemented.

For example, if

$$f=(A.\bar{B})+(C.[A+\bar{D}])+(A.C)$$

then
$$\bar{f}=(\bar{A}+B).(\bar{C}+[\bar{A}.D]).(\bar{A}+\bar{C})$$

When using this technique, it is advisable to collect groups of terms inside brackets in the manner shown to avoid errors.

7.10   Minimization techniques using the Karnaugh map

In an engineering sense, the usual meaning of the word minimization implies a solution to a problem requiring the minimum number of logic elements together with the minimum number of interconnections between the elements. The advent of integrated logic circuits has also introduced a modified meaning to minimization, insomuch that we can talk in terms of the minimum number of IC packages required by the circuit.

Any given problem may have several apparently similar solutions in terms of logical equations, but when each solution is realized as electronic hardware it is often the case that one emerges which is the best. The Karnaugh map may provide alternative solutions to a problem, and it is then up to the designer to select the most economical solution.

Consider the case of a problem which can be expressed in the following logical form

$$f=\bar{A}.B.\bar{C}+\bar{A}.B.C+A.B.C+A.B.\bar{C}$$

and we wish to obtain the simplest logical statement which satisfies this equation. The Karnaugh map of the function is shown in Fig. 7.11. The process of simplifying the logical expression is one of grouping together *adjacent cells* on the Karnaugh map. Adjacent cells are cells which differ in only one variable in the logical product terms which describe the cells. For example, the upper lefthand cell in Fig. 7.11 is the cell $\bar{A}.B.\bar{C}$ and that immediately to its right is the cell $A.B.\bar{C}$, which differ from one another only in terms of the variable A since one contains A and the other $\bar{A}$. This

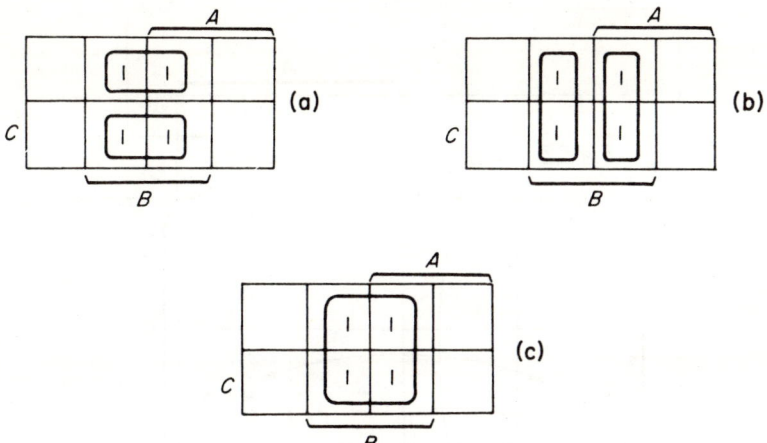

Fig. 7.11   Methods of grouping cells.

pair of cells can be grouped together in the manner shown in Fig. 7.11(a) and comprise the area which is the intersection of B AND $\bar{C}$. Similarly, the two lower cells in Fig. 7.11(a) are adjacent and are defined by B AND C, so that the whole map is defined by the expression

$$f = B.\bar{C} + B.C$$

Alternatively, we may group cells in the vertical plane, in the manner of Fig. 7.11(b), giving

$$f = A.B + \bar{A}.B$$

Or, since the four cells form a single group as in Fig. 7.11(c), we may group them together to give

$$f = B$$

which is the simplest statement representing the original equation.

Cells may be grouped in binary combinations, i.e., groups of 1, 2, 4, 8, etc. Where other combinations of adjacent cells are mapped, i.e., groups of 3, 5, 6, 7, etc., it is more convenient to combine them in the form of a number of binary combinations. For example, the three cells in Fig. 7.12 can be

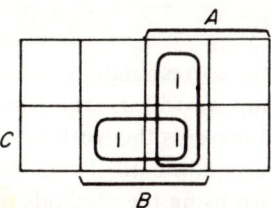

Fig. 7.12   A simplifying procedure on the Karnaugh map.

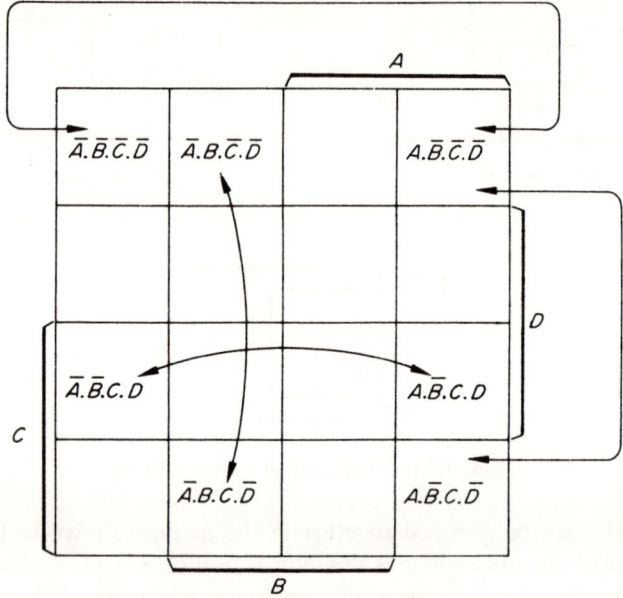

Fig. 7.13   End-to-end and top-to-bottom adjacency on the Karnaugh map.

grouped in the form of two groups of two cells to give the logical expression

$$f = A.B + B.C = B.(A + C)$$

Although the Karnaugh map appears as a two-dimensional surface it can, in fact, be regarded as having three dimensions. For example, the top row of cells in Fig. 7.13 is adjacent to those in the bottom row, as we can see by inspecting the defining equations of the cells. Thus, we may fold the top of the map down to meet the bottom to form a cylinder. Also, since the end cells are also adjacent, our cylinder can be bent round until the ends meet to give a three-dimensional toroidal shape, much like a tyre inner-tube. In this way, it is found that the four cells in the corners of Fig. 7.13 are adjacent in the same manner as the four cells in Fig. 7.11 are also adjacent to one another.

7.11   A design study

Table 7.4 is a truth table corresponding to, say, an industrial control system, and in the following we will use the Karnaugh map to deduce the logical expression which represents the truth table. The first step is to draw the Karnaugh map corresponding to Table 7.4, which is illustrated in Fig. 7.14. The map is drawn using the methods outlined above, and a brief description of the procedure is given here. The logical function correspond-

Table 7.4

	Inputs			Output
A	B	C	D	f
0	0	0	0	1
0	0	0	1	1
0	0	1	0	0
0	0	1	1	0
0	1	0	0	0
0	1	0	1	0
0	1	1	0	0
0	1	1	1	0
1	0	0	0	1
1	0	0	1	1
1	0	1	0	1
1	0	1	1	1
1	1	0	0	0
1	1	0	1	0
1	1	1	0	1
1	1	1	1	1

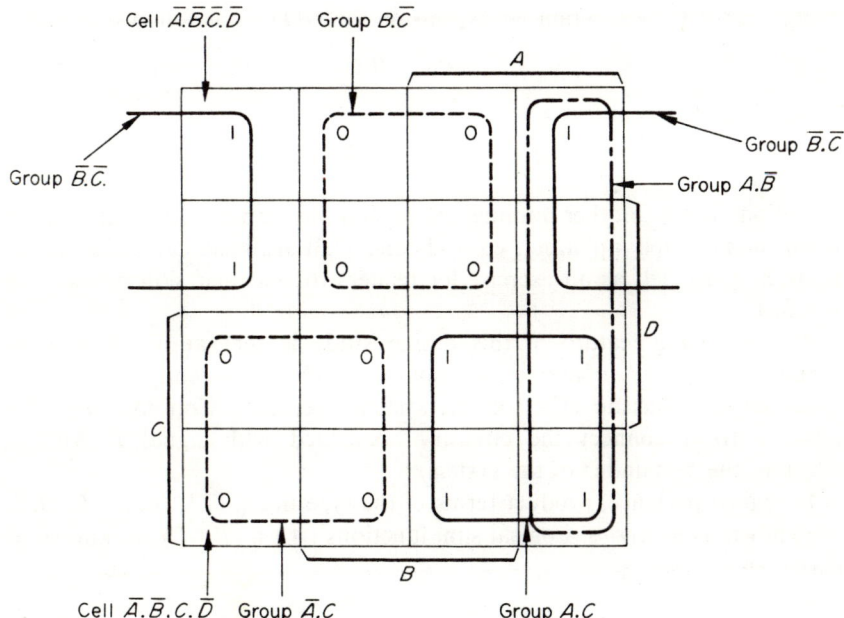

Fig. 7.14   Karnaugh map of Table 7.4.

ing to the first row of the truth table is $\bar{A}.\bar{B}.\bar{C}.\bar{D}$, which defined the cell in the upper left-hand corner of the Karnaugh map. Since the network must generate a '1' when this combination is satisfied, we place a '1' in that cell on the map. The third row of the truth table defines the cell $\bar{A}.\bar{B}.C.\bar{D}$ (the lower left-hand corner cell) which, according to the truth table, must have a '0' placed in it. The remaining cells are filled in this fashion.

The next step is to group all the 1's on the map in the *minimum* number of groups. The four cells in the lower right-hand corner of the map, group $A.C$, are an obvious choice. Also, using the concept of end-to-end adjacency, the four cells representing the group $\bar{B}.\bar{C}$ are also selected. The two groups cover all the 1s, so that one *minimal* form of solution is

$$f = A.C + \bar{B}.\bar{C} \tag{7.1}$$

As a point of interest, the four cells shown as the group $A.\bar{B}$ can also be included in the expression for f as follows

$$f = A.C + \bar{B}.\bar{C} + A.\bar{B} \tag{7.2}$$

However, the term $A.\bar{B}$ is *redundant* since all the cells in that group have already been defined in eq. (7.1) and, although eq. (7.2) is one form of solution, it is not a minimal logical form.

We can, alternatively, use a technique described earlier and say that the areas which include 0's are NOT the areas which include 1's. Thus, by grouping the 0's we obtain an expression for NOT f, as follows:

$$\bar{f} = \bar{A}.C + B.\bar{C}$$

hence, from theorem 9

$$f = \overline{\bar{A}.C + B.\bar{C}} \tag{7.3}$$

Equation (7.3) is another minimal expression for output f. Also, although it is not immediately apparent, eq. (7.1) and (7.3) are identical logical equations, and it is left as an exercise for readers to test their skill in verifying this fact.

An interesting feature of this design study is that at no point does variable D appear in any of the solutions! Quite evidently, input D *is redundant* and has no effect on the system operation. Consequently, it is possible to disconnect the circuitry associated with signal D without affecting the remainder of the system.

In logical parlance, product terms of the type in eq. (7.1) (i.e., $A.C$, $\bar{B}.\bar{C}$) are known as *minterms*. Logical sum functions ($\bar{A}+C$, $B+C$) are known as *maxterms*.

7.12 NOR and NAND functions

The name NOR has been coined to describe the operation NOT OR (or more correctly OR NOT), and is illustrated in Fig. 7.15. Here, the output from the OR gate is inverted by the NOT gate to give an output of $\overline{A+B}$

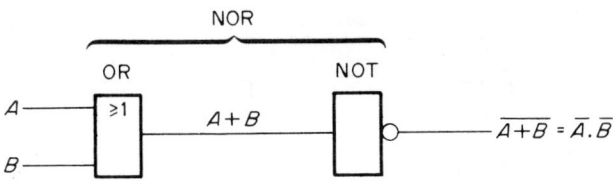

Fig. 7.15 The basis of the NOR gate.

which, by De Morgan's theorem, is equivalent to $\bar{A}.\bar{B}$. If the OR section of the gate has a fan-in of N, then the output is in the form

$$\overline{A+B+ \ ... \ +M+N} = \bar{A}.\bar{B}. \ ... \ .\bar{M}.\bar{N}$$

The truth table for a NOR gate with a fan-in of two is given in Table 7.5, and is seen to be the complement of the OR gate truth table. The truth table of the NOR gate may be summarized as follows:

When any input line is activated with a '1' signal the output is '0'; only when all the inputs are 0's is the output 1.

Table 7.5

Truth tables for two-input NOR and NAND gates

Inputs		Outputs	
		NOR	NAND
A	B	$\overline{A+B} = \bar{A}.\bar{B}$	$\overline{A.B} = \bar{A}+\bar{B}$
0	0	1	1
0	1	0	1
1	0	0	1
1	1	0	0

The name NAND is a contraction of the NOT AND operation as follows

$$\text{NAND} = \overline{\text{AND}}$$

A block diagram of a network generating the NAND function is illustrated in Fig. 7.16. The output from the circuit is the inverted AND function, $\overline{A.B}$, giving the truth table in Table 7.5. The NAND gate truth table may be summarized as follows:

When any input line has a '0' signal applied to it, the output is '1'; only when all input lines are activated by 1's is the output '0'.

An advantage of both NOR and NAND gates over other types of gate is

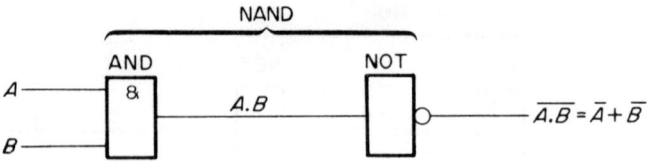

Fig. 7.16 The basis of the NAND gate.

Circuit		Function generated	
		NOR gates	NAND gates
(a)	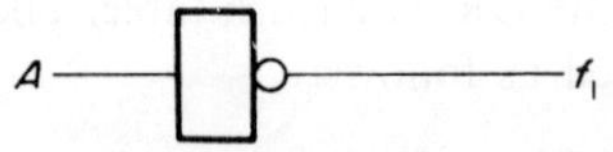 A———$\triangleright\!\circ$———f_1	$f_1 = \overline{A}$	$f_1 = \overline{A}$
(b)	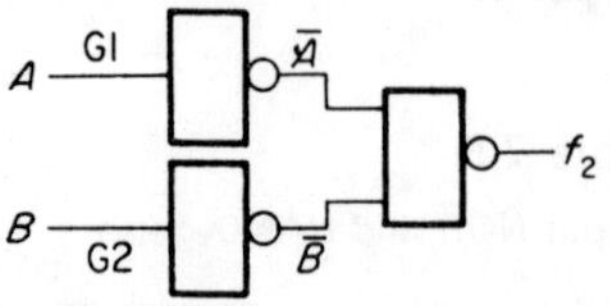 G1, G2 circuit ——f_2	$f_2 = A.B$	$f_2 = A+B$
(c)	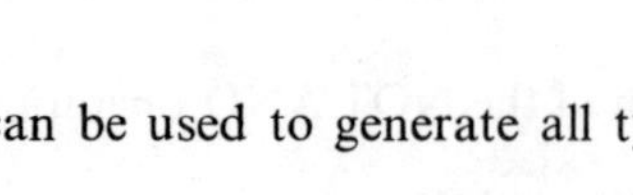 G3, G4 circuit ——f_3	$f_3 = A+B$	$f_3 = A.B$

Fig. 7.17 Basic configurations of NOR and NAND gates.

that they can be used to generate all types of logic functions as shown in Fig. 7.17.

The analysis of each configuration is given below.

NOR gates

CIRCUIT(a): In this case, only one signal is applied to the gate so that the output is

$$f_1 = \overline{A}$$

that is, it acts as a NOT gate.

CIRCUIT(b): Gates G1 and G2 act as invertors in the manner of circuit

(a) so that

$$f_2 = \overline{A} + \overline{B} = A.B$$

that is, it generates the AND function of the two inputs.

CIRCUIT (c): The output from G3 is $\overline{A+B}$ and, since G4 acts as an invertor, then

$$f_3 = \overline{\overline{A+B}} = A+B$$

which is the OR function of the inputs.

NAND gates

CIRCUIT (a): As before, only one input is used and the gate acts as an invertor so that

$$f_1 = \overline{A}$$

CIRCUIT (b): Gates G1 and G2 again act as invertors, so that

$$f_2 = \overline{\overline{A}.\overline{B}} = A+B$$

which is the OR function of the inputs.

CIRCUIT (c): The output from NAND gate G3 is $\overline{A.B}$, and from G4 is

$$f_3 = \overline{\overline{A.B}} = A.B$$

giving the AND function of the inputs.

7.13 Design of NAND systems using the Karnaugh map

The majority of modern logic systems is designed using only one type of gate, which may either be NAND or NOR gates. In this section of the book, we consider a method of designing logic systems employing only NAND elements.

As we have already seen, when we group the 1's on the Karnaugh map

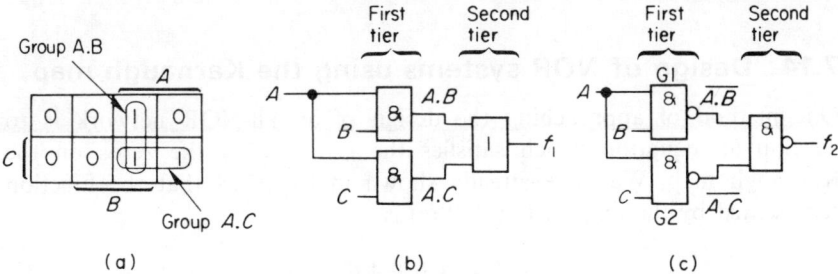

Fig. 7.18 Networks corresponding to Karnaugh map (a) are given in (b) and (c).

we obtain an AND-OR type of solution (see, for example, eq. (7.1)). It is shown below that this type of system can be directly replaced by an all-NAND network. Let us consider the design of a logic circuit to satisfy the Karnaugh map in Fig. 7.18(a), for which the logical equation is

$$f_1 = A.B + A.C$$

This equation is satisfied by the *two-tier* or *two-layer* AND-OR network in Fig. 7.18(b). Now let us consider the logic function generated by the all-NAND network in Fig. 7.18(c), which is

$$f_2 = \overline{(A.B).(A.C)} = \overline{(\overline{A}+\overline{B}).(\overline{A}+\overline{C})}$$

$$= (\overline{A}+\overline{B}) + (\overline{A}+\overline{C}) = A.B + A.C$$

that is, $f_1 = f_2$ and the two networks generate identical logical functions. The procedure for developing two-tier NAND networks is summarized in the following:

(a) Draw the Karnaugh map of the function and group the 1's in the manner outlined earlier, each loop being defined in the logical product form, e.g., in the form $A.B$, etc.

(b) Draw a two-tier NAND network having as many NAND gates at the input as there are loops on the Karnaugh map. The outputs from these gates are connected to the inputs of the NAND gate in the second tier.

(c) The variables defined in each of the individual loops are connected to the inputs of the first tier NAND gates in the manner shown in Fig. 7.18(c) (A and B in the case of G1, and A and C in G2).

The procedure outlined above does not necessarily lead to a minimal network, but it provides a basis for a working system.

Note: Using the above procedure it is necessary to construct a two-tier system for all Karnaugh map configurations, even if only one cell is defined.

7.14   Design of NOR systems using the Karnaugh map

One method of approaching the design of an all-NOR network is to develop an equation which satisfies the grouping of the zeros on the Karnaugh map. We have already shown in Fig. 7.18 that the function represented by the map in Fig. 7.19(a) is

$$f_1 = A.B + A.C$$

By grouping the 0's in the manner shown in Fig. 7.19(b), we obtain the

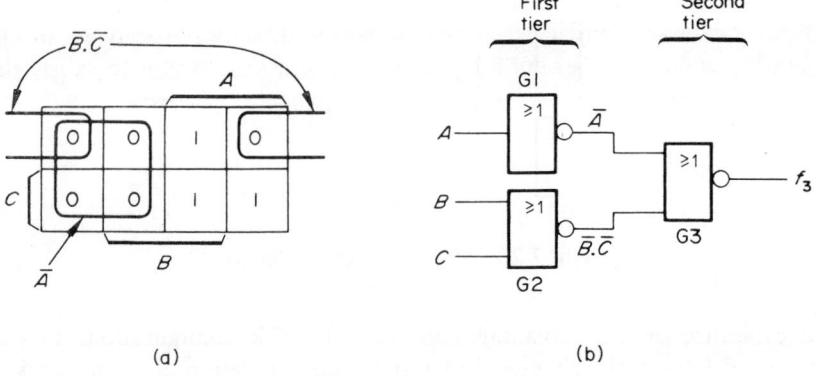

(a)   (b)

Fig. 7.19   Design in an all-NOR network.

equation

$$\bar{f}_1 = \bar{A} + \bar{B}.\bar{C}$$

hence $$f_1 = \overline{\bar{A} + \bar{B}.\bar{C}}$$ (7.4)

Equation (7.4) indicates that, in order to generate the function f_1, we need a NOR gate with two input signals, one being $\bar{A}$ and the other being the function $\bar{B}.\bar{C}$. That is, the inputs are equivalent to the groups of 0's on the Karnaugh map. The functions $\bar{A}$ and $B.C$ are generated by gates G1 and G2, respectively, in the first tier of the system. The design procedure based on this study is summarized as follows:

(a) Draw the Karnaugh map of the function and group the 0's, each loop being defined in the logical product from (e.g., $\bar{B}.\bar{C}$ in Fig. 7.19(a)).

(b) Draw a two-tier NOR network having as many NOR gates in the first tier as there are loops on the Karnaugh map. The outputs from these gates are used as inputs to the second tier gate.

(c) The **logical complements** of the variables defined in each of the individual loops are used as inputs to the first tier NOR gates in the manner shown in Fig. 7.19(b) (i.e., signal A in the case of G1, and signals B and C in the case of G2).

7.15   The wired-OR function

The wired-OR function is a name applied to gates used with their output terminals directly coupled together in the manner shown in Fig. 7.20.

The name wired-OR is derived from the fact that if the output of G1 is zero OR the output of G2 is zero, then the output from the circuit itself is zero. This statement is, in fact, not true of certain types of TTL NAND gates (see section 7.19) and modified types of gates have to be used in order

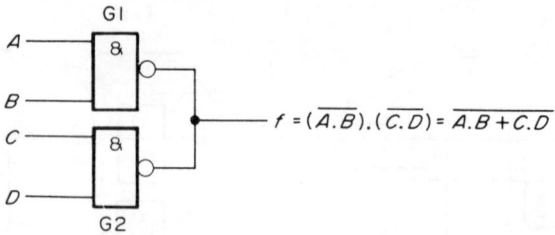

Fig. 7.20 The wired-OR function.

to capitalize on the advantages of the wired-OR configuration. This is discussed later in the chapter. When the actual operation of the network is considered, it is found that connecting the outputs of G1 and G2 provides the AND function of the individual outputs. That is, in Fig. 7.20

$$f = (\text{output of G1}) \text{ AND } (\text{output of G2}) = (\overline{A.B}).(\overline{C.D}) \qquad (7.5)$$

The design of logic circuits using the wired-OR configuration can be carried out as follows. It we consider the problem outlined in Fig. 7.19(a) and its solution in eq. (7.4), we see that eq. (7.4) and (7.5) are similar in general form.

If we apply inputs to NAND gates connected in the wired-OR configuration corresponding to *the terms defining the zeros* in Fig. 7.19(a), in the manner shown in Fig. 7.21, then the function generated is

$$f_4 = (\overline{A}).(\overline{B.C}) = \overline{A} + \overline{B.C}$$

the final form of the equation describing the Karnaugh map in Fig. 7.19(a) (see also eq. (7.4)).

The design procedure for the wired-OR configuration is summarized as

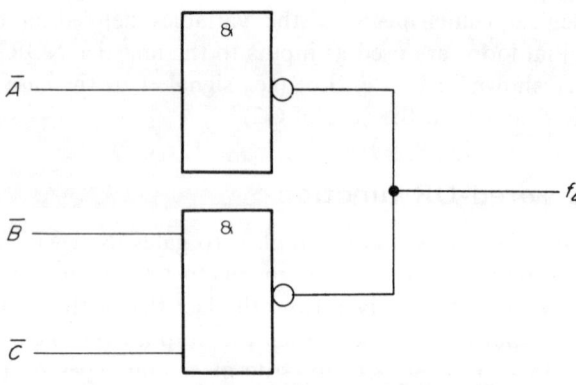

Fig. 7.21 Design of a wired-OR system.

follows:

(a) Draw the Karnaugh map of the function and group the 0's, each loop being defined in the logical product form.
(b) Draw a wired-OR network having as many NAND gates as there are loops on the Karnaugh map. The output terminals of all the gates are connected together as in Fig. 7.21.
(c) The variables defined in each of the individual loops are used as inputs to the separate NAND gates (one gate per loop), in the manner shown in Fig. 7.21.

NOR gates can also be used with their outputs connected together, the effect being simply to generate the NOR function of all the inputs. That is, the circuit merely acts as a NOR gate but with a greater fan-in capability as we shall show in the following. If gates G1 and G2 in Fig. 7.20 are replaced by NOR gates, the function generated is

$$\overline{(A+B)}.\overline{(C+D)}=\overline{A+B+C+D}$$

Thus, if two NOR gates each with a fan-in of ten are used in this

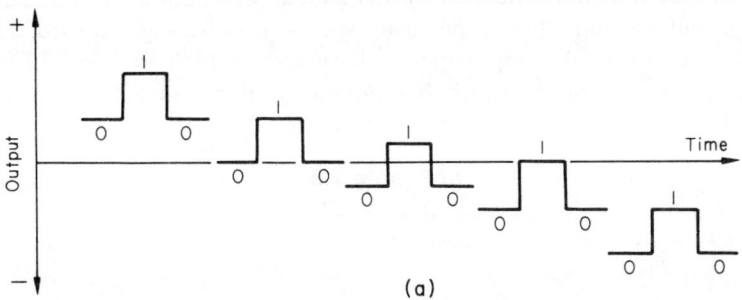

(a)

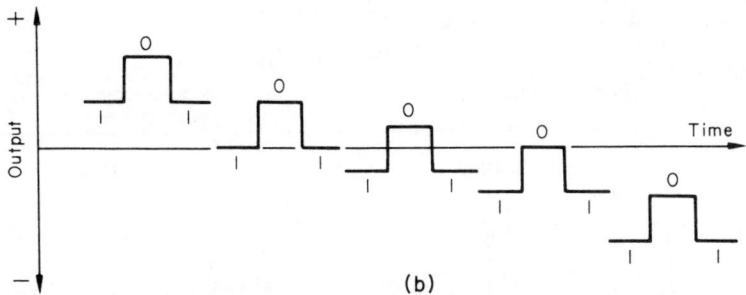

(b)

Fig. 7.22   (a) Positive logic levels and (b) negative logic levels.

configuration, then the result is an equivalent single NOR gate with a fan-in of twenty.

7.16 Logic signal levels

Up to this point in the chapter we have assumed that zero voltage has represented the logic '0' level and that a positive voltage represents the logic '1' level. This is conveneint with n-p-n transistors which have a positive collector supply voltage. In some instances other polarities may be used, as in the case of p-n-p bipolar transistors and also of p-channel MOSFETs, which use negative supply potentials. In other systems, 'floating' voltage levels are used in which neither of the two levels has zero value.

It is convenient to define two types of logic levels, namely positive logic and negative logic. In *positive logic* systems, the more positive (or less negative) of the two signal levels is taken to be logic '1', the less positive (or more negative) of the two levels having the logical value '0'. Illustrative examples of positive logic levels are given in Fig. 7.22(a).

Negative logic is the converse of positive logic, typical negative logic levels being illustrated in Fig. 7.22(b).

The logic function generated by any gate depends not only on the circuit design but also on whether positive logic or negative logic is used. This is illustrated in the following. Suppose that the truth table in Table 7.6 is that of an electronic gate in which H represents a high (more positive) voltage and L represents a low (less positive) voltage. In positive logic, $H = 1$ and

Table 7.6

Inputs		Output
A	B	
L	L	L
L	H	H
H	L	H
H	H	H

Table 7.7

Positive logic			Negative logic		
Inputs		Output	Inputs		Output
A	B		A	B	
0	0	0	1	1	1
0	1	1	1	0	0
1	0	1	0	1	0
1	1	1	0	0	0

Table 7.8

Positive function	Negative logic function
AND	OR
OR	AND
NOR	NAND
NAND	NOR

$L=0$, while in negative logic $H=0$ and $L=1$. Inserting these values in Table 7.6 we obtain the truth tables in Table 7.7. Investigating the results in Table 7.7 we see that the gate generates the OR function if positive logic levels are used, and generates the AND function if negative logic is used. A more complete list is given in Table 7.8.

Fortunately, it is usual to use only one logic level within any given system so that once the logic levels have been selected, the logic function performed by the gate is dependent only on its circuitry. However, if a gate is transferred from a system with logic levels of one kind to one with logic levels of the opposite kind, then its logic function is changed in accordance with Table 7.8.

With the exception of a few isolated cases, positive logic levels are assumed in the majority of the work in this book.

7.17 Resistor-transistor logic (RTL) gates

Three types of RTL gates are illustrated in Fig. 7.23, (a) and (b) being NOR gates and (c) is a NAND gate. The gates shown in Fig. 7.23(b) and (c) are

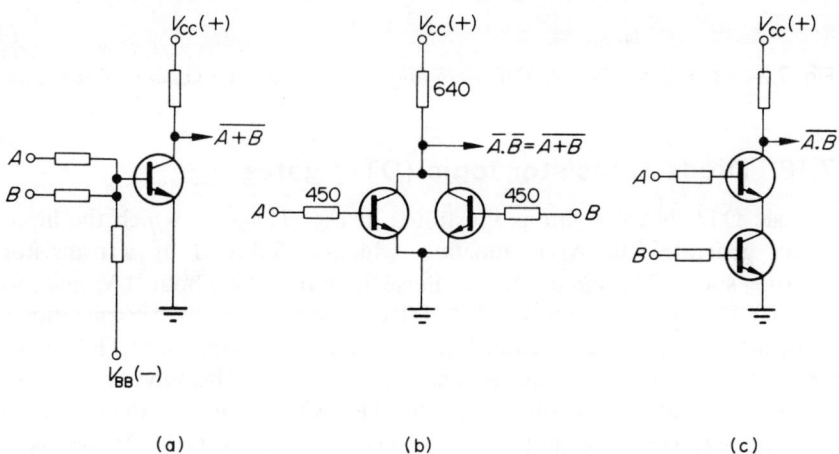

(a) (b) (c)

Fig. 7.23 (a) RTL NOR gate, (b) RTL (or DCTL) NOR gate, and (c) RTL (or DCTL) NAND gate.

also known as *direct coupled transistor logic* (DCTL) gates. The NOR function in circuit 7.23(b) is generated from a wired-OR connection of two invertors. RTL circuits were the first to be introduced in monolithic form, but have been superseded by other circuits which are superior in terms of switching speed, fan-out, and noise immunity as well as being more economical.

The gates shown in Fig. 7.23 are sometimes described as *current sourcing logic gates,* since the circuits act as a source of current for the driven gates. The NAND gates described in the following two sections are described as *current sinking logic gates* since the output transistor (TR2 in Fig. 7.24(b)) acts as a current 'sink' when it is turned on for the current flowing in the input lines of the connected gates.

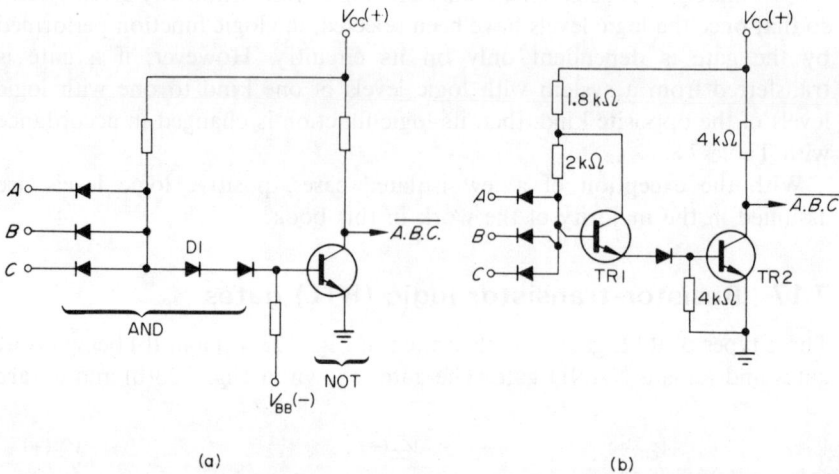

Fig. 7.24   (a) Basic DTL NAND gate and (b) an improved version of the gate.

7.18   Diode-transistor logic (DTL) gates

A basic DTL NAND gate is illustrated in Fig. 7.24(a) in which the input circuit generates the AND function, which is followed by a transistor invertor stage. The circuit has a noise immunity of about 1 V and, to improve the noise immunity, diode D1 can be replaced by a Zener diode.

An integrated circuit version, Fig. 7.24(b), includes transistor TR1 in the input circuit and only requires one power supply. The modification increases the available turn-on current for TR2 which improves the switching performance. The propagation time of DTL gates is typically 25–100 ns.

DTL gates of the types shown in Fig. 7.24 are suitable for use in the wired-OR configuration since, when TR2 is saturated it will hold the outputs of other gates connected to it at zero potential. Essential requirements of gates which are used in wired-OR networks is that their output resistance must be low when the output voltage is low, and their output resistance must be high when the output voltage is high.

7.19   Transistor-transistor logic (TTL) gates

A typical TTL NAND gate is shown in Fig. 7.25. The design represents a departure from conventional circuits in the use of the multiple-emitter transistor TR1, which has several emitter diffusions introduced during manufacture. The switching speed of TR1 is extremely fast since the base current I_1 is more or less constant irrespective of the states of the inputs. When one or more of the input lines are at zero potential, current I_1 flows from the base to those input terminals. When all inputs are at the '1' level, current I_1 flows from the base of TR1 into the base of TR2 via the collector of TR1. As a result, the rise, fall, and storage times associated with TR1 are virtually eliminated. From the above, we can see that the signal at the collector of TR1 is the AND function of the input signals.

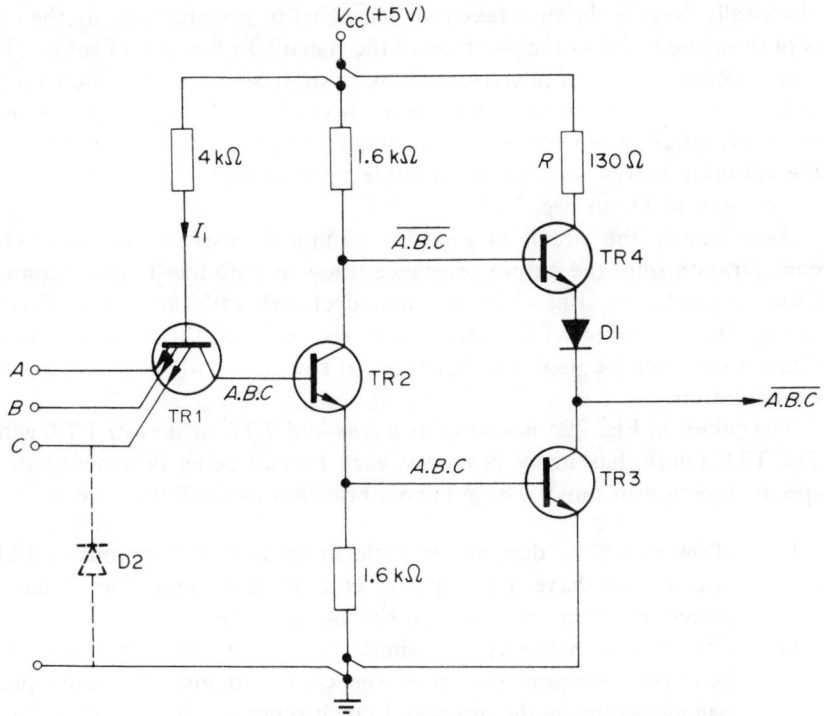

Fig. 7.25   A TTL NAND gate.

Transistor TR2 in Fig. 7.25 operates as a phase-splitting amplifier so that its collector signal is the NAND function of the inputs, and the emitter output is the AND function. Since the voltage levels at the collector and emitter of TR2 are complementary, then when TR3 is saturated then TR4 is cut off and vice versa. A feature of this arrangement is that the output resistance is very low for both '1' and '0' outputs, and is approximately equal to that of a saturated transistor.

During the very short transition period when the output changes from one logic level to the other, both TR3 and TR4 conduct simultaneously. To prevent an excessive surge of current through the output circuit when this happens, a resistor is included in series with the collector of TR4. Diode D1 also serves to reduce the current surge and to improve the noise immunity to about 1 V. The propagation delay of this type of gate is, typically, 10 ns with a further reduction to about 3 ns by the use of Schottky clamping diodes.

A problem associated with TTL gates is the effect on the circuit perfor-mance of its short transition time or rise time which, in a TTL gate with a 6 ns propagation delay, may be as small as 1 ns. The rapid change in output voltage can give rise to oscillations in output voltage due to reflections when signals are applied to lines which are electrically long. [A line is electrically 'long' if the time taken for the signal to propagate along the line is of the same order as the rise time of the signal.] In the case of some TTL gates, it corresponds to interconnections as short as 0.2 m. The oscillations arise from reflections of the voltage wave, firstly from the receiving end and, secondly, re-reflections at the transmitting end. One method of minimizimg the reflection waves is to include a diode between each input and earth, in the manner of D2 in Fig. 7.25.

As it stands, the circuit in Fig. 7.25 cannot be used in the wired-OR configuration since the output resistance is low in both the '1' and '0' states. Circuits similar to Fig. 7.25 are manufactured with an *open-collector* arrangement in which TR4, D1, and *R* are omitted; this type of circuit element can then be used with an external resistive load in the wired-OR configuration.

The circuit in Fig. 7.25 is known as a *standard TTL* or *normal TTL* gate. The TTL family has many branches, each branch being designed with a specific function in mind. The principal branches of the family are

L   (Low power) — designed to perform the normal functions of TTL circuits, but have a low power consumption. This branch has a larger propagation time than the 'normal' type.

LS   (low power Schottky) — similar to the L type, but includes Schottky clamping diodes in the circuit to give the same pro-pagation time as the 'normal' branch types.

ALS (advanced low power Schottky) — similar to LS, but a more advanced design to give higher speed.

7.20 Emitter-coupled logic (ECL or E^{2}CL) gates

The basis of the ECL gate is the emitter-coupled amplifier described in chapter 2. One type of circuit is shown in Fig. 7.26 in which the emitter-coupled amplifier includes transistors TR1-3 in the left-hand half of the circuit and TR4 in the right-hand half. Transistors TR5 and TR6 operate in the emitter follower mode to provide a low output resistance.

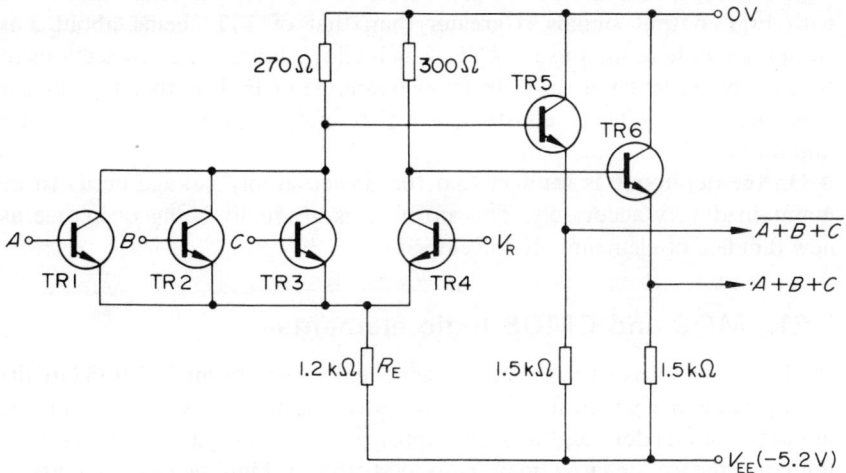

Fig. 7.26   One form of ECL OR/NOR gate.

The transistors operate in a non-saturated mode so that they carry current at all times. This is achieved by the use of a stabilized reference voltage V_R connected to the base of TR4. The potential difference between the two logic levels is typically 0.8 V, the reference voltage of about -1.17 V lying about midway between the two values. Thus, logic '0' (positive logic) is represented by a voltage of about -1.57 V, and logic '1' by about -0.77 V.

Since the input circuit is part of a long-tailed pair amplifier then, when all the inputs are at logic '0', transistor TR4 carries its maximum current and its collector voltage is 'low'. At this instant of time, the collector voltage of transistors TR1-3 is 'high'. When any input is at the logic '1' level, the current through that particular transistor increases, causing the current through TR4 to decrease. This action causes the collector voltage of TR4 to be 'high' and that at the collectors of TR1-3 to be 'low'. In this way, the logic function at the base of TR5 is the NOR function of the inputs, and that at the base of TR6 is the OR function.

Since the transistors operate in the unsaturated mode, the problems associated with the storage times of saturated transistors are eliminated. Moreover, since the voltage swing is small, propagation times of 1-3 ns are obtainable with ECL.

This type of gate is also known as *current mode logic* (CML) or *current steering logic* since the logic signal 'steers' the current flowing in R_E through the appropriate transistor.

ECL gates have a number of advantages over other families of logic circuits. Included in these are the potentially higher operating speed due to the smaller propagation delay and the fact that the output signal and its logical complement are available simultaneously. The fall time associated with ECL output signals is greater than that of TTL, being about 3 ns compared with about 1 ns for TTL, which allows longer interconnections to be used between elements. A further advantage of ECL is that the current consumed per gate is constant, so that power supply transients are minimized.

On the debit side is the fact that the power supply voltage needs to be maintained very accurately. This problem is gradually being overcome as new families of elements are developed.

7.21 MOS and CMOS logic elements

In chapter 1, it was shown that p-channel enhancement-mode MOS circuits were particularly suited for use in integrated circuits. Consequent upon the use of p-channel devices, the drain supply potential is negative; for engineering convenience, negative logic is used so that a 'high' negative voltage is taken to represent logic '1'.

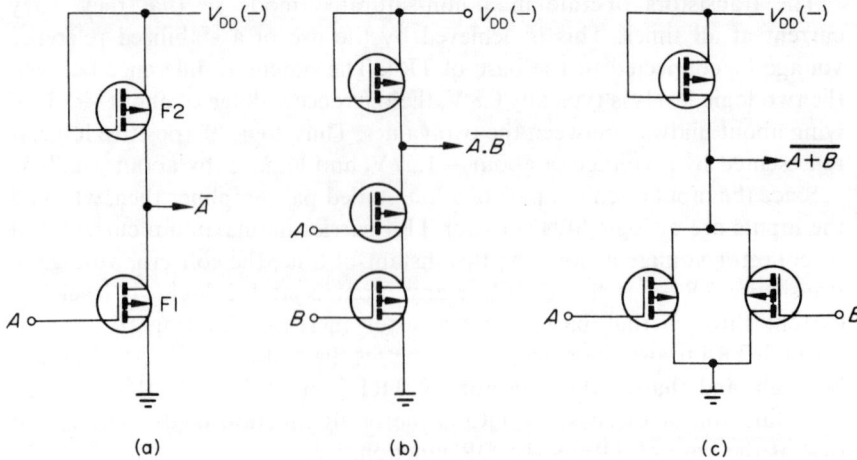

Fig. 7.27 Negative logic MOS gates: (a) INVERTOR, (b) NAND, and (c) NOR.

Typical p-channel MOS circuits are illustrated in Fig. 7.27. In Fig. 7.27(a), F1 operates as the active transistor and F2 as a pinch-effect resistor. The complex techniques required to isolate devices in the case of bipolar integrated circuits are not necessary in the case of p-MOS circuits, and all that is necessary is to ensure that the substrate-to-channel junction is reverse biased. This is brought about in p-channel devices by connecting the substrate to the most positive point in the circuit (earth in this case).

Supply voltages for this type of circuit usually lie in the range $-15\,\text{V}$ to $-20\,\text{V}$, with logic '1' having a value in the range $-9\,\text{V}$ to $-14\,\text{V}$. Logic '0' lies in the range $-2.5\,\text{V}$ to $-5\,\text{V}$. The propagation delay of p-MOS gates is, typically, 50–150 ns. The effective resistance of the transistors employed in the circuit is dependent on the ratio of channel width to channel length; for the invertor transistor, F1 in Fig. 7.27(a), this ratio is about 15 times greater than for the load transistor (F2) so as to ensure a low logic '0' voltage.

Most forms of MOS logic systems can operate with pulsed power supplies, and this has the effect of reducing the average power consumed by the system. However, where gates are cascaded, it is necessary to have more than one pulsed supply since it requires a finite time for the output voltage to reach its steady value. The so-called *two-phase* and *four-phase* systems use pulsed supply voltages which are applied sequentially to the system. It can be shown that multi-phase systems are capable of achieving a power consumption lower than p-channel systems and can offer a speed improvement of two to three times over them.

Since electrons have a higher mobility than holes, n-channel devices have a potentially higher operating speed than do p-channel devices. This fact

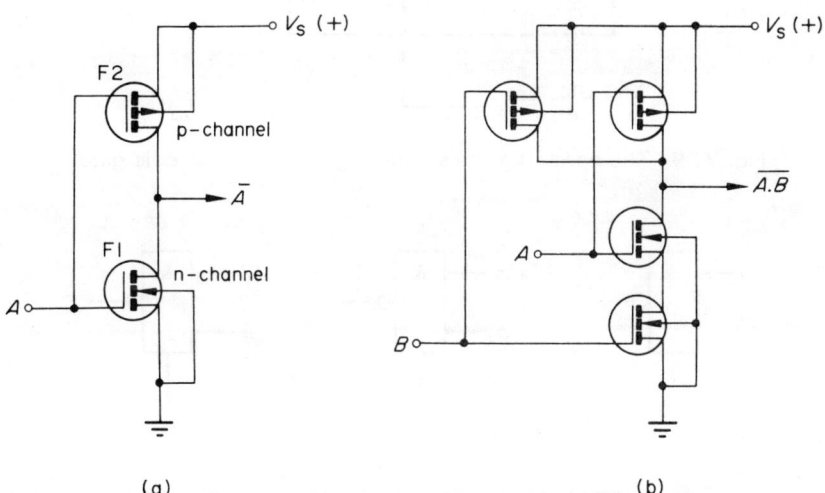

(a)   (b)

Fig. 7.28   Positive logic CMOS gates: (a) INVERTOR and (b) NAND.

is capitalized in complementary MOS (CMOS) circuits which utilize both n-channel and p-channel devices. Two circuits are shown in Fig. 7.28. In both cases, the supply voltage is positive and, for convenience, positive logic levels are used. The use of n-channel devices results in a two- to three-fold improvement in speed when compared with completely p-channel systems, but only at the cost of increased manufacturing complexity.

An added advantage of CMOS circuits is that the quiescent power consumption per gate can be as low as a few nanowatts. The reason for this is that when, in Fig. 7.28(a), F1 is ON then F2 is OFF and vice versa. Since the gate current of the MOSFETs is very small, the input power per gate is only a few nanowatts.

7.22   Tri-state logic or three-state logic gates

The advent of microprocessors has resulted in a completely new generation of logic devices, namely three-state logic or Tri-state logic (the latter being a trade name). A conventional logic gate provides a two-state output signal, that is, the output is either logic '0' or is logic '1'. A three-state gate has a third output condition, namely the output from the gate can be an open-circuit.

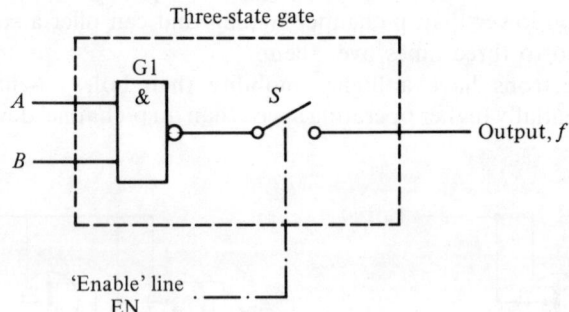

Fig. 7.29   The basis of a three-state logic or Tri-state logic gate.

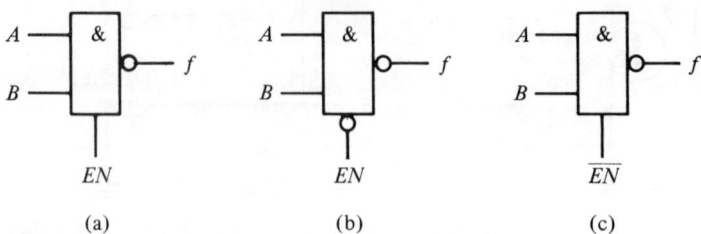

Fig. 7.30   Representation of 'enable' signals (these are also known as chip enable (CE), device enable (DE), device select (DS) and chip select (CS)).

The basic mechanics of its operation are illustrated in Fig. 7.29. The basic logic function associated with the chip is produced by gate G1 (in this case it is the NAND function). Switch S is an electronic switch which is controlled by a logic signal on the *chip enable line*, *EN*. Let us assume that when $EN = 1$, switch S is closed. Thus, when $EN = 1$, the output from the gate is given by the expression $f = \overline{A.B}$ and, depending on the signals A and B, the output from the three-state gate is either logic '0' or it is logic '1'. In this state the chip is said to be 'enabled' or 'selected'.

When $EN = 0$, switch S is open and the logic signal generated by gate G is isolated from the output terminal. In this case the chip is said to be 'disabled' or 'deselected'.

In some cases the chip is enabled by means of a logic '1' (see Fig. 7.29), and this is represented in the form in Fig. 7.30(a). However, certain chips are 'enabled', i.e., the internal switch S in Fig. 7.29 is closed, by the application of a logic '0' to the 'enable' pin of the chip (in this case the chip is disabled when a logic '1' is applied to the 'enable' line). Two methods of illustrating the latter type of gate are shown in Figs. 7.30(b) and (c). One method (diagram (b)) is to show a logic inversion 'bubble' together with an *EN* on the 'enable' line; alternatively, the 'enabling' line may enter the chip directly (diagram (c)), but the 'enabling' line has $\overline{EN}$ written by it.

The reason why Tri-state logic has been widely adopted in microprocessor-based systems is that these systems use a common set of busbars (known as *buses*) to connect the microprocessor to all its supporting devices or *peripherals* such as keyboards, video display units, etc. Since there may be many hundreds of these peripherals, only one peripheral at a time is given access to the data bus system, otherwise the possibility for confusion is very large if every peripheral is allowed to 'talk' to the bus system at the same time. The peripherals are therefore connected to the bus system through Tri-state logic chips, only one chip at a time being enabled by the system (see also chapter 9). In this way the central processing unit can control data transfer both to and from support chips.

Problems

Note: In the following the symbols 'v' and '+' represent the OR function.

7.1 Simplify the following Boolean expressions, showing all working.

(a) $f = A.B \vee \bar{A}.\bar{B}$
(b) $f = X \vee \bar{X}.Y$
(c) $f = (X \vee \bar{Y}).(\bar{X} \vee Y).(\bar{X} \vee \bar{Y})$
(d) $f = A \vee \bar{A}$

7.2

A	B	C	Output
0	1	1	0
0	1	0	0
1	0	1	1
1	1	0	1
0	0	0	0
1	0	0	0
1	1	1	1
0	0	1	0

(a) From the above truth table, give a Boolean algebraic expression for the output.

(b) Using Boolean algebra, simplify the expression in (a).

(c) Draw a logic diagram for the simplified expression given in (b).

7.3   Repeat part (c) of problem 7.2, but using (i) NOR gates, (ii) NAND gates.

7.4   (a) If $f_1 = (\bar{A}.B \vee C) \vee (A \vee \bar{B}.C)$, write down the simplified expression for $\bar{f}_1$.

    (b) (i) Minimize the following Boolean expression

$$f_2 = B.\bar{C} \vee \bar{A}.B.C \vee A.\bar{B}.\bar{C} \vee A.C$$

      (ii) Show how this minimized expression could be represented using NOR logic elements only.

7.5   The permissible states in a safety system are expressed in the logic function

$$f = \bar{A}.\bar{C} + B.\bar{C} + \bar{A}.B.C + A.B.C$$

By means of a Karnaugh map, or other suitable means, simplify this expression and draw a block diagram to realize it using (a) a combination of AND, OR and NOT gates, and (b) NAND gates only.

7.6   (a) Minimize the following Boolean expression.

$$f = \bar{A}.C.D.E \vee \bar{A}.B.C \vee A.\bar{B}.C.E \vee \bar{A}.\bar{B}.C.\bar{D} \vee \bar{A}.\bar{B}.C.D.\bar{E} \vee A.\bar{C}.\bar{D}.\bar{E}$$

    (b) Construct a logic diagram from the minimized expression using (i) AND, OR, and NOT elements, (ii) NAND elements only.

7.7   Obtain a Boolean expression for the function represented by the truth table below, and plot it on a Karnaugh map. Implement the equation using (i) NAND gates only, and (ii) NOR gates only. What is the relevance of input A?

A	B	C	f
0	0	0	0
0	0	1	1
0	1	0	1
0	1	1	0
1	0	0	0
1	0	1	1
1	1	0	1
1	1	1	0

7.8 (a) Using Karnaugh maps, verify the Boolean expression

$$A + B.C = (A + B).(A + C)$$

(b) Define the logic terms OR and NOT, and draw RTL circuits which produce these functions.

(c) Show how simple logic gates may be used to satisfy the following truth table:

Inputs		Output
A	B	S
0	0	0
1	0	1
0	1	1
1	1	0

7.9 Draw a non-equivalence (EXCLUSIVE-OR) logic diagram using

(a) NAND logic elements only
(b) NOR logic elements only.

Build up the Boolean expressions at each stage of the diagram, finally providing that both diagrams produce the same Boolean expression.

7.10 An integrated circuit element in a digital computer has four binary input channels A, B, C and D and a single output channel E, and conforms to the truth table below. Draw the logic diagram of this circuit element, assuming that the designer has produced the simplest (minimal) form.

A	B	C	D	E		A	B	C	D	E
0	0	0	0	0		1	0	0	0	0
0	0	0	1	0		1	0	0	1	0
0	0	1	0	1		1	0	1	0	0
0	0	1	1	1		1	0	1	1	0
0	1	0	0	1		1	1	0	0	1
0	1	0	1	1		1	1	0	1	1
0	1	1	0	1		1	1	1	0	0
0	1	1	1	1		1	1	1	1	0

7.11 (a) With the aid of logic drawings for the following two circuits, construct a truth table and show that they are equivalent.

Circuit 1: inputs B and C to an OR gate whose output is AND gated with input A to give output X.
Circuit 2: two AND gates with separate inputs B and C share input A. The outputs of these gates enter an OR gate to give output Y.

(b) Draw the logic diagram of the following expression and, using Boolean algebra with all working shown, simplify the expression and then draw the single logic element it represents.

$$Z = A.[\overline{\bar{B} \vee \bar{A}.(C \vee \bar{B}.\bar{C})}]$$

7.12 (a) Design a logic network consisting of NAND elements to translate the following input digits into the given digits at one time:

Input digits	Output digits
0	0
1	6
2	4
3	7
4	0
5	3
5	7
7	7

The input and output digits are represented in binary code using three input lines A, B, C and three output lines X, Y, Z, where C and Z are the most significant bits of the binary codes.

(b) Simplify your logic using Karnaugh maps.

7.13 Discuss the terms 'inhibit' and 'enable' in connection with input signals to logic gates. State which of the logic levels '1' and '0' are 'inhibit' and 'enable' signals when applied to (i) AND, (ii) OR, (iii) NOR, (iv) NAND gates. State in each case the output from the gate when it is 'inhibited'.

7.14 Draw a circuit diagram of a NAND element for use with positive logic. Include in your diagram the following components

(a) anti-saturation diodes
(b) clamping diodes.

Explain the operation of the circuit with particular reference to (a) and (b) above.

8. Counting circuits, timing circuits, and IC memories

8.1 Basic memory elements

All forms of *counting networks* and *sequential logic networks* incorporate memory elements to record the state of the problem at given instants of time. Memory elements or flip-flops are of two main types:

(a) Static memories
(b) Dynamic memories.

In a *static memory* element, such as the *S–R* flip-flop described in chapter 6, the element retains the stored information so long as its power supply is maintained. They are also known in computer technology as *volatile memories*. In *dynamic memories*, the information is stored in the form of an electrical charge held by the gate dielectric of a MOSFET. The capacitance of the gate dielectric is typically a few picofarads and, since the leakage current is very small, the charge leaks away very slowly. If, for example, the gate capacitance is 2 pF and the leakage current is 10^{-15} A, then the decay in gate voltage is

$$\mathrm{d}v/\mathrm{d}t = i/C = 10^{-15}/2 \times 10^{-12} = 0.5 \times 10^{-3} \ \mathrm{V/s}$$

that is 0.5 mV/s or about 1 V in 33 minutes. This time decay is normally very long compared with the usual time scale associated with, say, computer circuitry. In many dynamic memory circuits, the memory is 'refreshed' periodically by a recharging circuit so that the information is not lost.

8.2 Contact bounce elimination

All electrical switches which depend on metal-to-metal contact produce electrical noise when the contacts close due to the effects of contact 'bounce'. Where switches are used as input devices to counting circuits which use high-speed electronic circuit elements this effect can cause problems since, when the contacts close, the counter may count a large number of pulses due to the existence of the high frequency 'bounce' signals rather than a single pulse. Contact bounce elimination circuits using NOR and NAND elements are shown in Fig. 8.1(a) and (b), respectively.

The circuits are, in essence, versions of the basic S–R flip-flop. The output from both circuits is '0' when the switch is in position A, and is '1' when the switch is in position B. Using 5 V TTL NAND logic, the value of R would be about 1 kΩ.

8.3   Master-slave flip-flop

Prior to the advent of monolithic integrated circuits, the S–R flip-flop and its variations were used as the principal types of memory elements. Monolithic IC's have led to a completely new concept in circuit design, that is the two-stage master-slave flip-flop.

When flip-flops are used in *synchronous* or *parallel* counting systems, all the changes in the system occur simultaneously under the control of a clock signal. Should the circuit contain feedback links between the outputs and inputs of flip-flops, it is possible for oscillations to occur and it may not be possible to determine the true output from the system. By using two-stage master-slave flip-flops this problem is eliminated, since the feedback loop is never completely closed.

A block diagram illustrating the principle of master-slave flip-flops is shown in Fig. 8.2. Here the flip-flop is connected to the input line via the normally-open switch S1, and the master and slave sections are connected by normally-closed switch S2. When a '1' signal is applied to the control line, S1 is closed and S2 is opened, the complete sequence of events being described below.

(a) Clock signal '0'; the master stage is isolated from the input line and its output is connected to the slave. The output from the network is equal to the 'slave' output, which is equal to the 'master' output.

(b) Clock signal '1'; the slave stage is isolated from the master, and the

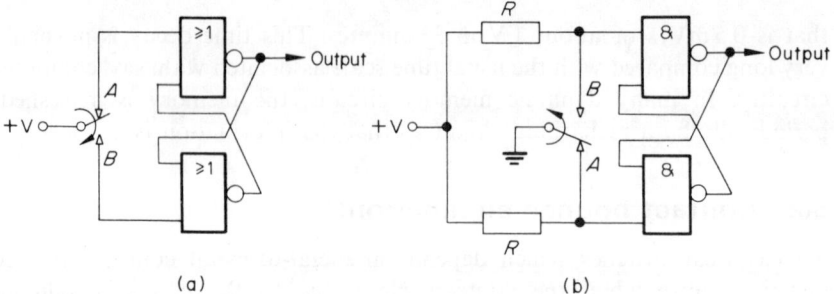

(a)   (b)

Fig. 8.1   Contact bounce elimination circuits.

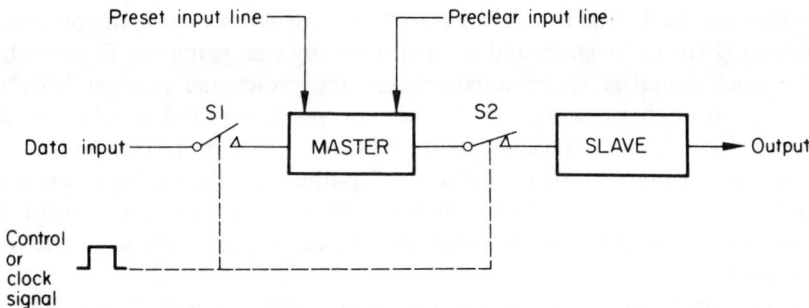

Fig. 8.2   A block diagram of a master-slave flip-flop.

master is connected to the input line. The master flip-flop is set to the
state of the input, and the slave output remains equal to the previous
state of the master stage.

(c) Clock signal '0'; the master stage is isolated from the input line and
its output is connected to the slave. The output from the slave then
becomes equal to the new state of the master stage.

Readers will note that at no stage is the slave flip-flop ever connected
directly to the input line. Consequent upon the method of operation, the
new input data is transmitted to the output at the instant the *trailing edge*
of the clock pulse occurs.

One form of master-slave *S–R* flip-flop is shown in Fig. 8.3 in which gates
G1 and G2 are equivalent to S1 in Fig. 8.2, G6 and G7 function as S2, while
G4 and G5 function as the master flip-flop and G8 and G9 as the slave.
Gate G3 ensures that when G1 and G2 are 'open' then G6 and G7 are
'closed' and vice versa.

Presetting the flip-flop is a process of setting Q to the '1' state whatever
the states of the input lines. It is carried out when the clock signal is '0' by

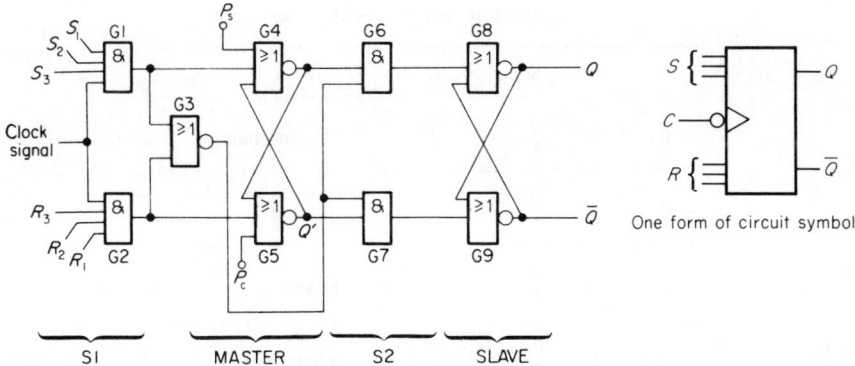

Fig. 8.3   One form of master-slave *S–R* flip-flop.

energizing the P_s line by a '1' signal. *Clearing* or *preclearing* is the process of setting Q to the '0' state, and is carried out by energizing the P_c line when the clock signal is '0'. In some circuits, the preset and preclear lines are driven through inverting stages, so that presetting and preclearing are carried out by the application of '0' signals to the appropriate lines.

Gates G1 and G2 also provide the flip-flop with the facility of carrying out logical operations. As we shall see later, these two gates permit the development of circuits with characteristics different to those of the *S–R* flip-flop.

Owing to the fact that the timing sequences within the flip-flop in Fig. 8.3 are controlled by the voltage level of the clock signal, this type of flip-flop is described as a *level-triggered master-slave S–R flip-flop*; its symbol is shown in Fig. 8.3.

The 'arrow' at the point where the clock line enters the flip-flop symbol is significant in two respects as follows. Firstly, the 'triangle' inside the flip-flop implies that changes inside the flip-flop are controlled by a change in the clock level applied to the clock input line; secondly, the inversion 'bubble' outside the flip-flop implies that it is a logic '1' to logic '0' change of the clock level which results in a change at the output of the flip-flop.

8.4   Flip-flop truth tables

Truth tables of combinational logic circuits do not change, but in the case of memory elements the truth table is dependent on the 'electrical' history of the input signal sequence. That is, it depends on the sequence of events prior to the moment we are considering. The truth tables of flip-flops are therefore *dynamic* in nature and are subject to change. The truth tables of the more popular types of flip-flops are listed in Tables 8.1 to 8.5. In the tables, the column headed Q_n lists the state of output Q *prior* to the

Table 8.1

Truth table of the *S–R* flip-flop

S	R	Q_n	Q_{n+1}	
0	0	0 1	0 1	Q_n (no change)
0	1	0 1	0 0	0 (reset)
1	0	0 1	1 1	1 (set)
1	1	0 1	? ?	Indeterminate

Table 8.2
Truth table of the T flip-flop

T	Q_n	Q_{n+1}	
0, 1 or 0→1	0	0	Q_n
	1	1	
1→0	0	1	$\overline{Q}_n$
	1	0	

Table 8.3
Truth table of the S–R–T flip-flop

S	R	T	Q_{n+1}	
0	0	0, 1 or 0→1	Q_n	No change in output
0	0	1→0	$\overline{Q}_n$	Trigger or toggle
0	1	0, 1 or 0→1	0	Reset output to '0'
0	1	1→0	?	Indeterminate
1	0	0, 1 or 0→1	1	Set output to '1'
1	0	1→0	?	Indeterminate
1	1	0, 1 or 0→1	?	Indeterminate
1	1	1→0	?	Indeterminate

Table 8.4
Truth table of the J–K flip-flop

J	K	Q_n	Q_{n+1}	comment
0	0	0	0	Q_n (no change)
		1	1	
0	1	0	0	0 (reset)
		1	0	
1	0	0	1	1 (set)
		1	1	
1	1	0	1	$\overline{Q}_n$ (toggle or trigger)
		1	0	

Table 8.5
Truth table of the D flip-flop

D	Q_n	Q_{n+1}
0	0	0
	1	
1	0	1
	1	

application of the clock pulse. The column headed Q_{n+1} lists the state of output Q *after* the application of the clock pulse.

In the case of the S–R flip-flop, Table 8.1, the outputs are unchanged when $S=0=R$, so that $Q_{n+1}=Q_n$. It was shown in chapter 6 that a '1' applied to the R line resets Q to '0', and a '1' applied to the S line sets Q to '1'. When both inputs are at the '1' level the operating condition is indeterminate.

The *trigger flip-flop* or T flip-flop, Table 8.2, has a single input line, the T line, and when it is energized at either '0' or '1' levels the output remains unchanged, so that $Q_{n+1}=Q_n$. Also, when the input changes from '0' to '1' (shown as $0\rightarrow1$ in the truth table) the output remains unchanged. However, when the input changes from '1' to '0' ($1\rightarrow0$) the output condition is inverted, so that $Q_{n+1}=\bar{Q}_n$. When this happens the flip-flop is said to *toggle* or *trigger*.

Table 8.3 shows the truth table of an S–R–T flip-flop, which is equivalent to a T flip-flop with additional set and reset overriding inputs.

Perhaps the most versatile of flip-flops is the master-slave J–K element whose truth table is listed in Table 8.4. Its truth table is generally similar to that of the S–R element (Table 8.1) if the J line is taken to be equivalent to the S line and the K line equivalent to the R line. The difference between the S–R and J–K memory elements lies in the final lines of their respective truth tables. In the J–K flip-flop the state of the output is clearly defined when $J=1=K$, whereas in the S–R flip-flop the output is indeterminate when $S=1=R$. Since the first three rows of the J–K element truth table satisfy the S–R truth table, we see that this element can be used to replace an S–R flip-flop since the respective truth tables are identical in this respect. Also, from the final line of the J–K truth table, the J–K element can be used to replace the T flip-flop.

The D flip-flop or *data latch* has a single input line (the D line) and performs much the same function as the circuit described in section 8.2, and has the truth table given in Table 8.5. This type of element is particularly useful as a buffer store between a counter and a digital readout device to eliminate readout 'flicker' while the counter is operating.

8.5 The master-slave J–K flip-flop

A basic level-triggered master-slave J–K flip-flop circuit is shown in Fig. 8.4 and comprises a flip-flop of the type shown in Fig. 8.3 with feedback links between $\bar{Q}$ and G1 and between Q and G2.

In this circuit, when $J=0=K$, both G1 and G2 are inhibited so that the output is unchanged after the clock pulse has been applied. This satisfies the first set of conditions in Table 8.4. When $J=0$, $K=1$, output Q becomes zero when a clock pulse is applied irrespective of the previous state of the output; the operation is described in the following. Suppose initially that

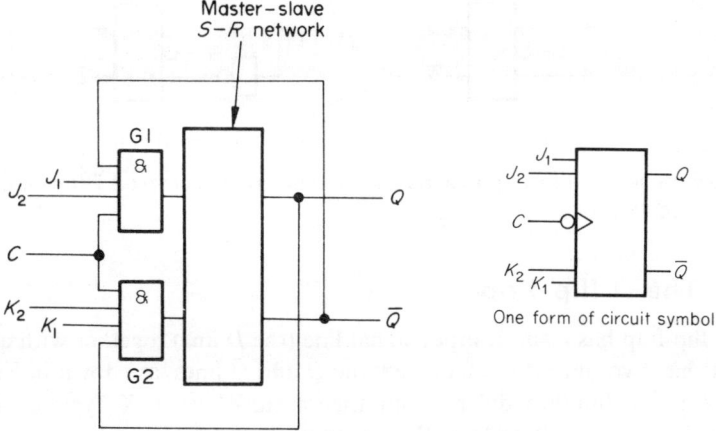

Fig. 8.4 A master-slave J–K flip-flop.

$Q = 1$. This signal is fed back to G2, which allows the K signal (which is logic '1') to be gated into the master stage of the flip-flop by the next clock pulse. On the incidence of the trailing edge of the clock pulse, the state of the master is gated into the slave to force $\bar{Q}$ to be '1', i.e., it forces Q to be '0'. If on the other hand Q was initially '0', this signal would close G2 and, at the same time, G1 would be opened to allow the J signal (logical '0') to be gated into the flip-flop. Since $J = 0$, output Q remains at zero at the end of the clock pulse.

Similarly, when $J = 1$, $K = 0$, the J signal opens G1 when a clock pulse is applied allowing, on the incidence of the trailing edge of the clock pulse, a '1' to be gated through to output Q.

When both J and K input lines are at the logic '1' level, both G1 and G2 are opened when the clock pulse is applied. Due to the feedback links, a signal equivalent to $\bar{Q}$ is gated into the upper input of the slave flip-flop and a signal equivalent to Q is gated into the lower input. When the clock signal falls to zero, the state of output Q becomes equivalent to the previous state of $\bar{Q}$, i.e., $Q_{n+1} = \bar{Q}_n$. This mode of operation is equivalent to the trigger flip-flop operation.

8.6 J–K flip-flop functions

As stated above, the J–K flip-flop can be used to perform other flip-flop functions, a number of which are described below.

The J–K element can be used as a clocked S–R flip-flop merely by using the J inputs as S inputs, and the K inputs as R inputs. It can also be used as a clocked T flip-flop in the manner shown in Fig. 8.3(a) and, when used with an invertor, it can be used as a D flip-flop as shown in Fig. 8.5(b).

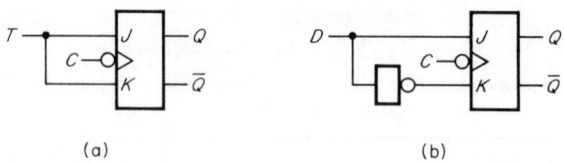

(a) (b)

Fig. 8.5 The J–K flip-flop may be used either (a) as a clocked T flip-flop or (b) as a D flip-flop.

8.7 The D flip-flop

A 'D' flip-flop has a single input signal line (the D line) together with a clock line; it has two output lines, namely the Q and $\bar{Q}$ lines, as shown in Fig. 8.6. This type of flip-flop differs from the master-slave $J–K$ type in that it responds to a logic '0' to logic '1' change in the clock signal (whereas the $J–K$ master-slave type responds to a logic '1' to a logic '0' change in the clock signal).

The D flip-flop can generally be used in applications where the $J–K$ type can be used, and has the advantage that only one signal line (the D line) is needed.

For the reason given in the final sentence of the first paragraph of this section, this type of flip-flop is described as an *edge-triggered flip-flop* since it responds to the leading edge (the '1' → '0' edge) of the clock pulse.

8.8 Binary code sequences

Forms of numbering systems in common usage include the *binary system* and *decimal system*. The systems differ in their *base* or *radix*, and in the decimal system it is ten which utilizes the digits 0, 1, 2, ..., 7, 8, 9. The radix of the binary system is two and the digits 0 and 1 are used. In the *natural binary code* or *pure binary code*, each digit is given a *weight* or value which increases by a factor of two for each digit. Thus

$$\text{Binary } 1011 = (1 \times 2^3) + (0 \times 2^2) + (1 \times 2^1) + (1 \times 2^0)$$
$$= \text{Decimal } (8 + 0 + 2 + 1) = \text{decimal } 11$$

A shorthand method of indicating the radix of the system is to write the radix as a suffix, viz:

$$1011_2 = 11_{10}$$

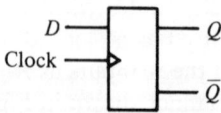

Fig. 8.6 An edge-triggered D flip-flop.

From the above we see that the weights of the pure binary code increase in the order 2^0, 2^1, 2^2, etc., and, for the convenience of the reader, the first twenty-four numbers of the binary code sequence are listed in Table 8.6, together with a number of other code sequences.

Table 8.6

Number sequences

Decimal	Pure binary code 2^4 2^3 2^2 2^1 2^0 16 8 4 2 1					8421 BCD 80 40 20 10 8 4 2 1								Aiken 2421 BCD 2 4 2 1				Johnson code
0	0	0	0	0	0	0	0	0	0	0	0	0	0	0	0	0	0	00000
1	0	0	0	0	1	0	0	0	0	0	0	0	1	0	0	0	1	00001
2	0	0	0	1	0	0	0	0	0	0	0	1	0	0	0	1	0	00011
3	0	0	0	1	1	0	0	0	0	0	0	1	1	0	0	1	1	00111
4	0	0	1	0	0	0	0	0	0	0	1	0	0	0	1	0	0	01111
5	0	0	1	0	1	0	0	0	0	0	1	0	1	0	1	0	1	11111
6	0	0	1	1	0	0	0	0	0	0	1	1	0	0	1	1	0	11110
7	0	0	1	1	1	0	0	0	0	0	1	1	1	0	1	1	1	11100
8	0	1	0	0	0	0	0	0	0	1	0	0	0	1	1	1	0	11000
9	0	1	0	0	1	0	0	0	0	1	0	0	1	1	1	1	1	10000
10	0	1	0	1	0	0	0	0	1	0	0	0	0					
11	0	1	0	1	1	0	0	0	1	0	0	0	1					
12	0	1	1	0	0	0	0	0	1	0	0	1	0					
13	0	1	1	0	1	0	0	0	1	0	0	1	1					
14	0	1	1	1	0	0	0	0	1	0	1	0	0					
15	0	1	1	1	1	0	0	0	1	0	1	0	1					
16	1	0	0	0	0	0	0	0	1	0	1	1	0					
17	1	0	0	0	1	0	0	0	1	0	1	1	1					
18	1	0	0	1	0	0	0	0	1	1	0	0	0					
19	1	0	0	1	1	0	0	0	1	1	0	0	1					
20	1	0	1	0	0	0	0	1	0	0	0	0	0					
21	1	0	1	0	1	0	0	1	0	0	0	0	1					
22	1	0	1	1	0	0	0	1	0	0	0	1	0					
23	1	0	1	1	1	0	0	1	0	0	0	1	1					
24	1	1	0	0	0	0	0	1	0	0	1	0	0					

The reader should note that whilst the decimal value 11 is pronounced 'eleven', the binary value 11_2 is pronounced 'binary one, one'; the binary value 1011_2 is pronounced 'binary one, zero, one, one'.

Where a decimal readout of a binary number is required, it is sometimes convenient to use a modified version of the binary code in which the *length* of the code sequence is restricted to ten groups of digits before it is repeated. Systems which code decimal numbers in pure binary form require at least four flip-flops and, since four flip-flops produce a natural code length of sixteen ($2^4 = 16$), when used in *binary–coded-decimal* (BCD) code generators six of the code groups are redundant and are not used.

Two examples of BCD codes are given in Table 8.6, and are the 8421

BCD code (usually referred to simply as the BCD code) and the 2421 BCD code or Aiken code. In the 8421 BCD code, the weight of the least significant digit is 1 and that of the most significant digit in the first group of four is 8. In this code, the code groups corresponding to decimal values between zero and nine are the same as those for the natural binary code. On the count of ten, the first four stages in the counter are reset to zero and a *carry* pulse is transmitted to the first stage of the next higher decade. As shown in Table 8.7, the higher decade has weights of 80, 40, 20, and 10, so that eight **bi**nary digits (*bits*) are required to store any number in the range zero to $(99)_{10}$.

The *Aiken code* is one form of 2421 BCD code in which the least significant digit has a weight of 1 and the most significant digit has a weight of 2. The next higher decade of a counter using this code would have weights of 20, 40, 20, and 10. There are many possible BCD code sequences, the relative advantages of one over the other depending on such factors as ease of circuit construction, operating speed, ease of decoding for readout and printout purposes, etc. A number of codes which include a *negative weight* in the code sequence are possible as, for example, in the $(-2)841$ BCD code the number $(6)_{10}$ is represented by the code group 1100.

In other codes the digits cannot be given weights, as in the case of the *Johnson code* in Table 8.6. One advantage of this code over some other codes is the relative simplicity with which it can be decoded into decimal. The Johnson code is also known as a *twisted-ring code* and as a *walking code*.

The widespread use of microprocessor-based systems has made the *hexadecimal code* (or *hex* code) very popular. This code has a radix of sixteen (see Table 8.7).

The first ten values (zero to nine) in the hex code are given their corresponding decimal values, the remaining six values in the code sequence corresponding to decimal ten to decimal fifteen are each given an alphabetical character in the range A to F (see Table 8.7). Thus

$$9_{10} = 9_{16}$$
$$10_{10} = A_{16}$$

and

$$15_{10} = F_{16}$$

Since the radix of the system is sixteen, the value 16_{10} is represented in hex as follows.

$$16_{10} = (1 \times 16^1) + (0 \times 16^0) = 10_{16}$$

The value 28_{10} is represented as follows

$$28_{10} = 16_{10} + 12_{10} = 1C_{16}$$

Table 8.7

The hexadecimal code

Decimal	Pure binary code	Hexadecimal code
0	000000	0
1	000001	1
2	000010	2
3	000011	3
4	000100	4
5	000101	5
6	000110	6
7	000111	7
8	001000	8
9	001001	9
10	001010	A
11	001011	B
12	001100	C
13	001101	D
14	001110	E
15	001111	F
16	010000	10
17	010001	11
18	010010	12
.	.	.
.	.	.
28	011100	1C
.	.	.
.	.	.
37	100101	25
.	.	.

The reason why the hexadecimal code is widely used in microprocessor-based systems (see also chapter 9) is that each hexadecimal character represents four binary digits. For example

$$0_{16} = 0000_2$$
$$9_{16} = 1001_2$$
$$A_{16} = 1010_2$$
$$F_{16} = 1111_2$$

The number of bits that can be handled in one operation by a microcomputer is known as the *word length* of the microcomputer. The majority of popular microcomputers have a word length of eight bits (also known as a *byte* — see also section 9.2). Since two hexadecimal characters can represent an eight-bit computer word, the hex code is clearly more economic to use than the binary code. Thus the eight-bit word 00001010_2 can be represented by OA_{16} and the word 11110110_2 by $F6_{16}$. The hex code is clearly a more convenient form of code than is binary in allowing the user to communicate with the computer. The application of the hexadecimal code is liberally illustrated in chapter 9.

It is not possible to give a full treatment of code sequences here, and for a more detailed treatment readers are referred to more specialized texts. [See, for example, *Logic Circuits* by N. M. Morris (McGraw-Hill).]

8.9 Pure binary counters

Counters can be divided into *asynchronous* and *synchronous* systems, the essential difference between them being that the clock pulse is used in synchronous systems to coordinate operations, whereas in asynchronous counting systems the operations are initiated by the incoming pulses. Synchronous systems are generally faster in operation than asynchronous systems.

An asynchronous counter working in the pure binary code using J–K flip-flops connected as T flip-flops (that is, $J = 1$ and $K = 1$ in all cases, and the input pulse is connected to the clock input line) is illustrated in Fig 8.7(a); the associated waveforms are also shown in diagram (b). A synchronous counter operating in the same code is shown in Fig. 8.7(c). In both circuits, output A is the most significant digit of the code group and has a weight of 8, and D is the least significant digit and has a weight of 1. The code sequence generated is listed in Table 8.8. The operation of the asynchronous counter is described below.

Table 8.8

Code sequence generated by the circuits in Fig. 8.7

Pulse number	States of the Q outputs A B C D 8 4 2 1				States of the $\bar{Q}$ outputs $\bar{A}$ $\bar{B}$ $\bar{C}$ $\bar{D}$			
Initial condition	0	0	0	0	1	1	1	1
1	0	0	0	1	1	1	1	0
2	0	0	1	0	1	1	0	1
3	0	0	1	1	1	1	0	0
4	0	1	0	0	1	0	1	1
5	0	1	0	1	1	0	1	0
6	0	1	1	0	1	0	0	1
7	0	1	1	1	1	0	0	0
8	1	0	0	0	0	1	1	1
9	1	0	0	1	0	1	1	0
10	1	0	1	0	0	1	0	1
11	1	0	1	1	0	1	0	0
12	1	1	0	0	0	0	1	1
13	1	1	0	1	0	0	1	0
14	1	1	1	0	0	0	0	1
15	1	1	1	1	0	0	0	0
16	0	0	0	0	1	1	1	1

The counter outputs of Fig. 8.7(a) are initially set to zero by applying a pulse to the *clear* line. When the first pulse is applied, its leading edge ($0 \rightarrow 1$ edge) has no effect on the state of the counter (see Table 8.2) and the outputs remain unchanged. However, when the input voltage is reduced to zero ($1 \rightarrow 0$ edge), output D changes from '0' to '1'. This transition is applied to FFC, but it has no effect on the state of output.

The trailing edge of the second pulse again causes FFD to change state, this time from '1' to '0'. As we have already explained, a transition of this type when applied to the T input of a flip-flop causes it to toggle. Consequently, the output of FFC changes from '0' to '1'. This process is continued until, after 15 pulses have been applied, each output is at the '1'

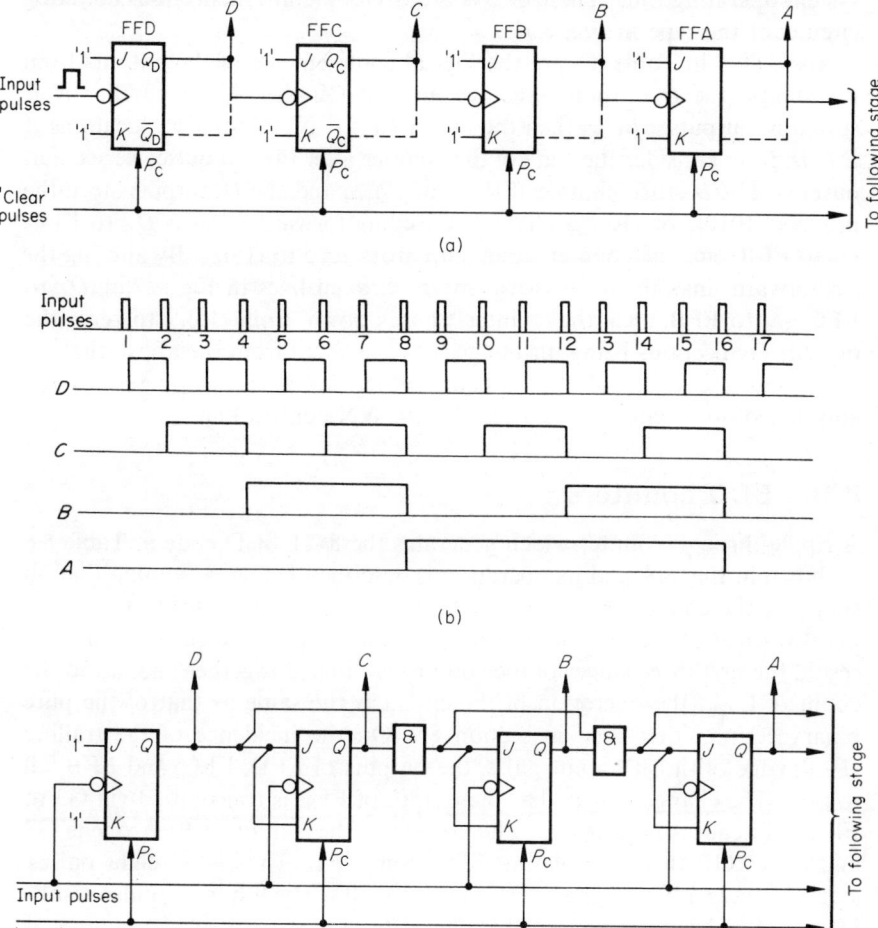

Fig. 8.7 Basic forms of pure binary counter; (a) asynchronous, (b) wave forms, (c) synchronous.

level. The trailing edge of the 16th pulse causes the output of FFD to change from '1' to '0', which results in the output of FFC changing from '1' to '0' and so on. This change *ripples through* the counter until, after a very short time, all the outputs are again zero. The $1 \rightarrow 0$ transition at the output of FFA can be used to generate a *carry* signal for the next stage of the counter.

As we have already said, the changes in the outputs of the flip-flops ripple through the counter from the input to the output, and it is for this reason that this type of counter is also known as a *ripple-through counter*. In cascading states of this type, the 'ripple-through' delays are cumulative and, in large systems, the time delay can represent a significant proportion of the system operating time. These delays are overcome in synchronous counting circuits of the type in Fig. 8.7(c).

Also listed in Table 8.8 are the logical complements of the outputs from the stages (the $\bar{Q}$ outputs), and we note that if the binary number N is stored in outputs $ABCD$, then the number $(15 - N)$ is stored in locations $\bar{A}$ $\bar{B} \ \bar{C} \ \bar{D}$. If we consider the state of the counter after the 5th pulse, we see that outputs $ABCD$ store number $(0101)_2$ or $(5)_{10}$, and the $\bar{Q}$ outputs store the number $(1010)_2$ or $(10)_{10}$. That is by feeding forward outputs Q_D to FFC, Q_C to FFB, etc., the counter 'counts up' from zero to $(15)_{10}$. By altering the feedforward links to those shown by the broken lines in Fig. 8.7(a) ($\bar{Q}_D$ to FFC, $\bar{Q}_C$ to FFB, etc.), the counter 'counts down' from $(15)_{10}$ to zero (the outputs connections being unchanged). It is evident from the above that we can design a counter which can count either 'up' or 'down' merely by the addition of logic circuitry and an UP/DOWN control line.

8.10   BCD counters

A ripple-through counter which generates the 8421 BCD code in Table 8.6 is shown in Fig. 8.8, and its operation is described below. Assuming that all stages of the counter have initially been set to zero then, since $\bar{A} = 1$, gate G1 is opened to the flow of information, and G2 is closed since $A = 0$. As a result, the first three stages of the counter are linked together and, up to the count of $(7)_{10}$, the operation of the circuit is the same as that of the pure binary counter described in section 8.9. On the incidence of the trailing $(1 \rightarrow 0)$ edge of the 8th input pulse, the outputs of FFD, FFC, and FFB fall to zero. The change in signal at the output of FFB is transmitted via G3 to FFA, and causes it to toggle. The change in the output of FFA opens G2 and closes G1, thereby isolating FFC from FFD. Thus, after eight pulses, stage A stores a '1' and the other stages store 0's. The 9th input pulse results in output D becoming '1', so that the code group stored in the counter is $(1001)_2 = (9)_{10}$. The 10th pulse causes FFD to toggle so that $D = 0$ which, in turn, toggles FFA via gates G2 and G3 so that output $A = 0$. The binary

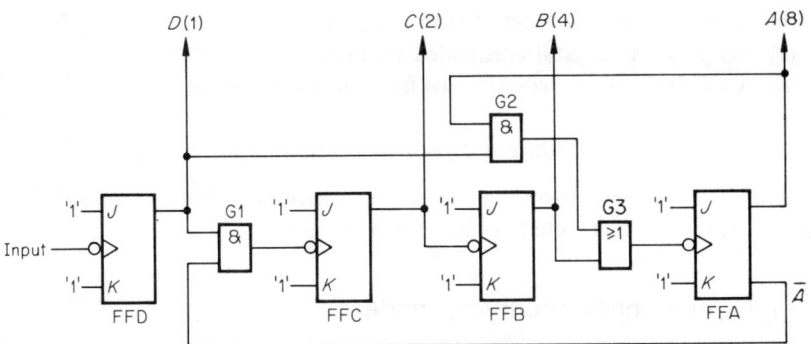

Fig. 8.8   A ripple-through 8421 BCD counter.

sequence stored by the counter is now 0000 and the circuit is once more in its initial state.

One form of synchronous 8421 BCD counter using *J–K* flip-flops is shown in Fig. 8.9. The clock pulses are, in effect, the pulses to be counted and all the changes in the outputs are synchronized with the trailing edge of the clock pulse waveform. The design of synchronous counters requires specialized techniques and is not dealt with in this book. Readers interested in this type of problem are referred to specialized texts on the topic. [See, for example, *Logic Circuits* by N. M. Morris (McGraw-Hill).]

8.11   Digital instruments for measuring frequency and time

General-purpose counter-timer instruments contain a number of essential sections which are:

(a) Voltage comparators for converting input waveforms into rectangular pulses

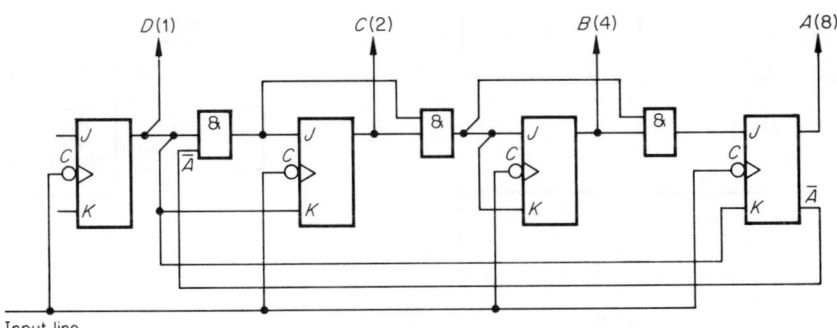

Fig. 8.9   A synchronous 8421 BCD counter.

(b) A scaling or frequency dividing section
(c) An accurate crystal-controlled oscillator
(d) A number of electronic gates for control purposes
(e) A counter
(f) A means of providing a digital readout.

The function performed by the instrument depends on the way in which the sections are interconnected, as illustrated below.

8.11.1 Frequency (counter) mode

A simplified block diagram of an electronic pulse counter is shown in Fig. 8.10. The input waveform is converted into a series of pulses by comparator C, these pulses being gated into the counter by G1 which is controlled by the output from the scaling section. The accuracy of the instrument is largely dependent on the accuracy of the timing pulse generated by the scaler; for the highest possible accuracy a crystal-controlled oscillator is used as the controlling element.

Many instruments of this kind include data latch devices (D flip-flops) as buffers between the counter and the display in order to overcome the problem of 'flicker' of the readout devices while the counter is operating.

The mode of operation described above is best used where the signal frequency is relatively high. For low-frequency signals, the period operating mode (see section 8.11.2) is best employed, in which the periodic time of the input signal is measured.

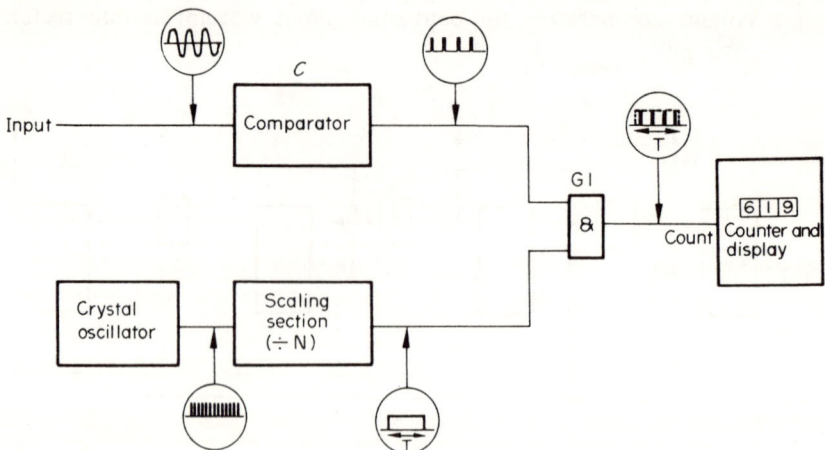

Fig. 8.10   A block diagram of a frequency measuring instrument.

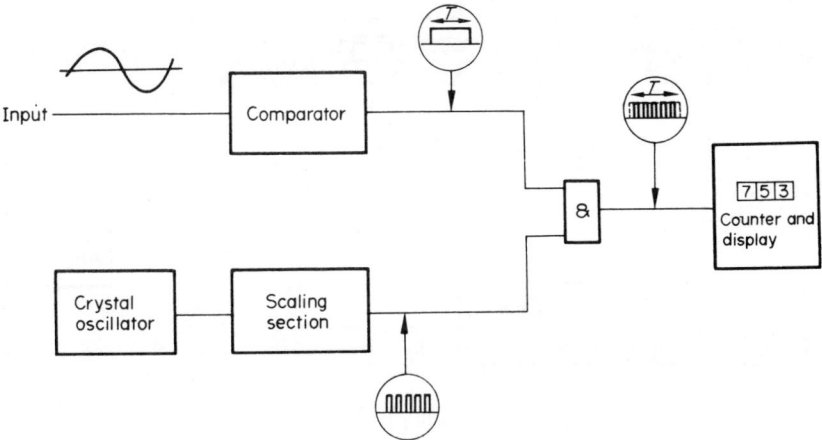

Fig. 8.11 A block diagram of a period measuring instrument.

8.11.2 Period (timer) mode

In this mode, Fig. 8.11, the input signal is used to gate pulses from the crystal oscillator into the counter. If, for example, the incoming frequency is 1 Hz and the internal oscillator frequency is 1 MHz then, with a 1 MHz counter, it is possible to measure the periodic time to within $\pm 1\,\mu s$. As a result, periodic times of low frequency signals can be measured to a high degree of accuracy.

8.11.3 Events counter mode

The circuit in Fig. 8.12 enables us to determine the number of events at input B which occur in time interval A. This could be used, for example, in determining the number of components passing a given point on a production line in a given length of time.

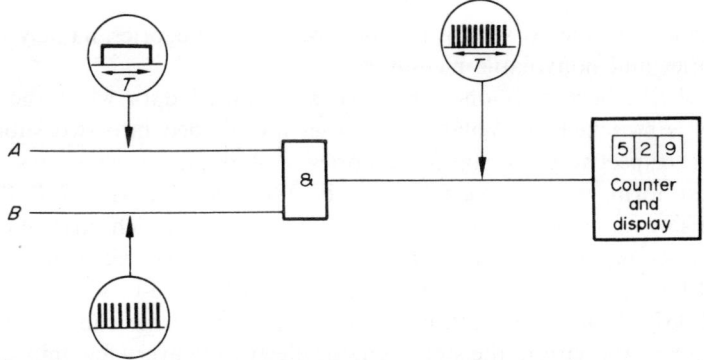

Fig. 8.12 A block diagram of an events counter.

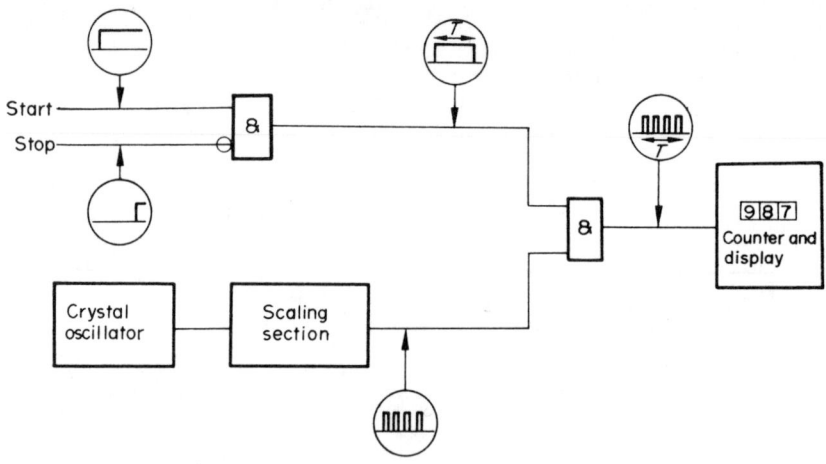

Fig. 8.13 A time interval meter.

8.11.4 Time interval mode

In Fig. 8.13 the start and stop signals provide a gating pulse which controls the application of clock pulses to the counter. This type of circuit is used in applications where the time interval between two consecutive events is to be measured.

A variation of this circuit can be used to measure the width of pulses by allowing, say, the positive-going edge to start the operation and the negative-going edge to stop the operation of the counter. A modification of this technique will allow us to measure the mark-to-space ratio of a rectangular wave train.

8.12 Types of semiconductor memory

Semiconductor memory chips fall into two main categories, namely volatile memories and nonvolatile memories.

A *volatile memory* is one which loses its stored data when the power supply is switched off. Volatile memories are divided into two subgroups known respectively as static memories and dynamic memories. In its simplest form, a *static memory* can be thought of as an *S–R* flip-flop comprising a cross-connected pair of transistors (which may either be bipolar transistors or FETs). A *dynamic memory* element contains MOSFETs, and it relies on the capacitance of the gate-to-source region of the MOSFETs to store electrical charge; since this charge can leak away, it is necessary to 'refresh' the stored charge electrically every few milliseconds. Volatile memories are described as *random access memories* (RAMs) or *read-write memories*. The name read-write memory is sometimes used because it is possible either to read data from the memory or write data to it.

A *nonvolatile memory* does not lose its stored data when the power is switched off. However, the stored data cannot be changed during normal operation and for this reason they are also known as *read-only memories* (ROM). A wide variety of ROMs are manufactured, the following being the most popular.

ROM

This type has its memory contents fixed during the process of manufacture by a photomasking and diffusion process, and the stored data cannot be changed by the user at a later stage. This type gives the lowest cost per chip if the production volume is very high.

PROM (Programmable ROM)

This type is available in an unprogrammed form and is 'field' programmed by an engineer using a PROM programming device. The memory elements are frequently semiconductor fusible links which can either be left intact or can be 'blown' by passing a high current pulse through selected fuses during the programming process. Once a memory has been programmed, it cannot be reprogrammed. This type of ROM is used where the program has been fully tested and 'debugged', that is, all errors have been removed during the testing period.

EPROM (Erasable PROM)

This type leaves the factory in an unprogrammed form, and is programmed by applying suitable voltages to pins on the memory chip. The memory element is covered by a transparent material; if necessary, all the data can be erased at a later time by exposing the chip to very strong ultraviolet radiation. After erasure, the chip can be reprogrammed once more.

EAROM (Electrically Alterable ROM)

This type of memory accepts data in much the same way as a RAM, but is very much slower in accepting data.

Problems

8.1 Draw the circuit diagram of a bistable multivibrator suitable for counting in binary code from a pulse source. Describe, with the aid of waveform diagrams, its principle of operation; give reasons for the use of steering diodes and pulse differentiating circuits. Explain with the aid of a block diagram, how a number of such stages can be connected to count in decades.

8.2 With the aid of a circuit diagram, describe the operation of a bistable circuit using two p-n-p transistors. Show how the circuit could be triggered for use as

(a) a register element
(b) a binary counter.

State the approximate operational frequency limitation of your circuit, and indicate which part of the circuit causes this limitation.

How can the circuit be modified to increase the operational frequency?

8.3 (a) Draw a logic diagram of a J–K master-slave bistable multivibrator for use in a synchronized digital system.

(b) Describe its operation fully for all combinations of JK input conditions.

(c) What principal advantage does such a device have over the gates $S–R$ bistable?

8.4 Design a ripple-through counter which works in the following code, and state the 'weight' of each digit

$$
\begin{array}{cccc}
0 & 0 & 0 & 0 \\
0 & 0 & 0 & 1 \\
0 & 0 & 1 & 1 \\
0 & 1 & 0 & 0 \\
0 & 1 & 0 & 1 \\
0 & 1 & 1 & 1 \\
1 & 1 & 0 & 0 \\
1 & 1 & 0 & 1 \\
1 & 1 & 1 & 1 \\
0 & 0 & 0 & 0 \quad \text{etc.}
\end{array}
$$

8.5 Design a counter which works in the following code, and state the 'weights' of the digits

$$
\begin{array}{cccc}
0 & 0 & 0 & 0 \\
0 & 0 & 0 & 1 \\
0 & 0 & 1 & 0 \\
0 & 0 & 1 & 1 \\
0 & 1 & 1 & 1 \\
1 & 0 & 0 & 0 \\
1 & 0 & 0 & 1 \\
1 & 0 & 1 & 0 \\
1 & 0 & 1 & 1 \\
1 & 1 & 1 & 1 \\
0 & 0 & 0 & 0 \quad \text{etc.}
\end{array}
$$

8.6 An encoder generates the following code sequence. Design a convertor using (i) NOR gates, (ii) NAND gates, to convert the code into a pure binary code.

$$
\begin{array}{cccc}
0 & 1 & 0 & 1 \\
0 & 0 & 0 & 1 \\
0 & 0 & 1 & 1 \\
0 & 0 & 1 & 0 \\
0 & 1 & 1 & 0 \\
1 & 1 & 1 & 0 \\
1 & 0 & 1 & 0 \\
1 & 0 & 1 & 1 \\
1 & 0 & 0 & 1 \\
1 & 1 & 0 & 1 \\
0 & 1 & 0 & 1 \quad \text{etc.}
\end{array}
$$

9. Microprocessor-based systems

9.1 Introduction to microcomputer systems

A block diagram of a typical microcomputer is shown in Fig. 9.1 and comprises

- (a) The *central processing unit* (CPU) or *microprocessor chip* which houses the *arithmetic and logic unit* (ALU) and the *control unit.*
- (b) The *read-only memory* (ROM) which contains the permanent operating progam of the microcomputer system.
- (c) The *random access memory* (RAM) or read-write memory which is used for the temporary storage of data.

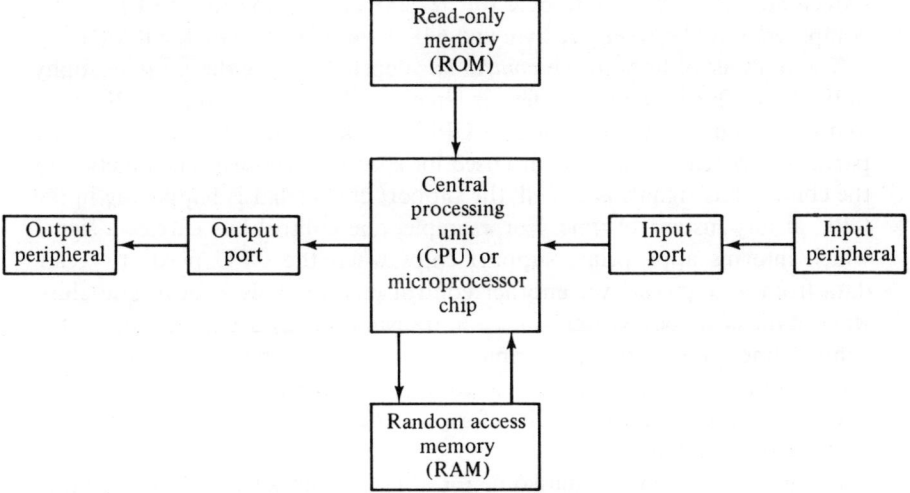

Fig. 9.1   Basic microcomputer structure.

283

(d) A number of *input peripherals*, e.g., switches, keyboards, etc., which are connected to CPU by *input ports*, which are special-purpose IC chips.

(e) a number of *output peripherals*, e.g., lights, printers, cathode-ray tubes, etc., which are connected to the CPU by *output ports* which are special IC chips.

A number of peripherals combine the function of an input peripheral with an output peripheral; an example of this type is a computer *terminal* which contains a keyboard and either a cathode-ray tube or a printer. Many applications require a 'port' that can function both as an input port and an output port; special-purpose *programmable I/O ports* (PIO) are widely used in these applications (see also section 9.19).

Microprocessor-based systems are organized around a 3-*bus system* (see Fig. 9.2) as follows.

The *data bus* carries data between the CPU and its support chips (and thence, if appropriate, to peripherals). This bus carries the data in 'parallel', so that there are as many lines in the bus as there are bits in the word length of the CPU. There are eight lines in the data bus of an 8-bit CPU (such as the Intel 8085, the Zilog Z80, and the Rockwell 6502), and sixteen lines in the case of a 16-bit CPU (such as the Intel 8086, the Motorola 68 000, and the Zilog Z8000). Moreover, the data bus is *bidirectional* so that it can carry data in either direction.

The *address bus* usually has sixteen lines in it, allowing it to address any one of $2^{16} = 65\,536$ locations. In computer jargon, this is described as 64K locations, where $1K = 2^{10} = 1024_{10}$. If each location can handle eight bits or one byte of data, then the theoretical storage capacity of a computer having sixteen address lines is 64K byte (or 524 288 bits) (many low-cost personal computers have between 8K byte and 64K byte of user-available RAM).

The number of lines in the *control bus* depends on the design philosophy of the CPU. There may be as few as ten control lines in a simple CPU or as many as twenty in a sophisticated CPU (not all of which are used in any particular system since many are used for special purposes). The function of the control bus signals is to 'tell' the support chips what is happening in the CPU at any instant of time. For example, one control bus carries a signal which informs appropriate support chips when the CPU needs to 'read' data from a chip, and yet another control bus line tells appropriate chips when data is to be 'written' into (or transmitted to) a chip. Some of the control lines are dedicated output lines (examples being the 'read' and 'write' control bus lines described above), and others are dedicated input lines (these are concerned with informing the CPU about events in other parts of the computer).

It is generally the case that some (or all) of the lines from the three buses are connected to each support chip (see Fig. 9.2). To illustrate the operation

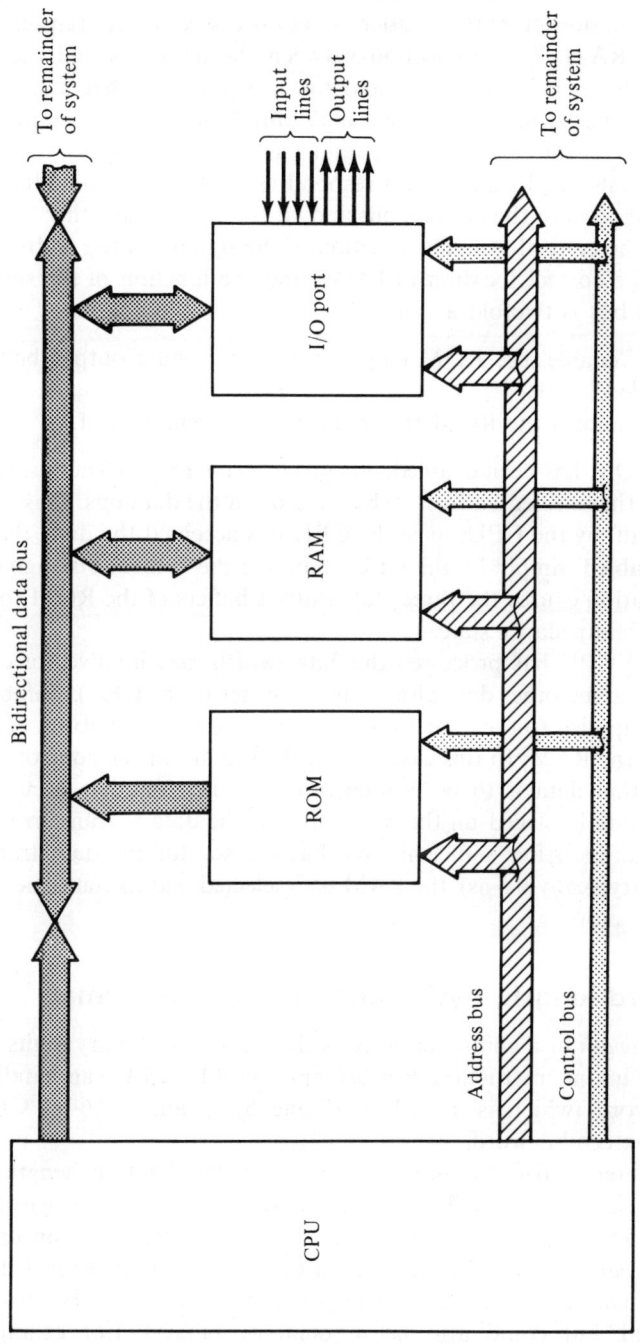

Fig. 9.2 A block diagram of a microprocessor-based system.

of the three-bus system, consider the case in which data stored in a particular location in ROM is, after being processed, to be transferred to a location in RAM. The connection between the data bus and the output lines from the ROM and RAM (and the I/O ports for that matter) is via three-state buffers which are in their high impedance mode when not in use, i.e., they are 'tri-stated' leaving the data bus lines 'floating'.

Logic signals, i.e., 1's and 0's, corresponding to the required address in the ROM are applied to the address bus and, at the same time, other signals are applied to the control bus. The function of the signals on the address bus is to 'select' the correct location in ROM, and the function of the signals on the control bus is twofold as follows:

(a) They 'enable' the ROM chip, i.e., the three-state output buffers are enabled.

(b) They inform the ROM that data is to be read from it.

Once the ROM has been enabled, it is given access to the data bus, allowing the data in the selected address to be 'placed' on the data bus. This data can then be 'read' by the CPU; once the CPU has accepted the data, the ROM chip is 'disabled' simply by the CPU changing the control bus signals. The latter operation causes the three-state output buffers of the ROM to return to their high impedance state.

When the CPU has processed the data (which may involve, for example adding it to some other data already in a register in the CPU), suitable logic signals are applied to the address bus and control bus to enable a particular location in the RAM. In this case, one of the signals on the control bus tells the RAM that data is to be written into it. When this has occurred, the processed data is placed on the data bus and the data is transferred to the selected address. After sufficient time has elapsed for the data transfer to take place (typically 0.5 μs), the RAM is deselected and its data lines are tri-stated once more.

9.2 Word length, bytes, and instruction format

The *word length* of a computer word is the number of binary digits (*bits*) it can handle in one instruction. For example, an 8-bit CPU can handle of an eight-bit word (which is described as one *byte*), and a 16-bit CPU can handle a sixteen-bit word.

A computer *instruction* is simply a group of bits (the 'length' of an instruction in an 8-bit CPU may be 1-byte, 2-bytes, or 3-bytes) which defines a computer operating sequence. Each instruction has an *operation code* or *opcode*, which is the first byte of the instruction. If the instruction is written in *machine code*, the opcode is expressed as a number (usually is hexadecimal, but could also be in octal or binary). For example, the instruction which causes the contents of an 8-bit register known as the accumulator or A-register to be added to the contents of another register known as the B-register in an Intel 8085 CPU, has the hexadecimal opcode 80 (usually written as 80_{16}) which corresponds to the binary opcode 1000 0000, this instruction is described in section 9.3. Alternatively, the instruction may be written down as an *assembly language instruction*; this is in the form of three or more alphabetical characters which are combined in some cases with a number. For example, the 1-byte instruction with the opcode 80_{16} (see above) can be expressed in the form of the assembly language instruction ADD B; that is, 'ADD the contents of register B to the contents of the accumulator'. The assembly language form of the in-struction, i.e., the mnemonic form is easier to follow than the machine code form.

In a 1-byte instruction, the CPU only needs the one byte of information in order to understand and execute the instruction. In a 2-byte instruction (occupying two 8-bit memory locations), the first byte is the opcode (which defines the operation to be carried out) and the second byte could, for example, be the data which is to be operated on. The first byte of a 3-byte instruction is, once again, the opcode; the second and third bytes may either be data or an address of a location in the memory of the computer. A 3-byte instruction occupies three consecutive 8-bit memory locations.

9.3 CPU architecture

The *architecture* or internal structure of a typical 8-bit CPU is shown in Fig. 9.3 (this is very loosely based on the Intel 8085 CPU).

The CPU contains a number of general-purpose 8-bit registers including the *accumulator* or A-register and registers B, C, D, E, H, and L (note: the number in parentheses in each register in Fig. 9.3 indicates the number of bits used in that register). Other registers include the *instruction register*, IR, which holds a copy of the instruction currently being executed, and the *flag register*,* FR, which is eight bits wide, of which only five bits are used as follows.

Sign flag, S: this is 'set' to logic '1' if the arithmetic sign of a result of a computation is negative.

Carry flag, C: this is 'set' if the carry-out from an arithmetical operation is '1'.

Parity flag, P: this is 'set' if the result of an arithmetic or logic operation results in an even number of 1's in the solution.

---

*A 'flag' is a flip-flop which can either be 'set' or 'reset' to indicate the occurrence or otherwise of a particular condition in the CPU.

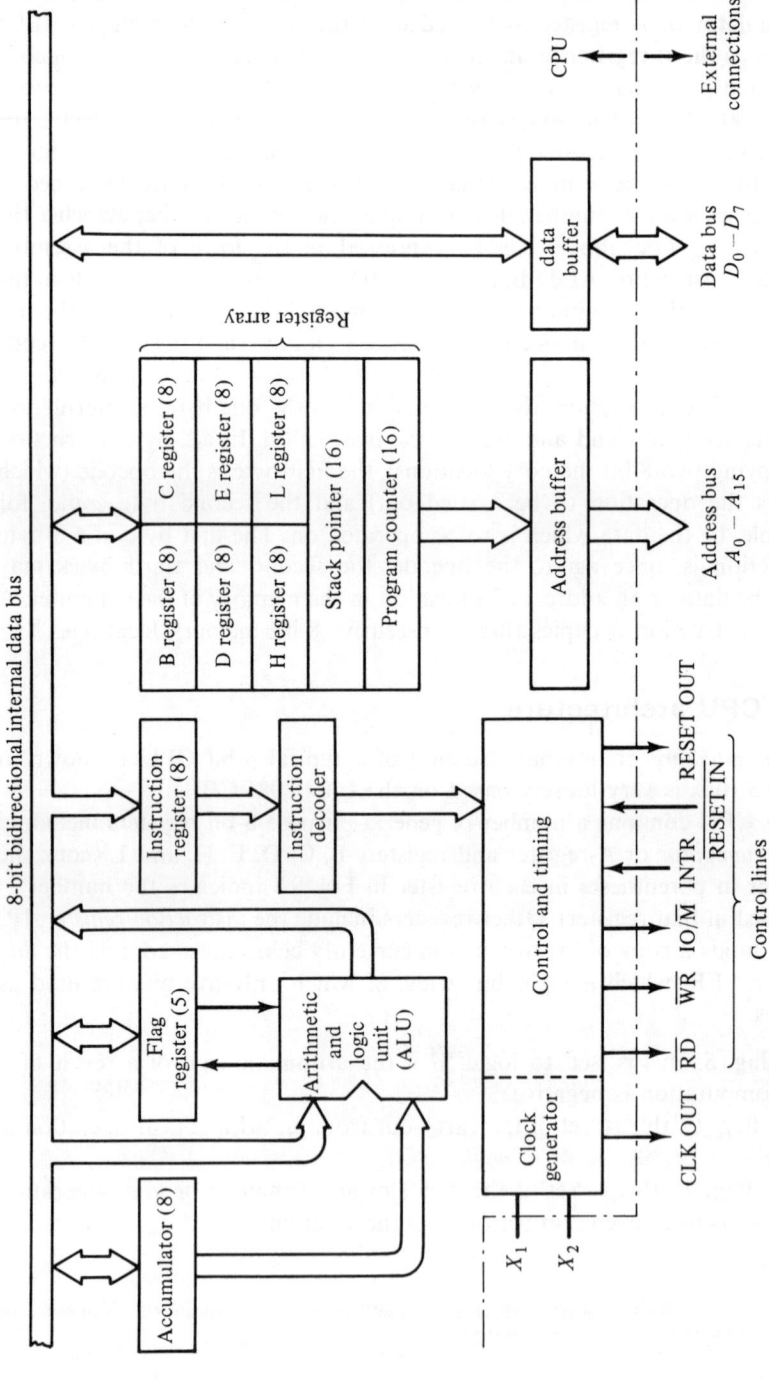

Fig. 9.3 The architecture of a typical 8-bit CPU (loosely based on the Intel 8085).

Auxiliary carry flag, A_C: this is 'set' if a carry of '1' is transferred from bit 3 to bit 4 of the accumulator during an arithmetic operation.

Zero flag, Z: this is 'set' if an arithmetic operation gives a result of zero.

The *address buffer* functions as a buffer amplifier between the 16-bit *program counter*, PC, and the external address bus of the system; the PC stores the 16-bit address of the location currently being accessed by the CPU. The *data buffer* is a buffer amplifier between the internal data bus of the CPU and the external data bus lines of the microcomputer.

The *instruction decoder* performs the function of decoding the opcode of the instruction held in the IR; the decoded instruction results in signals being generated by the *control section* of the CPU. The function of the control signals in Fig. 9.3 are described below.

$\overline{\text{RD}}$ (an output signal) – this line has a logic '0' on it when data is being 'read' from a support chip.

$\overline{\text{WR}}$ (an output signal) – this line has a logic '0' on it when data is 'written' into a support chip.

$\text{IO}/\overline{\text{M}}$ (an output signal) – a logic '1' on this line implies that an I/O device is being addressed; a logic '0' on this line implies that a memory location is being accessed.

INTR (an input line) – a logic '1' on this line means that an external device wishes to 'interrupt' the normal operation of the program.

$\overline{\text{RESET IN}}$ (an input line) – a logic '0' on this line resets not only the program counter to zero but also the interrupt enable flag.

RESET OUT (an output line) – a logic '1' on this line indicates that the CPU is being reset. This signal is generally used as a RESET signal for the remainder of the computer.

The CPU in Fig. 9.3 has an oscillator built into the chip whose frequency determining element (usually a crystal) is connected externally to terminals X_1 and X_2 of the control and timing section. The clock signal from the CLOCK OUT terminal is used as a clock signal for some of the support chips, e.g., chips containing timers.

Finally, the *arithmetic and logic unit*, ALU, can accept data either from the internal data bus or from the accumulator, and can perform a limited range of arithmetic and logic functions on the data. These functions usually include

  addition
  subtraction
  comparison of data
  data shifting
  AND operations
  OR operations
  EXCLUSIVE-OR operations

Mathematical operations such as multiplication and division are achieved by the appropriate use of addition, subtraction, and shifting instructions. Mathematical and logical operations may affect the state of flags in the flag register, an example of each kind being given below.

Consider the execution of an ADD B instruction (this being a 1-byte instruction — opcode 80_{16}), which causes the data in register B to be added to the data in the accumulator, the result being left in the accumulator. This is expressed either in the form

$$(\text{accumulator}) \leftarrow (\text{accumulator}) + (\text{register B})$$

or in the form

$$(A) \leftarrow (A) + (B)$$

The parentheses in the above expressions are a shorthand way of saying 'the contents of'; thus (register B) is interpreted as 'the contents of register B'. Suppose that the accumulator contains the hexadecimal value $A3_{16}$ and register B contains BO_{16}; the ADD B instruction has the following result

original contents of the accumulator $= A3_{16} = 1010\ 0011_2$
contents of register B $= BO_{16} = 1011\ 0000_2$

final binary value in the accumulator $0101\ 0011$ $\leftarrow$Non-zero result sets $Z=0$.
Carry-out of '1' from bit_7 sets $C=1$.   Four 1's in result sets $P=1$.
$\text{Bit}_7 = 0$ sets $S=0$.   No carry from bit_3 to bit_4 sets $A_C = 0$.

An ANA C instruction (which is a 1-byte instruction — opcode $A1_{16}$) causes the eight bits in register C to be ANDed on a bit-by-bit basis with the contents of the accumulator, the result being left in the accumulator. Suppose that the accumulator originally contains $A3_{16}$ and that register C contains $B1_{16}$. The effect of an ANA C instruction is as follows

original contents of the accumulator $= A3_{16} = 1010\ 0011_2$
contents of register C $= B1_{16} = 1011\ 0001_2$
final contents of the accumulator $1010\ 0001$ $\leftarrow$Non-zero result sets $Z=0$.
$\text{Bit}_7 = 1$ sets $S=1$.   Three 1's in result sets $P=0$.

Note: the ANA C instruction causes the C-flag to be reset to '0' and the A_C flag to be set to '1'.

9.4 Instruction sets

The *instruction set* of a CPU is the range of general-purpose instructions available to the CPU user; each instruction in the set produces a known response in the CPU. A relatively simple CPU may have only fifty or sixty instructions, while a sophisticated CPU may have more than two hundred instructions. A number of instructions are described in detail in this chapter, and the full range of 8085 CPU instructions is given in appendix A. The reader is referred to specialized texts* for details of other instruction sets.

9.5   Addressing modes

The addressing modes of the CPU are the methods which are available to specify the address to be used in an instruction. The number and variety of addressing modes available in a given CPU provide an indication of the effectiveness of the CPU in manipulating instructions and data. Some CPUs have as few as four addressing modes, whilst others have fifteen or more addressing modes.

The 8085 has six addressing modes, namely direct addressing, register addressing, register indirect addressing, immediate addressing, implied or inherent addressing, and stack addressing. These modes are described in this chapter.

9.6   Direct addressing

Each direct addressing mode instruction is a 3-byte instruction, the first byte being the opcode whilst bytes two and three contain the 16-bit address of the data or operand referred to in the instruction (byte two contains the low-order byte and byte three the high-order of the address). The LDA (LoaD the Accumulator) direct addressing instruction may appear in a program as shown in Table 9.1.

Table 9.1

Memory location (hex)	Instruction or data (hex)	Instruction mnemonic	Comment
2000	3A	LDA	LoaD the Accumulator from
2001	80		low byte of address
2002	20		high byte of address

The first byte of the instruction (the opcode), stored in memory location 2000_{16} (this was chosen at random), is $3A_{16}$. This opcode causes the contents of the address specified in bytes two and three of the instruction to be copied into or 'loaded' into the accumulator.

The low byte of the address, i.e., 80_{16} or $1000\,0000_2$, at which the data is to be found is stored in the second byte of the instruction (which is stored in

*See, for example, *Logic Circuits*, 3rd edn, N. M. Morris McGraw-Hill, London, 1984.

location 2001_{16}), and the high byte of the address, i.e., 20_{16} or $0010\,0000_2$ is stored in location 2002_{16}. If location 2080_{16} stores the hex value AB_{16}, then after the LDA 2080_{16} instruction has been executed, the accumulator contains AB_{16}.

9.7   Register addressing or register direct addressing

This is generally similar to direct addressing (see section 9.6) except that the address specified by the instruction is either a register or a register pair within the CPU rather than a memory location. The following 1-byte ADD D instruction is an example of this type.

Memory location (hex)	Instruction (hex)	Mnemonic	Comment
2010	82	ADD D	ADD the contents of register D to the accumulator.

The opcode, 82_{16}, of this 1-byte instruction is stored in memory location 2010_{16} (the location was chosen at random). When this instruction is decoded, it causes the contents of register D to be added to the contents of the accumulator, the result being left in the accumulator.

9.8   Register indirect addressing

This type of instruction specifies a register pair which contain the address of a memory where the data is to be found (the high-order byte of the address is in the first register of the pair, and the low-order byte is in the second register). The following is an example of register indirect addressing.

Memory location (hex)	Instruction (hex)	Mnemonic	Comment
2015	12	STAX D	STore Accumulator contents indirectly

The 1-byte STAX D instruction causes the contents of the accumulator to be stored in the memory location specified by the contents of the register pair D and E. Suppose that register D contained 20_{16}, register E contained $A2_{16}$ and that the accumulator stored 15_{16}. After the execution of the instruction, memory location $20A2_{16}$ contains 15_{16}.

9.9 Immediate addressing

This type of instruction contains the data itself either as the second byte of a 2-byte instruction or as the second and third bytes of a 3-byte instruction. The following is an example of a 2-byte immediate addressing instruction.

Memory location (hex)	Instruction or data (hex)	Mnemonic	Comment
2030	C6	ADI	ADd Immediate to the accumulator
2031	A2		data A2 (hex)

This instruction causes the hex value A2, which is stored in the second byte of the instruction, to be added to the contents of the accumulator, the result being left in the accumulator.

9.10 Implied addressing or inherent addressing

In this form of addressing, the operation code itself specifies the required address; in many cases the implied address is either the accumulator or the flag register. The following is an example of implied addressing.

Memory location (hex)	Instruction (hex)	Mnemonic	Comment
202A	2F	CMA	CoMplement the contents of the Accumulator

This instruction causes the contents of the accumulator to be logically complemented; for example, if the accumulator contains 19_{16} before the instruction is executed, it contains $E6_{16}$ after its execution.

9.11 Stack addressing

A 'stack' is a sequence of memory locations in which data is stored, and stack addressing is implemented by means of an address stored in the stack register of the CPU (see also section 9.14). Stack addressing is also known as *auto-indexed addressing*.

9.12 A sample program

The following program illustrates the use of a number of the above addressing modes. The program in Table 9.2 causes the decimal value XX (describ-

Table 9.2

A decimal addition program

Address	Byte			Instruction	Comment
	1	2	3	Mnemonic	
2000	XX				; decimal data$_1$
2001	YY				; decimal data$_2$
2002	ZZ				; decimal sum
					the contents of locations
					2003 to 201F are
					unimportant
2020	21	00	20	LXI H,2000	; 'point' to location 2000
2023	7E			MOV A,M	; move decimal data$_1$ to accumulator
2024	23			INX H	; 'point' to location 2001
2025	86			ADD M	; add decimal data$_2$
2026	27			DAA	; convert result to decimal
2027	23			INX H	; 'point' to location 2002
2028	77			MOV M,A	; store result in location 2002
2029	76			HLT	; halt

ed as data$_1$) stored in location 2000_{16} to be added to the decimal value YY (data$_2$) in location 2001_{16}, the result ZZ being stored in location 2002_{16}. The program itself commences at address 2020_{16}.

The column headed 'address' in Table 9.2 contains the range of hexadecimal addresses used in the program. Locations 2000_{16}, 2001_{16}, and 2002_{16} are used for the storage not only of the decimal values to be added together but also the result. Locations 2003_{16} to $201F_{16}$ are not used, and contain electronic 'garbage'. Addresses 2020_{16} to 2029_{16}, inclusive, are used to store the main program.

The second column headed 'byte' is subdivided into three parts; the byte column enables the programmer to write down any instruction up to three bytes in length. In this program, the instructions are either one byte long or three bytes long.

The column headed 'instruction mnemonic' is used to write down the mnemonic version or *assembly language* version of the instruction in the 'byte' column. Assembly language instructions are generally simpler to use than machine code instructions since they are easier to understand.

Finally, general comments are written down in the 'comment' column. The function of these comments is to assist the reader to understand the program.

The first instruction in the program is the LXI H,2000 instruction, which is stored in locations 2020 to 2022. The first byte of the instruction is its opcode (corresponding to the value 21_{16} in the 'byte 1' column) which corresponds to the LXI H part of the instruction. This opcode causes the CPU to load the register pair L and H with the hex value given in locations 2021_{16} and 2022_{16}, respectively. The purpose of the 16-bit value stored in the pair of registers H and L is to act as a 'pointer' which points at a particular location, M, in the memory of the computer (the low-order eight bits of the address is stored in register L [the eight bits being specified by the data in the 'byte 2' column of the instruction] and the high-order eight bits of the address being stored in register H (the eight bits being specified by the data in the 'byte 3' column of the instruction). When the LXI H,2000 instruction has been executed, register H stores 20_{16} and register L stores 00_{16}; the HL register pair therefore 'point' at memory location M whose address is 2000_{16} (which is the address where decimal data$_1$ is stored).

The next instruction, which is the 1-byte MOV A,M instruction (opcode $7E_{16}$), results in the contents of location M (that is, the contents of location 2000_{16}) being MOVed into the Accumulator, as follows

$$(\text{accumulator}) \leftarrow (\text{memory location M})$$

where the parentheses above are read as 'the contents of'. That is to say, the decimal value data$_1$, i.e., XX is 'moved' to the accumulator (strictly speaking XX is not moved, but *a copy of it is transferred to the accumulator*).

The 1-byte INX H instruction (opcode 23_{16}) causes the value stored in the register pair H and L to be incremented, i.e., increased by unity, so that the memory pointer now points at location 2001_{16} (where decimal data$_2$ is stored).

The next instruction, the 1-byte ADD M instruction (opcode 86_{16}) causes the contents of memory M, i.e., address 2001_{16}, to be ADDed to the accumulator contents, the sum remaining in the accumulator. That is, the accumulator now stores the sum of data$_1$ and data$_2$.

Since all microprocessors operate in binary, the addition described above is carried out in binary, the 'decimal' value of data$_1$ and data$_2$ being stored as binary-coded decimal (BCD) values. However, since we are dealing with decimal data, it is necessary to 'adjust' the binary sum to the correct decimal value. This is done by inserting a 1-byte DAA instruction (opcode 27_{16}) after the ADD M instruction; the function of the DAA instruction is to Decimal Adjust the Accumulator contents. On completion of this instruction, the accumulator stores the decimal sum of $data_1$ and $data_2$.

Next, the register pair H and L are incremented once more by means of a second INX H instruction, resulting in the memory pointer HL pointing to location $M = 2002_{16}$.

The 1-byte MOV M,A instruction (opcode 77_{16}) causes the contents of the accumulator (which is the decimal sum of $data_1$ and $data_2$) to be 'moved' into or copied into location M. That is

$$(\text{memory location M}) \leftarrow (\text{accumulator})$$

This is, the sum is stored in location 2003_{16}.

The final instruction is the 1-byte HLT (HaLT) instruction (opcode 76_{16}) which causes the CPU to suspend operations, thereby terminating the program. The addressing modes of the instructions used in the program in Table 9.2 are listed in Table 9.3.

Table 9.3

Instruction	Addressing mode
LXI H,2000	immediate
MOV A,M ⎫	
ADD M ⎬	register indirect
MOV M,A ⎭	
INX H	register
DAA	inherent

9.13 Subroutines

A subroutine is a subprogram which can be 'called' by the main program in order to perform a specific task; each subroutine may be called many times during the operation of the main program.

For example, consider a program which needs to perform multiplication at a number of points in the main program. The programmer can either write out a multiplication routine every time he needs it or he can write it out once in the form of a subroutine; the main program can 'call' for the subroutine every time it is needed by the main program. The latter method is more economic than the former not only in terms of the time taken to write the complete program, but also in terms of the amount of memory needed to store the complete program (including the subroutine).

The general principle is illustrated in Fig. 9.4 in which a subroutine is called twice during the operation of the main program. Suppose that the first instruction in the subroutine occurs at location 2090_{16}; the 3-byte CALL 2090 (shown as CALL SUB in the diagram) is inserted in the main program each time the subroutine is called. When this instruction is executed by the CPU, program control is transferred to address 2090_{16}, and

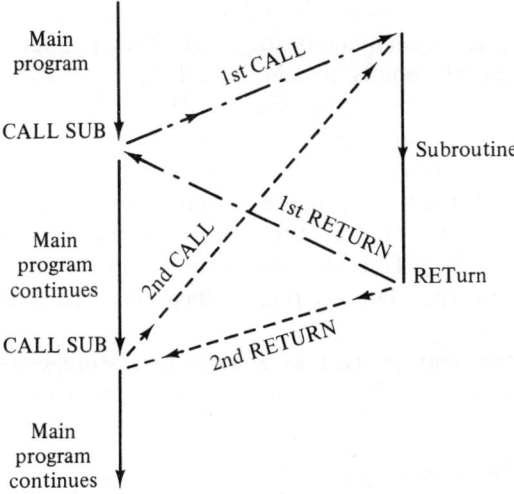

Fig. 9.4   Basic method of implementing a subroutine call.

the subroutine is executed. At the same time that control is transferred to the subroutine start address, the address to which a 'return' is to be made in the main program is stored in the area of memory known as the 'stack'. The operation of the stack is described in section 9.14.

The final instruction in the subroutine is a *return* instruction (whose mnemonic is RET – opcode $C9_{16}$). This instruction results in the return address in the main program being recovered from the stack and transferred to the program counter of the CPU. When this happens, an orderly return is made to the main program. The CALL and RETurn procedure described above is repeated every time that a subroutine is called.

9.14   Stack organization

A stack is a group of memory locations organized in a *last-in, first-out* (LIFO) manner; that is, the last byte of data to be stored on the stack is the first to be removed. The CPU contains a 16-bit *stack pointer* register, SP (see Fig. 9.3), which is used to point at the 'top' of the stack; the programmer must initialize the SP at the highest available address in RAM. If this address is $20FF_{16}$, then the stack pointer is initialized by means of a 3-byte LXI SP,20FF instruction.

When the CPU executes a subroutine CALL instruction, the CPU automatically 'pushes' onto the stack the two bytes (sixteen bits) of the address in the main program to which the return must be made *after* the subroutine has been completed. For example, if the top of the stack was initialized at $20FF_{16}$, the two bytes corresponding to the return address in the main program are stored in locations $20FE_{16}$ and $20FD_{16}$, respectively. At the same time, the contents of the 16-bit stack pointer are decremented twice so that the SP 'points' to address $20FD_{16}$.

The subroutine is completed when a RETurn instruction is executed; when this occurs, the two bytes in locations 20FD and 20FE are transferred to the program counter. This ensures that control is transferred to the correct address in the main program; simultaneously, the CPU increments the SP twice (once for each byte removed from the stack) so that it points once more to $20FF_{16}$. Thus the address to which the stack pointer points changes each time that data is either 'pushed' onto the stack or is 'popped' from it.

The stack can also be used as a temporary storage location for data either from any of the CPU general-purpose registers or the flag register.

9.15   Subroutine organization

The four general stages in the body of a subroutine are illustrated in Fig. 9.5. These are

(a) The 'save status' instructions which result in vital data being stored or 'saved' on the stack.
(b) The 'main body' of the subroutine which may be, for example, a multiplication routine.
(c) The 'restore status' routine which results in the vital data which was 'saved' on the stack in step (a) being restored to the correct locations.
(d) The 'return' routine, which is a single RET instruction.

The reason for the 'save status' section of the subroutine is described below. Since it is frequently the case that several of the general-purpose registers are used in the operation of the main body of the subroutine, the in-

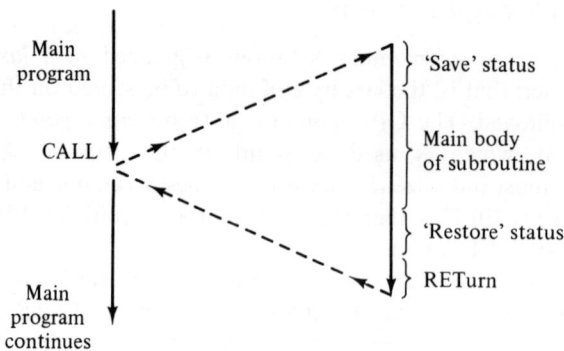

Fig. 9.5   Stages in a subroutine.

structions in the 'save status' part of the subroutine are dedicated to 'saving' the data in these registers by 'pushing' them onto the stack.

When the instructions in the main body of the subroutine have been executed, the status of the CPU is restored to its original condition by means of a series of 'pop' instructions which 'pop' the original contents of the registers from the stack into the registers. Where the main body of the subroutine does not use the general-purpose registers, stages (a) and (c) of the subroutine are omitted.

A return is made to the main program by means of a RETurn instruction.

9.16 Nested subroutines

The memory stack enables subroutines to be 'nested'; that is, it allows one subroutine to call a second subroutine, which in turn can call a third subroutine and so on. The general technique is illustrated in Fig. 9.6.

For example, an electronic time delay can be generated by means of a nested subroutine program in which the second subroutine in Fig. 9.6 generates a delay of one second which is 'called' sixty times to generate a time delay of one minute. The first subroutine in Fig. 9.6 provides a counting mechanism in which the number of minutes is counted, and if this counting routine is 'called' into use sixty times, then the program provides a delay of one hour.

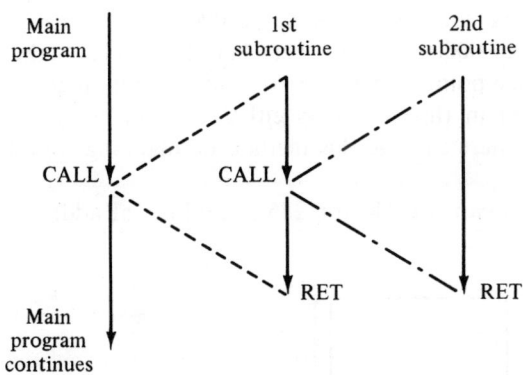

Fig. 9.6 Nested subroutines.

9.17 Interrupts

An *interrupt* is a signal that can be applied to the CPU which causes the normal sequence of operations to be suspended; program control is then handed to a special form of subroutine known as an *interrupt routine*.

There are two classes of interrupt, namely a nonmaskable interrupt and a maskable interrupt. The two types are illustrated by the following analogy. Suppose that an engineer is sitting in his office by his telephone. If the telephone rings he can either choose to answer the 'phone or he can ignore it; the telephone signal is a form of 'interrupt' signal which interrupts his normal train of thought. The telephone is an example of a maskable interrupt which he can effectively 'mask out' if he chooses to ignore it. On the other hand, if someone enters his office and asks him a question, it is equivalent to a nonmaskable interrupt which he cannot ignore and must respond to.

The 8085 CPU has one nonmaskable interrupt pin and four maskable interrupt pins of varying priority. That is, a high priority interrupt signal can interrupt the program of a lower priority interrupt, but not vice versa. The INTR interrupt line shown in Fig. 9.3 has the lowest priority of the four maskable interrupts on the 8085.

9.18 Memory mapped addresses and I/O mapped addresses

A location in a microcomputer system can be addressed by either of two methods (but not both), namely it can be memory mapped or it can be I/O mapped.

A *memory mapped address* requires the CPU to apply signals to all its sixteen address lines in order to define the location being addressed. That is, it can have any address in the range 0000_{16} to $FFFF_{16}$ (that is in the range $0000\,0000\,0000\,0000_2$ to $1111\,1111\,1111\,1111_2$); the address therefore lies at some point in the $2^{16} = 64K$ addressing range of the system. The address defined in this way may either be a memory or an I/O device; all microcomputers can use this method of 'mapping' and address (see Fig. 9.7(a)).

Many CPUs can use $\frac{1}{4}K$ (or 256_{10}) additional addresses using an *I/O*

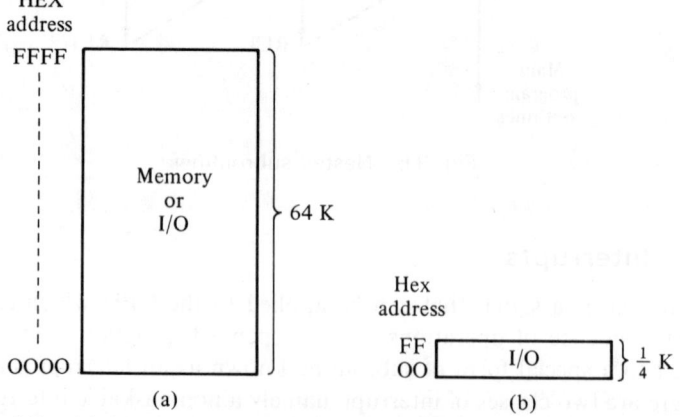

Fig. 9.7 (a) Memory mapping and (b) I/O mapping (not to scale).

mapping technique which, as its name implies, is used only with I/O devices. This method enables the system to use up to 256_{10} addresses relating to I/O devices without interfering with its 64K of memory mapped addresses (see Fig. 9.7(b)). This feature is not available on all microprocessors.

9.19 I/O ports

Input ports and output ports are IC chips which enable the CPU to 'listen to' and 'talk to' external circuits, respectively. In effect they comprise a set of synchronously operated switches which connect peripherals to the data bus (*note:* only one peripheral is given access to the data bus at any one instant of time). Certain I/O ports are dedicated for use as input ports and cannot perform any other function; yet other ports are dedicated output ports and cannot perform any other function.

There also exists a range of I/O ports which can be programmed by software to operate either as input ports or as output ports. These are known as *programmable I/O ports* (PIO); individual manufacturers use a range of other names for these ports including programmable peripheral interface (PPI), peripheral interface adaptor (PIA), versatilie interface adaptor (VIA), etc. PIO chips may also contain circuit elements which enable them to deal with interrupts and timer/counter operations. In the following, the attention of the reader is directed towards a basic PIO.

The PIO described below is based on the I/O facilities of the Intel 8155 chip (see Fig. 9.8). The chip has the usual chip select signal lines including an active-low write pin (driven by the $\overline{\text{WR}}$ control bus line of the CPU – see also Fig. 9.3.) and an active-low read pin (driven by the $\overline{\text{RD}}$ control bus line of the CPU). The chip enable pin on the PIO is driven from the address bus decoding network and the reset pin is activitated by the RESET OUT line of the CPU (see also Fig. 9.3). When the reset pin of the chip is activated, all registers in the PIO are reset to zero (this occurs when the CPU is reset).

The basic PIO chip described here contains four registers, namely the command/status register (C/S), the port A (PA) register, the port B (PB) register and the port C (PC) register. The registers have the I/O mapped addresses in Table 9.4.

Table 9.4

I/O mapped address (hex)	Function of register
20	C/S
21	PA
22	PB
23	PC

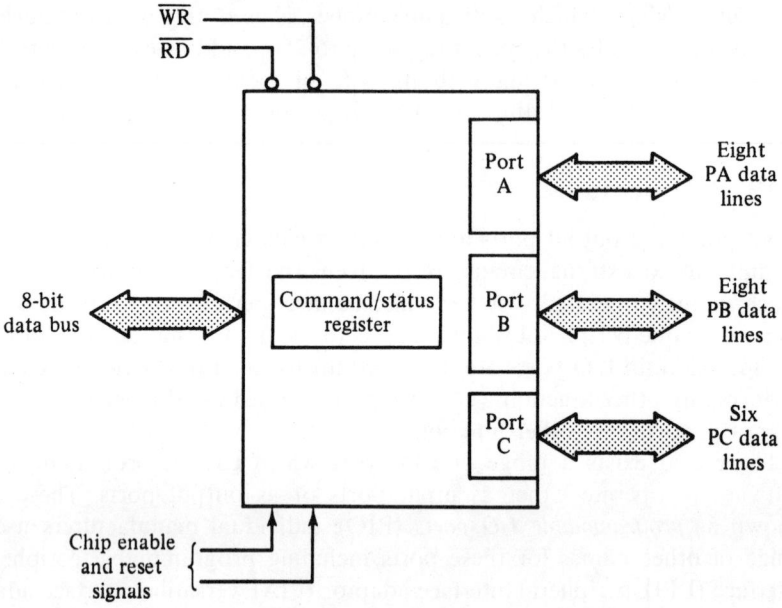

Fig. 9.8   A simplified programmable I/O port or PIO.

The port is 'initialized' by transferring a 'command byte' to the C/S register; the function of this byte is to specify the operating mode of the other registers in the chip. That is, it specifies whether the ports operate as input ports or as output ports. The function of each bit in the C/S register has a specific function and is shown in Fig. 9.9(a). The 8155 chip has a timer/counter built into it but, owing to restrictions of space the timer/counter is not discussed in this chapter. In the programs described in this chapter, the timer is turned off by writing logic 0's into bits six and seven of the C/S register. Also, as we shall not be using the interrupt facilities of the port, logic 0's are written into bits five and four of the C/S register (see Fig. 9.9(b)).

The operating state of port C (which can handle six lines – see Fig. 9.8) is specified by bit 3 and bit 2 of the C/S register. For the purpose of programs in this chapter, we specify that port C is to act as an input port by writing logic 0's in both of these bits (see Fig. 9.9(b)). The operating state of the 8-bit port B is specified by the condition of bit number one of the C/S register, and the operating state of the 8-bit port A by bit zero of the C/S register. In the programs in this chapter, a logic '1' is written into bit one of the C/S register (which defines port B as an output port) and a logic '0' is written in bit zero (which defines port A as an input port). The C/S register is therefore initialized by the two instructions below

MVI A,02

OUT 20

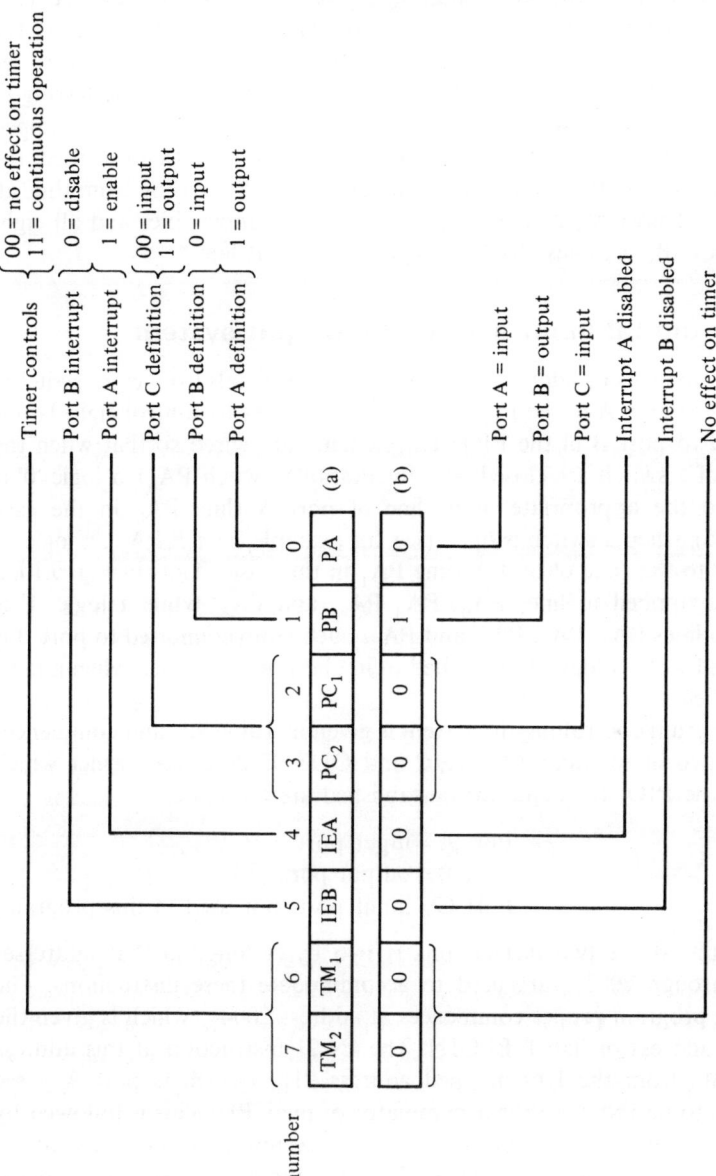

Fig. 9.9 The command/status (C/S) register of an Intel 8155 PIO and (b) a typical example of its use.

The first of these instructions, MV1 A,02, causes the hexadecimal value 02 (or 0000 0010_2) – which corresponds to the command word in Fig. 9.9(b) – to be moved into the accumulator of the CPU (*note:* MVI A is a mnemonic representing the instruction MoVe Immediate into the Accumulator). The second instruction, OUT 20, causes the data to be OUTput to the location having the I/O mapped address 20_{16}. Referring to Table 9.4, the reader will note that the C/S register has the I/O mapped address 20, so that after the above two instructions have been executed, the C/S register of the PIO stores the command word 20. Arising from this, all eight lines of port A, i.e., lines PA_7 to PA_0 act as input lines and all eight lines of port B, i.e., lines PB_7 to PB_0 act as output lines*.

9.20 A digital input and digital output system

In this section we consider a system in which the signals from eight switches connected to port A of the PIO in Fig. 9.10 are used to control eight lamps connected to port B of the PIO. The switches are wired so that when the contacts of a switch are closed (see, for example, switch PA_0), a logic '0' is applied to the appropriate input line of port A (line PA_0 in the case considered); when a switch is open (see, for example, switch PA_1), a logic '1' is applied to that line of port A (line PA_1 in this case). Thus in Fig. 9.10, a logic '0' is applied to lines PA_0, PA_3 PA_4, and PA_6, while a logic '1' is applied to lines PA_1, PA_2, PA_5, and PA_7. Each lamp connected to port B is illuminated when a logic '1' is applied to it, and is extinguished when a logic '0' is applied to it.

The program controlling the system is given in Table 9.5, and commences with the two instructions MVI A,02 and OUT 02 described earlier which initialize the PIO; the conditions established are therefore

Port A = input port
Port B = output port
Port C = input port (not used in this program)

Each of the above two instructions is two bytes long, so that addresses 2000_{16} through 2003_{16} are used to accommodate these instructions. The operating program proper commences at address 2004_{16} which is given the symbolic address or 'label' BEGIN. The IN 21 instruction at this address causes data from the I/O mapped address 21_{16} (which is port A – see Table 9.4) to be INput to the accumulator of the CPU. This is followed by an OUT 22 instruction which causes the accumulator contents to be OUTput to the I/O mapped address 22_{16}, i.e., the data is output to port B

---

*Note: Many PIOs allow any of their lines to act as input lines and any as output lines. It is therefore possible in these PIOs to have, say, six input lines and two output lines on port A, and four input lines and four output lines on port B. The 8155 only allows the user to select which complete port acts as input or output.

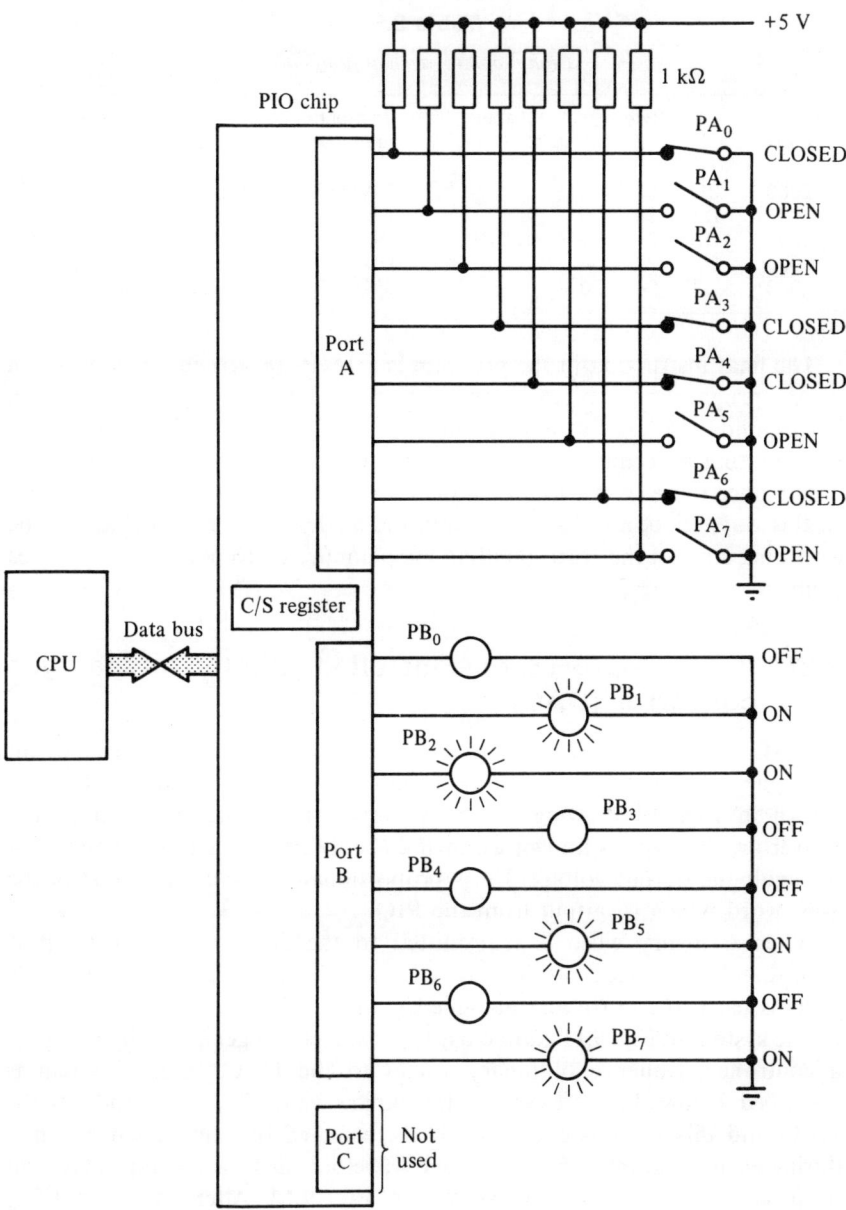

Fig. 9.10 A simple digital control scheme (control and address lines omitted for clarity).

(see also Table 9.4). That is, the state of switch PA_0 controls lamp PB_0, switch PA_1 controls lamp PB_1, etc. When the switches are in the positions shown in Fig. 9.10, lamps PB_0, PB_3, PB_4 and PB_6 are extinguished and the other lamps are illuminated.

Table 9.5

Digital control system program

Address (hex)	Byte 1	2	3	Label	Instruction mnemonic	Comment
2000	3E	02			MVI A,02	; Load command byte
2002	D3	20			OUT 20	; Initialize PIO
2004	DB	21		BEGIN:	IN 21	; Input data from port A
2006	D3	22			OUT 22	; Output data to port B
2008	C3	04	20		JMP BEGIN	; Repeat program loop

The final instruction in the program is a JMP BEGIN instruction which causes program control to be transferred back to address 2004_{16} (which corresponds to the symbolic address BEGIN). The INputting and OUTputting procedure is then repeated so that the state of the switches is 'input' once more to the accumulator, after which it is 'output' to the lamps. In this way, as soon as the position of the blade of any switch changes, the operating state of the corresponding lamp connected to port B also changes state.

9.21 A microprocessor-controlled digital-to-analogue convertor (DAC)

A DAC is a circuit which converts a binary word into an equivalent analogue voltage. The basis of one form of microprocessor-driven DAC is shown in Fig. 9.11: owing to the value of the resistances used in the convertor, this type is known as an R–$2R$ convertor. It can be shown that the analogue output voltage, V_A, is proportional to the digital value of the 8-bit word which is output from the PIO.

When a binary word is transmitted to the DAC via the PIO, it is immediately converted into its equivalent analogue voltage. In this way, the CPU is used to control an analogue signal.

The system in Fig. 9.11 can be used as a waveform generator by applying a controlled sequence of binary values to the DAC; such a system is described below. If, for example, the binary value 00_{16} is output to the DAC, and this value is continuously incremented or is increased by unity during each program step, the voltage waveform at V_A is a ramp waveform or a sawtooth waveform as shown in Fig. 9.12. After 255_{10} or FF_{16} incrementing steps, the analogue output voltage reaches its maximum value. The next incrementing step causes the total value to become zero again (it is assumed in this case that we are dealing with an 8-bit CPU). By rapidly repeating the incrementing process, the system produces a sawtooth output waveform at V_A.

A program which generates a sawtooth waveform in association with Fig.

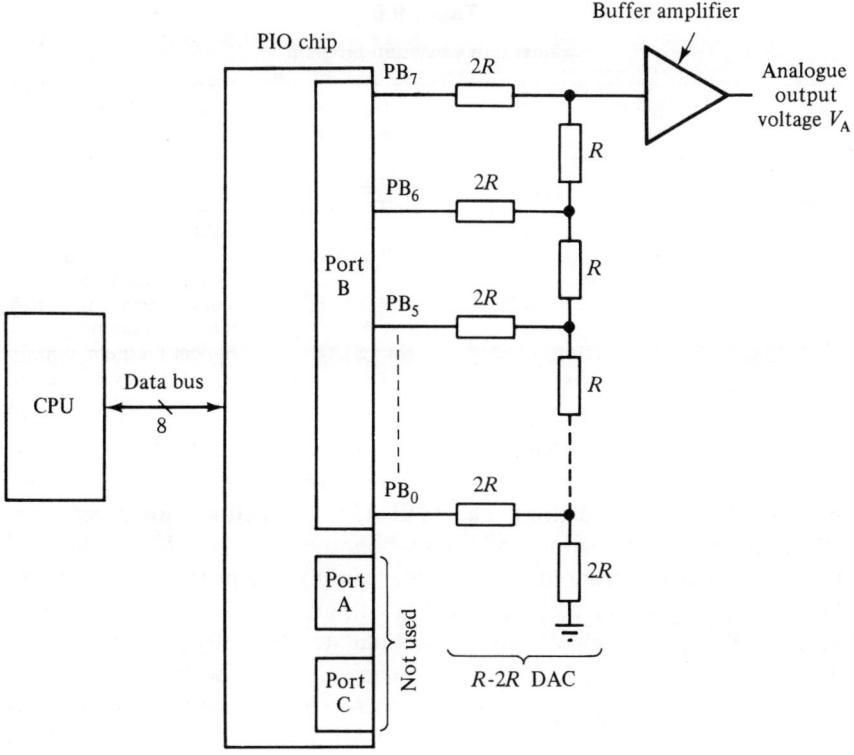

Fig. 9.11 An R-2R DAC which is interfaced to a microcomputer (control and address lines omitted).

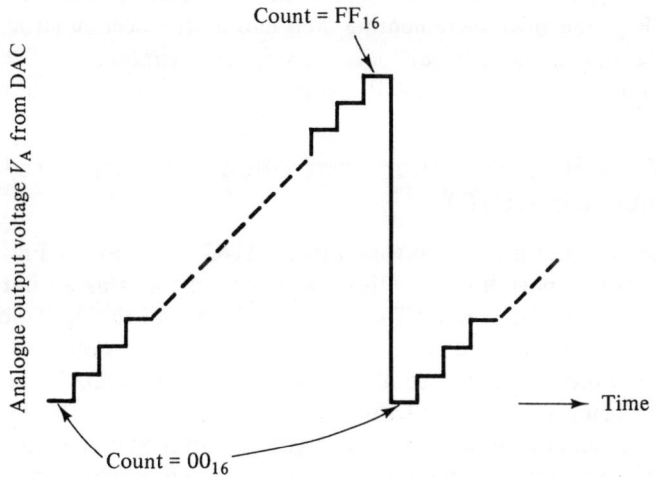

Fig. 9.12 A sawtooth waveform (see text).

Table 9.6

Sawtooth waveform program

Address (hex)	Byte 1	2	3	Label	Instruction mnemonic	Comment
2000	3E	02			MVI A,02	
2002	D3	20			OUT 20	; Initialize Port B
						; =output port
2004	3E	00			MVI A,00	; Set the contents of
						; accumulator to zero
2006	D3	22		START:	OUT 22	; Output 'count' to port B
2008	3C				INR A	; Increment count
2009	C3	06	20		JMP START	; Repeat program loop

9.11 is given in Table 9.6. The first two instructions (MVI A,02 and OUT 20) initialize port B of the PIO as an output port (ports A and C are not used in this application). The MVI A,00 instruction moves zero, i.e., 00_{16} into the accumulator, which initializes the data stored in a 'counter' (the accumulator storing the value of the 'count'). The program loop proper begins at the symbolic address START with an OUT 22 instruction which outputs the 'count' value from the accumulator (initially zero) to port B, where it is converted into its analogue equivalent by the DAC.

An INR A instruction increments or increases the count in the accumulator by unity, and this is followed by a JuMP back to START instruction. At this point, program control is transferred to address 2006_{16}, where the 'count' value in the accumulator is transmitted to port B once more. Each time the loop is executed, the value of the count is increased, and with it the analogue output voltage increases. The limit is reached when the counter stores FF_{16}; the next incrementing step causes the accumulator to store 00_{16} once more, and the sawtooth waveform begins again. The loop is cycled continuously until the CPU operation is stopped by the operator.

9.22 A microprocessor-controlled analogue-to-digital convertor (ADC)

The basis of one form of software-driven ADC is shown in Fig. 9.13. In this application, port B of the PIO once more operates as an output port and port A as an input port. The general principle of the operation of the ADC is as follows.

An 8-bit word is transmitted along the data bus to the DAC via the PIO and, after being converted to its analogue equivalent by the DAC, it is compared in an electronic comparator with the unknown analogue voltage, V_U, whose value is to be measured. In the following, it is assumed that the

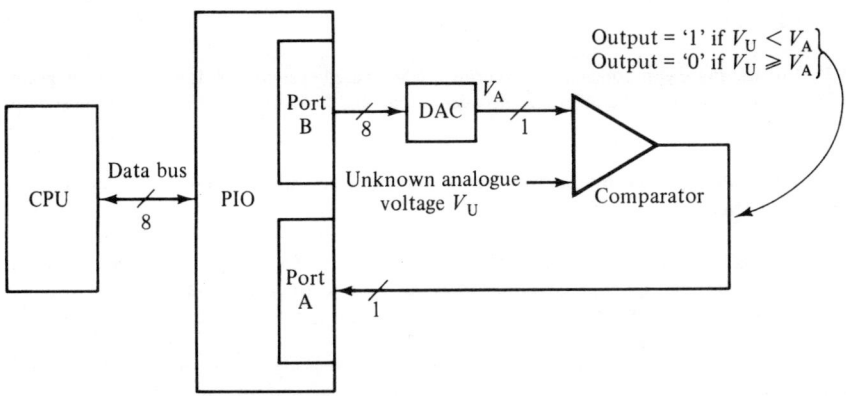

Fig. 9.13   Software-driven ADC.

comparator output is logic '1' if V_U is less than V_A (the output from the DAC), and is logic '0' if V_U is either equal to or is greater than V_A. The output voltage from the comparator is applied to one of the input lines of port A, and the logic level on this line is tested by the CPU to see if equality between V_A and V_U has been achieved. If equality has not been reached, the value of the binary word which is output by the CPU is increased, and the testing procedure is repeated. When equality is reached, the value of the binary word applied to port B represents to some scale the value of the unknown voltage V_U.

The name given to the software-driven ADC depends on the method used to generate the pattern of binary words applied to port B. If the bit-pattern is generated by a sawtooth program of the type in Table 9.6, the ADC is described as a *continuous balance ADC* (see also section 5.10). An alternative ADC (which also uses the hardware shown in Fig. 9.13) providing a faster method of reaching the balance point between V_A and V_U is known as a *successive approximation ADC**.

The ADCs described above are software driven; that is, the microprocessor software is in control of the conversion at all times, and the CPU is not free to perform any other function. Many commercially available ADCs have their own on-chip 'clock' which generates its own bit testing pattern; this leaves the CPU to perform other calculations while the ADC chip deals with the analogue-to-digital conversion. This type of ADC chip informs the CPU when it has completed its calculation by applying a signal to the CPU either through an input port or through one of the interrupt pins of the CPU.

---

*For further details the reader is referred to the third edition of *Logic Circuits* by N. M. Morris (McGraw-Hill, London).

Problems

9.1 List the main components in a microprocessor-based system and draw a block diagram of the system. Describe the difference between 'serial' and 'parallel' data in terms of the system operation.

9.2 What is understood by the 'word length' of a microcomputer? Describe the function of and the difference between RAM and ROM.

9.3 Discuss the three-bus system used by microprocessor-based systems. State the number of lines in each bus in a microprocessor-based system and hence explain the number of memory locations or I/O locations that may be addressed by the system. Discuss the function of two typical control bus lines.

9.4 What is meant by (i) a machine code instruction and (ii) an assembly language instruction. Give an example of each.

9.5 What is a 'flag' in a microprocessor? List the flags used in an 8085 CPU and state the condition which causes each of them to be 'set' and 'cleared'.

9.6 Discuss the function of the arithmetic and logic unit of a CPU, and describe typical operations carried out by it (hint: study the instruction set in appendix A).

9.7 Describe the following addressing methods and give an example of each type: direct addressing, register direct addressing, register indirect addressing, indexed addressing, relative addressing, immediate addressing, implied addressing, stack addressing.

9.8 The following information is stored in successive locations in memory, commencing at location 2010_{16}.

$$3A, 00, 20, 2F, 32, 01, 20, 3C, 32, 02, 20, 76$$

Convert the program into its assembly language form (this process is known as 'disassembling' the program). Describe the function performed by the program and give the hex address of the final byte in the program. If location 2000_{16} contains 01_{16}, state the hex value stored in locations 2001_{16} and 2002_{16} after the program has been executed, giving the reason for your solution.

9.9 The following assembly language program is stored in successive locations in memory, commencing at location 2010_{16}.

```
LXI   H,2000
MOV   A,M
INX   H
ADD   M
INX   H
MOV   M,A
HLT
```

Convert the program into its machine code form (this process is known as 'assembling' the program). Describe the function performed by the program and give the hex address of the final byte in the program. If the data in location 2000_{16} is 02_{16}, and in location 2001_{16} is $0A_{16}$, state (with a reasoned argument) the value of the date stored in location 2002_{16} after the program has been executed.

9.10 Write a program to initialize ports A, B, and C of the PIO described in section 9.19 as input ports and to read a data byte from each port in turn. The data from port C is to be stored in register C, the data from port B is to be stored in register B and the data from port A is to be stored in the accumulator.

9.11 Discuss the use of a 'stack' in the memory in connection with subroutines and interrupts. Give an example of a subroutine and an interrupt program.

9.12 Draw a block diagram and explain the operation of a microprocessor-driven digital-to-analogue convertor. Show how such a system can be used as a waveform generator.

9.13 Explain, with the aid of a diagram, the operation of a software-driven analogue-to-digital convertor.

10. Power electronics

10.1 Thyristors

Thyristors are bistable semiconductor switching devices which are triggered into conduction by the application of a signal to the *gate* electrode. The two most popular forms of thyristor are the *reverse blocking thyristor* (popularly referred to simply as a thyristor) and the bi-directional thyristor (known by its trade name of the *Triac*). In the following, we shall refer to the two types as thyristors and triacs, respectively.

The thyristor is a four-layer p-n-p-n device whose operation can be described by means of a *two-transistor analogy*, see Fig. 10.1. We assume that the centre p-n region can be divided diagonally as shown, so that the thyristor is split into a p-n-p transistor (T1) and an n-p-n transistor (T2). For the moment, we will ignore the effects of the n-gate electrode G2. First,

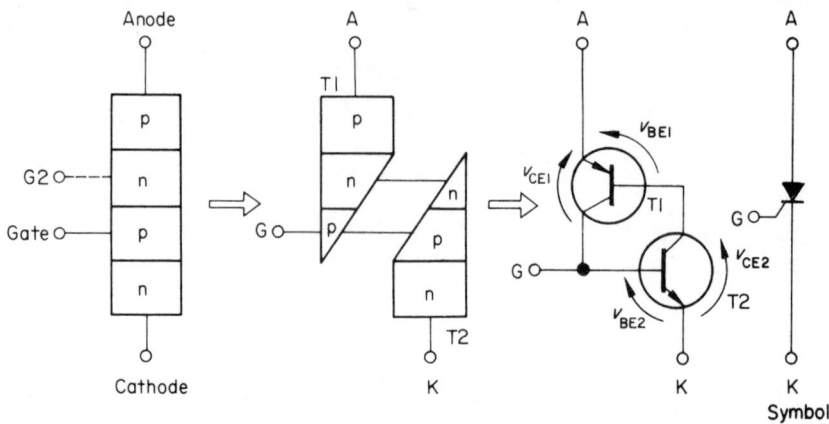

Fig. 10.1   A two-transistor analogy of the thyristor.

312

let us consider the operation of the device with the gate disconnected. The application of a negative potential to the anode causes the upper p-n junction to be reverse biased, and in this operating condition the thyristor *blocks* the flow of current. A positive anode potential forward biases the upper p-n junction but, with zero gate signal, the centre p-n junction is reverse biased. Again, the thyristor cannot conduct unless the anode voltage is raised to a value at which avalanche breakdown occurs. This method of turn-on is allowable in some circuits, but is not normally recommended since it can lead to the thyristor being damaged. Thus, provided that either forward or reverse breakdown voltages are not applied, when the gate voltage is zero the thyristor blocks the flow of current for both forward (positive) and reverse (negative) anode voltages.

If a positive potential is applied to the gate electrode (G) while the anode is positive, then transistor T2 is turned on and begins to conduct. Since T2 collector current is also the base current of T1, then T1 also begins to conduct. We also see from Fig. 10.1 that the collector current of T1 flows into the base of T2. Hence, so long as the anode is positive then each transistor maintains the other in a conducting state. Consequent upon this fact, during the period that the anode is positive the thyristor acts as a self-latching relay. From this we conclude that an impulsive gate signal of a few microseconds duration will be sufficient for triggering purposes.

The static characteristic of a thyristor is shown in Fig. 10.2. Once the thyristor has been triggered into conduction, the forward voltage across it falls to about 0.75–1.25 V, the actual value depending on the load current. The effect of initially applying progressively larger values of gate current has the result of turning the thyristor on more rapidly. Gate pulses applied while the anode is negative do not trigger the thyristor into conduction.

If a negative voltage is applied to the gate region, it cannot trigger the thyristor into conduction (whatever the anode potential) as it applies a reverse bias to the emitter junction of T2.

A reason for the low forward p.d. across the device can be deduced from the two-transistor analogy as follows. The voltage appearing between the anode and cathode when the thyristor is conducting is the sum of the saturated values of either v_{CE1} and v_{BE2} or v_{CE2} and v_{BE1}. From information given earlier in the book, we know that the sum of the p.d.s is about 1 V, which agrees quite well with the measured value.

Once the thyristor has been turned on it remains in a conducting state until the anode current has fallen below the *holding current*, which usually has a value in the range between a fraction of a milliampere in small thyristors and up to about 50 mA in large thyristors. The *turn-off time* of the thyristor is defined as the time required for the device to achieve its full blocking capability, and is the time taken for the free charge carriers to recombine and to disappear. The turn-off time lies between about 5 and $200\,\mu s$, and depends on the rating and construction of the thyristor.

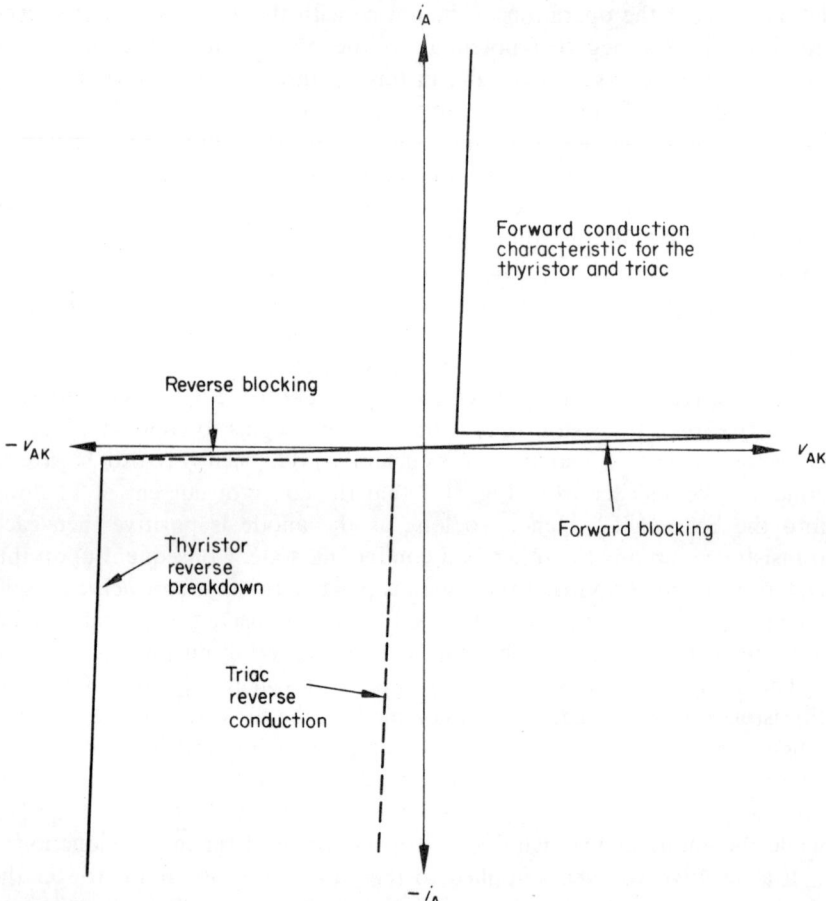

Fig. 10.2   The static characteristic of the thyristor is shown by the full line. The characteristic of the triac differs only in the third quadrant, and is shown by the broken line.

Thyristors with rapid turn-off times are constructed by reducing the lifetime of the charge carriers in the forward blocking junctions by *gold-doping* techniques. This, unfortunately, also reduces both the current rating and the forward blocking capability of a given type of device. As an illustration of the rating of a fast turn-off thyristor, a 400 A, 1.5 kV device can have a turn-off time in the range 30–40 μs.

Several forms of circuit symbol are in common usage, that shown in Fig. 10.3 being a popular form.

One form of structure of the triac is shown in Fig. 10.4, its static characteristic being generally similar to that of the thyristor, with the exception that it can be triggered into conduction when terminal A1 is

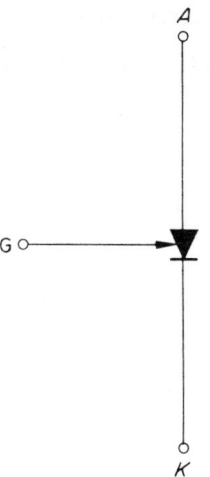

Fig. 10.3   An alternative symbol for the thyristor.

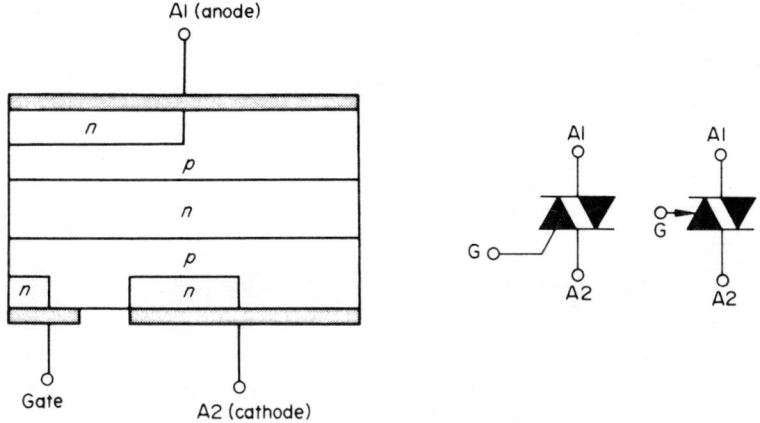

Fig. 10.4   The bi-directional thyristor (the *triac*).

either positive or negative with respect to terminal A2. The latter means that controlled conduction can take place in the third quadrant of its characteristic (see Fig. 10.2).

A further difference between the triac and the thyristor is that, in the case of the triac, it is possible to use either positive or negative gate voltages to trigger the device into conduction for either polarity signal at the A1 terminal. The triac generally has a lower gate triggering sensitivity than the thyristor.

10.2   The silicon controlled switch (SCS)

The silicon controlled switch is a device with a p-n-p-n structure similar to that of the thyristor, but has both p- and n-gate regions brought out to terminations. The additional gate connection leads to increased versatility, and the device is used in a wide range of applications.

One application is as a store and driver stage for a numerical indicator tube, shown in Fig. 10.5(b). In this application, a pulse is applied to the base of TR1 which, by invertor action, reduces the voltage applied to the anodes of the SCS's long enough for them to turn off. This causes the neon

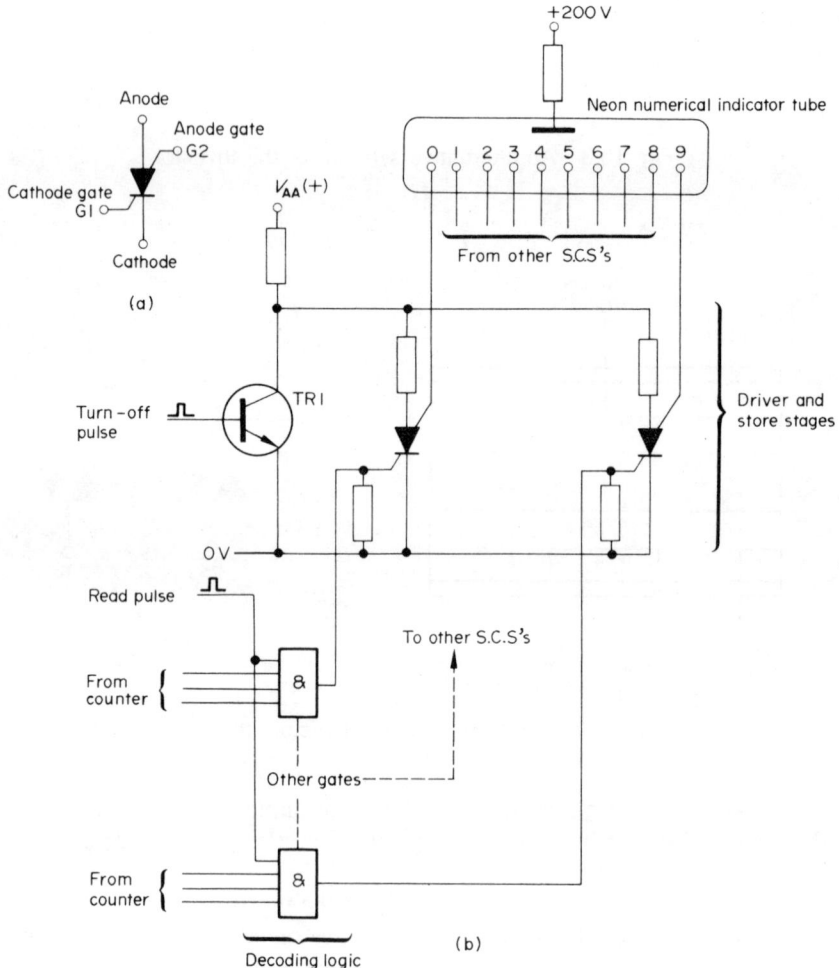

Fig. 10.5   (a) The silicon controlled switch and (b) its use in a numerical indicator circuit.

indication to be extinguished. In the meantime, the voltage level applied to the 'read' line is zero, so that the outputs from the AND gates are zero. After the turn-off pulse has fallen to zero, a positive potential is momentarily applied to the 'read' line, thereby opening the AND gates. Only one of the gates has all its input lines activated, so that only one of the SCS's has its turn-on gate activated. This SCS turns on very rapidly, illuminating the numerical indicator tube cathode connected to its anode gate.

In this circuit, the cathode gate is used as the turn-on electrode, the anode as the turn-off electrode, and the anode gate as the output electrode. Other configurations are possible, and this circuit merely serves to highlight the versatility of the device.

In some industrial applications, light-sensitive SCS's capable of switching several amperes and having turn-on and turn-off times of 50–100 μs are in use.

10.3 Basic single-phase circuits

A basic thyristor circuit for single-phase half-wave control is shown in Fig. 10.6(a). The thyristor circuit provides a 'chopped' half-wave output into the load; if the triac (shown in the inset) is used, the output is in the form of a 'chopped' full-wave output. The thyristor is triggered by a control circuit which is usually a pulse generator, suitable pulse generators having already been described in chapter 6. For smooth control, the turn-on pulse must be synchronized to occur at the same point in each cycle, one circuit of this type being described later in this section. With a resistive load, the current through the thyristor falls to zero when the anode voltage reaches zero. The thyristor then reverts to its blocking state for the negative half-cycle, and continues blocking for the first part of the following positive half-cycle up to the point at which it is triggered into conduction again. If the angle at which conduction commences (the *delay angle*) after the zero point in the cycle is α, and the angle at which conduction ceases is β, then the *mean* output voltage V_L in the case of the thyristor is

$$V_L = \frac{1}{2\pi} \int_\alpha^\beta (V_M \sin \omega t - V_A) \mathrm{d}(\omega t)$$

where V_M is the maximum value of the supply voltage and V_A is the p.d. across the thyristor when it is conducting. For most practical purposes, $\beta = \pi$ radians (180 degrees) and V_A has a value of about 1 V, and may be ignored in many cases. These assumptions result in the following simplified equation:

$$V_L = \frac{1}{2\pi} \int_\alpha^\pi V_M \sin \omega t \, \mathrm{d}(\omega t) \tag{10.1}$$

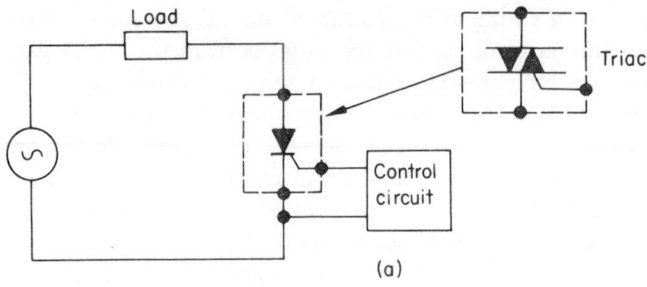

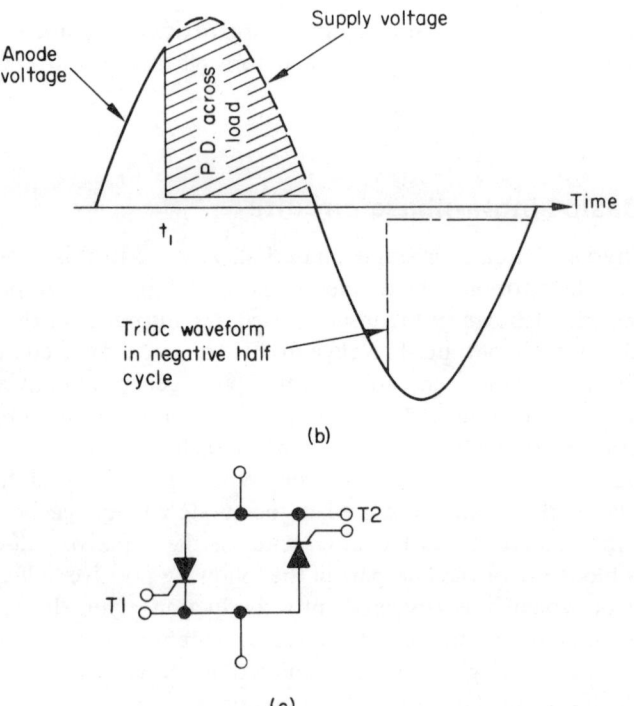

Fig. 10.6   (a) A single-phase controlled circuit and (b) voltage waveforms. An inverse-parallel connected pair of thyristors is shown in (c).

Completing the integration yields

$$V_L = \frac{V_M}{2\pi}(1 + \cos \alpha) \qquad (10.2)$$

$$= 0.159\, V_M (1 + \cos \alpha) \simeq 0.225\, V_{r.m.s.}(1 + \cos \alpha)$$

where $V_{r.m.s.}$ is the r.m.s. value of the supply voltage.

Thus, by phasing the gate pulses 'forward' so that $\alpha = 0$, the maximum output voltage of V_M/π is developed across the load. In this case, the thyristor merely acts as a rectifying element and the output voltage waveform is identical to that of a single-phase half-wave rectifier. By phasing the gate pulses 'back' so that $\alpha = \pi$ radian, the average output voltage is reduced to zero and the thyristor completely blocks the flow of current. Increasing α beyond π radian or 180 degrees has no further effect on the output voltage.

The form of control used here is known as *phase control*, since the load current and voltage are controlled by shifting the phase of the gate pulse relative to the supply (anode) voltage.

While the thyristor acts as a controlled rectifier and provides a d.c. output in the form of a series of unidirectional pulses, the triac is a bi-directional controlling element which provides a controlled a.c. output. A controlled a.c. output can also be developed by replacing the thyristor with the *inverse-parallel* connected pair of thyristors in Fig. 10.6(c). Inverse-parallel circuits can be triggered from a common pulse generator if the gates of the thyristors are driven via isolated secondary windings in the manner shown in Fig. 6.7.

A practical form of half-wave circuit is shown in Fig. 10.7, and can be regarded as consisting of three sections, namely the voltage limiting and synchronizing circuit, the pulse generator, and the load circuit. The load circuit is similar to that shown in Fig. 10.6, and we will now concentrate on

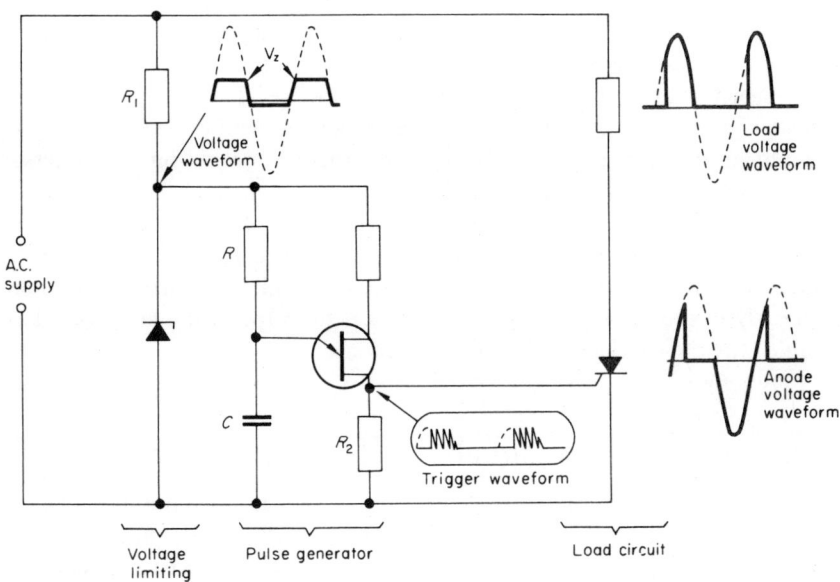

Fig. 10.7 A half-wave thyristor circuit.

the operation of the remainder of the circuit. The Zener diode acts as a voltage limiting circuit which, when the supply voltage is positive, applies a positive voltage V_Z to the pulse generator. When the supply voltage is negative, the voltage applied to the pulse generator is about $-0.7\,\text{V}$, corresponding to the voltage across a forward biased silicon diode. In order to ensure that the Zener diode is not damaged in normal operation, the value of R_1 must be such that it limits the maximum current through the Zener diode to a safe value when the peak voltage is applied.

Initially, capacitor C is discharged so that in the first positive half-cycle it begins to charge through R. When the voltage across C is equal to the peak-point voltage of the unijunction transistor, the UJT switches to the ON state and develops a voltage pulse across R_2, the time taken for the first pulse to be generated being computed on the basis of the work in section 6.7. The first pulse triggers the thyristor into its conducting state. Clearly, it is the timing of the first pulse which is of importance in this type of circuit.

For the remainder of the positive half-cycle a positive potential is applied to the pulse generator, and it continues to apply pulses to the thyristor. However, these pulses have no further effect on the thyristor since it is already conducting. The supply voltage falls to zero at the end of the positive half-cycle so that the thyristor current falls to zero and the device assumes a blocking state; it also causes the pulse generator to stop operating.

During the negative half-cycle, the thyristor is in its reverse blocking mode and does not pass load current. Any charge held by capacitor C at the end of the positive half-cycle is rapidly discharged through the Zener diode and resistor R, thereby ensuring that C is fully discharged at the commencement of the following positive half-cycle. This being the case, the delay time of the first gate pulse in each positive half-cycle relative to the commencement of the cycle is constant, thereby providing the necessary degree of synchronization.

Since the delay time is dependent on the value of the time constant RC, then alteration of the value of R controls the mean load voltage. A low value of R causes the capacitor to charge rapidly to turn the thyristor on early in the cycle, thereby providing a high output voltage. With a large value of R the average load voltage is reduced.

10.4   Single-phase circuits with inductive loads

When an attempt is made to reduce the current in an inductive load, the resulting flux change in the inductor core is such as to generate a 'back' e.m.f. which tends to maintain the flow of current. In thyristor circuits, the effect of this e.m.f. is to maintain the flow of current through the thyristor for a short time after the supply voltage has become negative. The current in the inductive circuit does not fall to zero until the stored energy in the load has fallen to zero.

Circuit waveforms for an inductive circuit controlled by a thyristor are shown in Fig. 10.8(a), and for a circuit controlled either by a triac or a pair of inverse-parallel connected thyristors is shown in Fig. 10.8(b).

In the case of the thyristor, Fig. 10.8(a), the device is triggered at t_1 in the first positive half-cycle, causing its anode voltage to fall to about 1 V. Conduction commences and current begins to build up in the load. The current waveshape can be derived from the solution of the differential equation of the circuit, and by this means it can be shown to have the typical 'lop-sided' appearance shown in the figure. When the anode voltage reaches zero, the anode current continues to flow for the reason given above until time t_2, when the thyristor current falls below the holding current. The thyristor then turns off and resumes a blocking state until triggered into conduction at t_3 in the second positive half-cycle. This process is repeated in each cycle.

In some instances, it is desirable to cause the thyristor to cut off at the commencement of the negative half-cycle, and this can be brought about by shunting the load with a diode. The function of the diode is to provide an alternative path for the flow of inductive current. In this type of application the diode is known as a *flywheel diode* or *spark quench diode*, and an example of its use is illustrated in Fig. 10.10.

In a single-phase triac circuit or an inverse-parallel thyristor circuit the wave-forms are as in Fig. 10.8(b). Here the triac is triggered at t_1 in the positive half-cycle and conduction continues until t_2 when the circuit current falls to zero. It is triggered in the negative half-cycle at time t_3 and conducts until t_4 when the current again falls to zero. The complete cycle of events is repeated again when the triac is triggered at t_5.

In triac and inverse-parallel circuits, a voltage spike may be generated at the leading edge of the triac voltage waveform (e.g., at t_2 or t_4 in Fig. 10.8(b)) if the trigger angle is greater than the load angle. Care has to be taken in the design of these circuits to ensure that the intial value of dV/dt (rate of change of applied voltage) does not exceed the maximum allowable value for the thyristor. More will be said about this later.

10.5 Other single-phase circuits

A range of typical single-phase circuits is shown in Fig. 10.9. The centre-tap circuit in Fig. 10.9(a) provides full wave d.c. control, the transformer allowing it to be used where a non-standard output voltage is required or where the load has to be electrically isolated from the power supply. The inverse-parallel circuit 10.9(b), provides full wave control of alternating currents, and the bridge circuits 10.9(c) and (d) can be used to control either

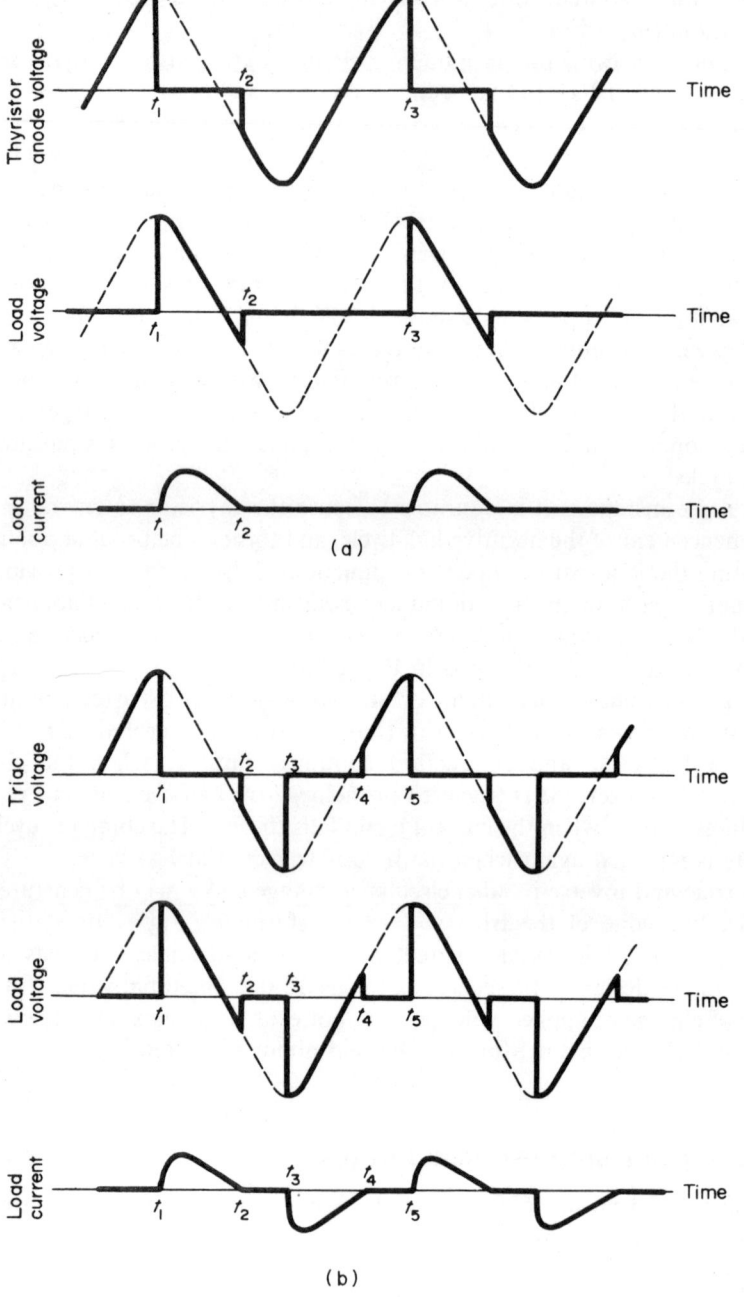

Fig. 10.8   (a) Voltage and current waveforms in a single-phase circuit with an inductive load and (b) waveforms for a triac or inverse-parallel thyristor circuit.

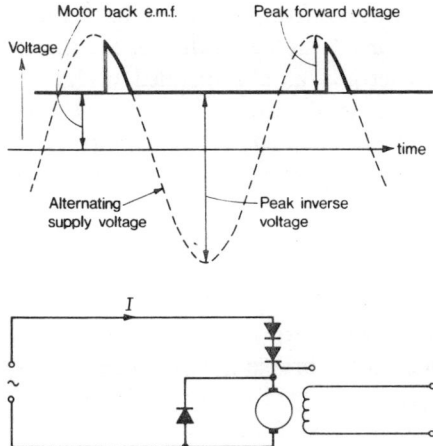

Fig. 10.9 Typical single-phase full-wave thyristor circuits: (a) centre-tap, (b) inverse-parallel, (c) and (d) bridge circuits.

a.c. or d.c. loads by placing the loads in the positions shown. Usually, only one of the two loads is included, the other being replaced by a short-circuit. Circuit 10.9(d) has, for a given type of thyristor, twice the current rating of circuit (c).

When the load is inductive it is advisable in circuit Fig. 10.9(c) to shunt the load with a flywheel diode to ensure that the thyristor current is reduced

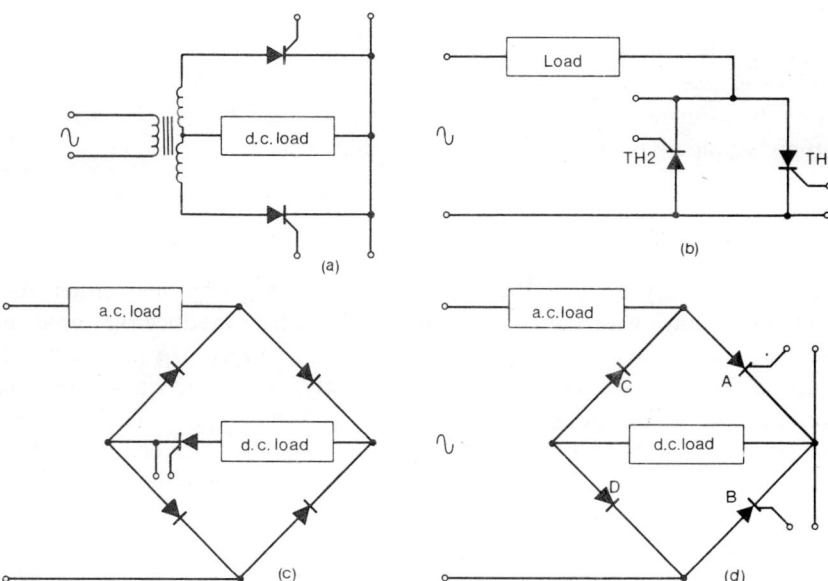

Fig. 10.10 A basic single-phase motor control scheme.

to zero at the end of the positive half-cycles. If, in Fig. 10.9(d), we were to use thyristors in positions A and C, with diodes in positions B and D, the circuit operates satisfactorily as a controlled bridge rectifier and has the added advantage that the diodes are correctly connected to act as a flywheel diode. The average output voltage from circuits 10.9(c) and (d) with a resistive load is twice that from the half-wave circuit, hence

$$V_{\mathrm{L}} = \frac{V_{\mathrm{M}}}{\pi}(1 + \cos \alpha) = 0.45 V_{\mathrm{s}} (1 + \cos \alpha)$$

where V_{s} is the r.m.s. supply voltage.

The circuit in Fig. 10.9(d) is known as a *half controlled bridge* since only one-half of the devices in the bridge are controlled. Bridge circuits in which all the elements in the bridge are thyristors are known as *fully controlled bridge* circuits; half controlled bridges are suitable for use in a.c.-to-d.c. (rectifier) applications, whereas fully controlled bridges can be used in d.c.-to-a.c. (invertor) applications as well as in rectifier circuits.

10.6   Control of d.c. motors

The inherently high controllability of d.c. machines makes them particularly useful in control systems. In this section, we will investigate the basic properties of systems. A basic relationship which holds for all d.c. machines is

$$E \propto \Phi \omega$$

where E is the (back) e.m.f. of rotation, Φ is the flux in the magnetic circuit, and ω is the speed of rotation of the armature. If we can neglect the armature voltage drop due to flow of armature current, then the armature supply voltage V is approximately equal to E, hence

$$V \propto \Phi \omega$$

or
$$\omega \propto V / \Phi$$

That is, we may control the armature speed by controlling either the armature voltage V or the magnetic flux Φ. In order to reduce the speed to zero, we must either reduce the armature voltage to zero or we must make the field flux infinitely large. The latter is, clearly, not a practical solution in conventional machines as it would require an infinitely large field current. An exception to this is the *split-field machine*, which is described in chapter 12. As a result, armature voltage control is a very popular method of controlling the speed of a d.c. machine when the speed variation demanded by the system lies in the range between zero (standstill) and the normal maximum speed in either direction of rotation. To increase the speed beyond the normal maximum, a practical solution is to reduce the field current below its normal value, thereby reducing the magnetic flux.

The control of electrical machines is dealt with in more detail in chapter 12, and in this chapter we will concentrate on some of the technical problems associated with armature voltage control.

A single-phase half-wave circuit is shown in Fig. 10.10, together with ideal waveform diagrams. In this circuit, the field current is maintained at a constant value by an independent supply source. When the motor is at standstill, the back e.m.f. is zero and the whole of each positive half-cycle of the supply voltage is available for control purposes. As the motor speeds up the back e.m.f. rises, so reducing the time available for the motor to draw current. The thyristor can only be triggered into conduction during the period when its anode is positive with respect to its cathode, that is when the supply voltage is greater than the back e.m.f. In this type of circuit, the peak repetitive inverse voltage can approach a value of twice the supply voltage, and a diode is sometimes connected in series with the thyristor in the manner shown to increase the reverse blocking capability of the circuit. The diode connected across the armature acts as a flywheel diode to divert the flow of armature inductive current after the point at which the supply voltage becomes negative.

10.6.1 The thyristor as an invertor

In reversing drives such as steel rolling mills, mine winders, etc., the work motor (the drive motor) has frequently to be brought to a standstill rapidly. A convenient way of rapidly reducing the motor speed is to convert the mechanical energy stored in the rotating load into electrical energy by using the work motor as a generator; the electrical energy is then returned to the supply system. This process, known as *regenerative braking*, requires the thyristor to operate in an *inverting mode* in which it converts a direct current into an alternating current.

Inverted operation can most easily be understood by reference to the single-phase system in Fig. 10.11. In order to define the 'generating' or power supply state, and the 'consuming' or motoring state, the normal (rectifying) mode of operation is shown in Fig. 10.11(a). For the thyristor to conduct it is necessary for its anode to be positive with respect to its cathode, and in Fig. 10.11(a) we assume that V_s is 102 V and that E is 100 V. The arrowheads on the 'potential' arrows point towards the more positive terminal, and the direction of the current arrows indicate the actual direction of flow of current in the circuit. We see from Fig. 10.11(a) that a generating state occurs when the voltage and current arrows act in the same direction, and a power consuming state is represented by one in which the arrows act in opposite directions.

By reversing the armature connections with respect to the transformer connections, as in Fig. 10.11(b), conduction can again take place in the

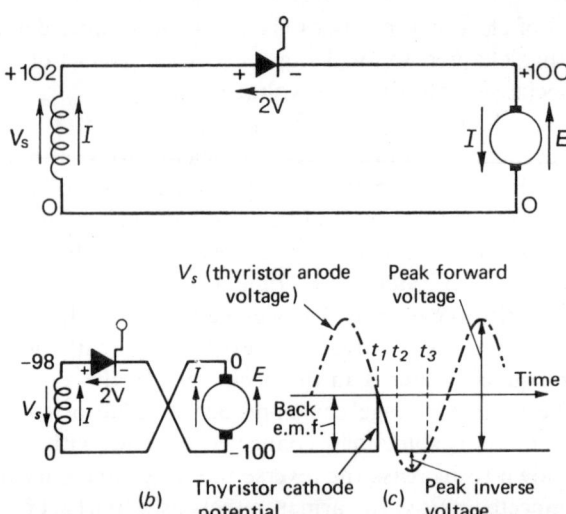

Fig. 10.11 The thyristor operating (a) in its normal (rectifying) mode and (b) in the inverted mode. Waveforms for inverted operation are shown in (c).

negative half-cycle of the supply voltage waveform so long as the cathode is maintained more negative than the anode. From a consideration of the voltage and current arrows in this mode of operation, we see that the rotating machine operates in a generating mode and the transformer winding operates in a consuming mode. Thus, the thyristor may be triggered at time t_1, Fig. 10.11(c), and if the circuit is non-inductive current flow ceases at t_2. If the armature circuit is inductive, current flow continues beyond t_2 in the manner described in section 10.4. In any event, conduction should not be allowed to continue beyond t_3 in Fig. 10.11(c), otherwise it will not be possible to turn the thyristor off as the anode voltage once more is positive with respect to the cathode. Should this happen, a short-circuit will develop across the power supply.

The three-phase half-wave fully controlled circuit in Fig. 10.12(a) can be used as an invertor when terminal Q is positive with respect to terminal P. For rectifying action to occur, gate pulses are applied to the R-phase thyristor at point L on the waveforms in Fig. 10.12(b), and for inversion the R-phase thyristor can be activated after point M. In the inverted mode, the Y-phase thyristor must be triggered before point N, otherwise the R-phase thyristor will continue conducting for the remainder of the cycle, and the supply will be short-circuited. Sufficient time must, of course, be allowed for the current to commutate from the R-phase thyristor to the Y-phase thyristor before point N, this period of time being increased by the effect of the reactance of the power supply system.

In the half-wave circuit in Fig. 10.12(a), for inversion the gate signal must

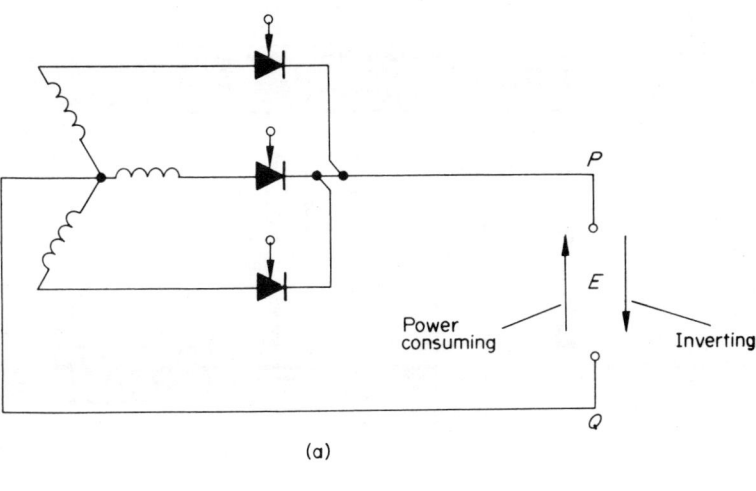

(a)

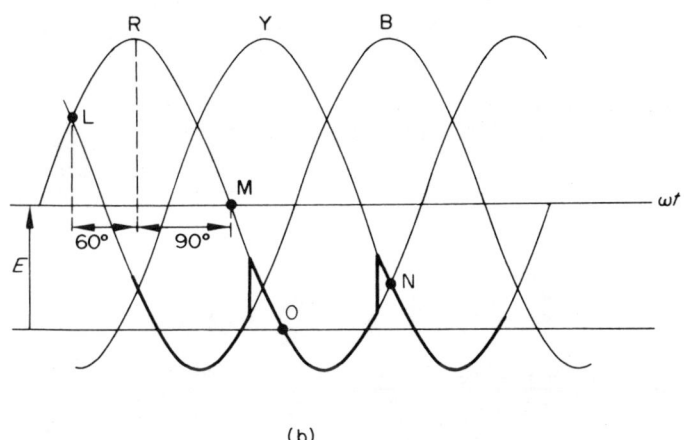

(b)

Fig. 10.12 Waveforms in a three-phase invertor.

be phased back by at least 150 degrees or $(\pi/m + \pi/2)$ radian, where m is the number of phases. There is, in fact, a limited range over which the thyristor can be fired, since triggering must occur before point 0 if the thyristor is to conduct.

10.7 Phase control of triac circuits

Two examples of triac circuits using phase control are shown in Fig. 10.13. Figure 10.13(a) uses a version of the unijunction transistor circuit described in section 10.3. In this case, the pulse generator is energized from a full-wave rectifier so that pulses are generated in both positive and negative half-

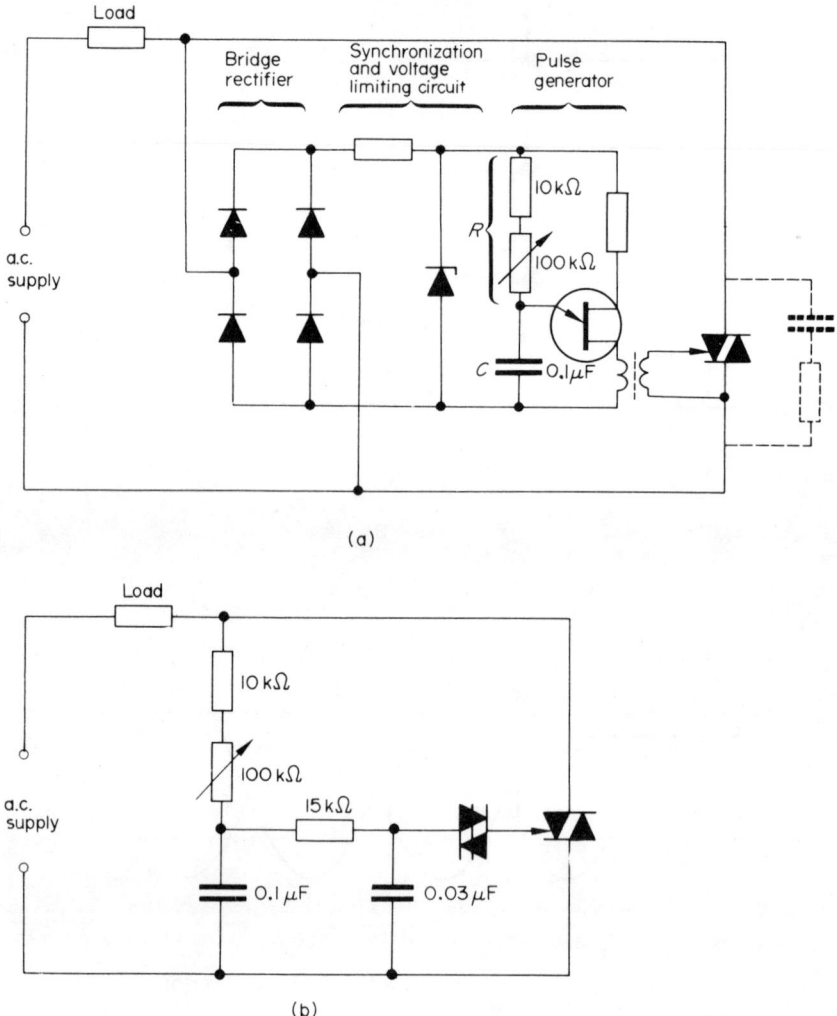

Fig. 10.13 Two types of triac circuit employing phase control.

cycles of the supply voltage waveform. This ensures that the triac is triggered in both half-cycles. The circuit in Fig. 10.13(a) also differs from that in Fig. 10.7 in that the pulse generator power supply is derived from the voltage across the triac, so that when the triac has been triggered the pulse generator supply voltage falls to zero for the remainder of the half-cycle.

As we have already mentioned, switching inductive loads can give rise to voltage spikes being generated across the triac with the attendant possibility of accidental dV/dt triggering. Triacs are generally more prone to spurious triggering from this cause than thyristors, and it is advisable in such cases to limit the possibility of dV/dt triggering by shunting the triac with the R–C network shown in broken line in Fig. 10.13(a). This type of protection is discussed in more detail later in the chapter.

Phase control is obtained in Fig. 10.13(a) by altering the value of R, thereby altering the charging time constant of the R–C circuit.

A circuit fulfilling the same function as that described above is shown in Fig. 10.13(b). This circuit employs a version of the diac relaxation oscillator previously described in section 6.8. In this case, a double R–C network is used to provide a smooth control over the load current.

10.8   Burst-firing control

A disadvantage of phase control techniques is that the rapid changes of voltage and current in the circuit generate harmonic frequencies in the medium- and short-wave radio-frequency bands, and where equipment includes long unscreened leads, audio-frequency interference may also be radiated. It is possible to minimize these problems either by using polyphase systems (i.e., 3-, 6-, or 12-phase systems) or by using L–C interference suppressors. In some instances polyphase supplies are not available, e.g., in domestic installations, and, where the power consumed is large, the cost of the suppressor may be a significant proportion of the initial capital cost of the equipment. Also, the power factor of the current drawn from the supply in phase control systems is related to the delay angle, so that if the output voltage is low (i.e., the delay angle is large) then the power factor is low.

An alternative method of control known as *burst-firing* (also known as *integral cycle firing*, *zero-point firing*, and *zero voltage firing*) can be used in some cases to overcome these defects. By arranging the firing circuit so that the thyristors or triacs are switched on at the zero voltage point in the cycle, and that load current flows for a burst of an integral number of half-cycles or cycles, the problem of radio-frequency interference is avoided.

Typical output voltage waveforms for a burst-firing system are shown in Fig. 10.14. In Fig. 10.14(a) the circuit is triggered into conduction for two successive half-cycles out of every six, and in Fig. 10.14(b) current flows for four successive half-cycles out of six. The output r.m.s. voltage V_0 from the circuit is

$$V_0 = V_r \sqrt{\frac{T_b}{T_p}}$$

where V_r = r.m.s. voltage of the supply

T_b = Period of the burst of conduction expressed in half-cycles

T_p = Period of the complete pattern expressed in half-cycles

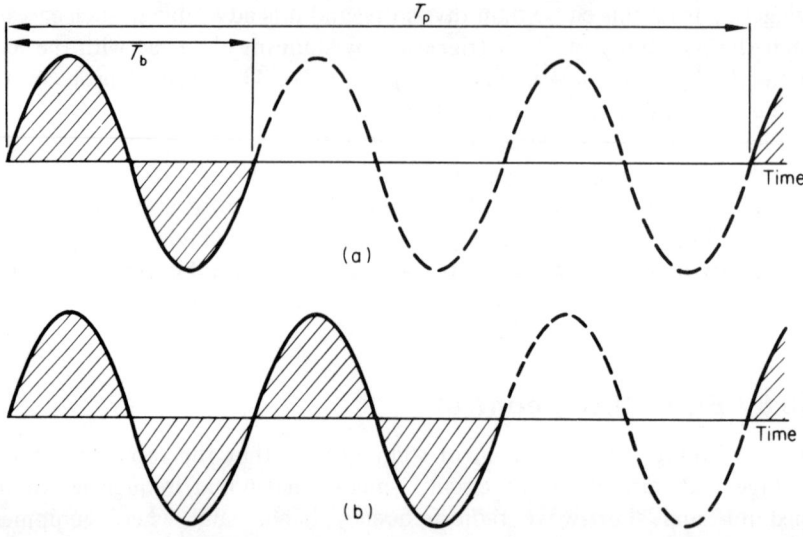

Fig. 10.14   Burst-firing waveforms.

In Fig. 10.14(a) the output voltage is

$$V_0 = V_r \sqrt{(2/6)} = 0.577\, V_r$$

and the power delivered into a resistive load is

$$P_0 = P_r T_b / T_p$$

where P_r is the power developed in the load when the full supply voltage is applied. In the case of Fig. 10.14(a), $P_0 = P_r/3$.

Loads such as electric heaters, with long time constants compared with the periodic time of one cycle of mains frequency, are well suited to burst-firing control methods. Burst-firing control is not without its disadvantages, as it is unsuitable for the control of lamp loads due to the flicker generated by the on/off pattern frequency at low output power. Nor is it suited to motor control, as low speeds can only be obtained by a small value of T_b/T_p. It is also unsuitable for the control of transformer-input loads.

The basis of one form of burst-firing scheme is shown in Fig. 10.15. In this circuit, gates G1 and G2 are NAND gates operating with positive potentials and positive logic, and are cross-connected to form a memory circuit. The output of G1 drives transistor TR, so that when the output from G1 is logic 1, TR is saturated and the capacitor is short-circuited. This inhibits the operation of the UJT pulse generator. When the output from G1 is logic '0', TR is cut off and the oscillator generates pulses at its maximum rate.

With zero volts applied to the control line, the output from G1 is logic '1'

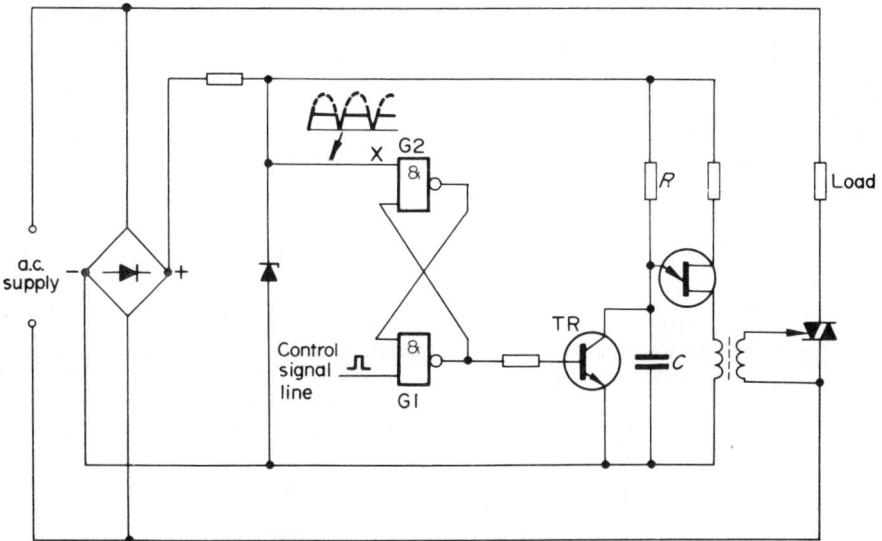

Fig. 10.15   A schematic diagram of one form of burst-firing control circuit.

so that TR short-circuits C and the pulse generator does not function. When the control signal line is raised to a positive potential it loses control over gate G1, and G1 acts simply as an invertor to the output of G2. With a positive voltage on the control line, the pulse generator remains inoperative until the voltage at point X falls to zero, which occurs at the commencement of the supply voltage waveform. This causes the output of G2 to rise to the logic '1' level, and the output of G1 to fall to logic '0'. As a result, pulses are applied to the triac gate and it is triggered into conduction. As soon as the supply voltage rises from zero volts, the voltage at X begins to rise and it has no further control over G2. But, due to the latching action of G1 and G2, the output from G1 remains at zero so long as the control voltage is positive.

When the control signal is again reduced to zero, the output of G1 drives TR into saturation and inhibits the operation of the pulse generator. The triac continues to carry current for the remainder of that half-cycle, after which it resumes its blocking state.

A simple method of controlling the magnitude of the load current in a burst-firing system would be to drive the control line by means of a pulse generator with a rectangular wave with a variable mark-to-space ratio. The circuit can then be converted into a closed-loop voltage regulator by driving the variable mark-to-space ratio oscillator by a comparator which compares the actual load voltage with the desired value of voltage.

The frequency of oscillation of the UJT pulse generator in Fig. 10.15 must be high compared with that of the supply frequency and, with a supply frequency in the range 50–100 Hz, the pulse repetition frequency should be several kilohertz.

10.9   Polyphase circuits

One form of three-phase circuit, the star connected half-wave circuit, was illustrated in Fig. 10.12(a), and other circuits are shown in Figs. 10.16 and 10.17. The circuits in Fig. 10.16 are half controlled circuits, that in Fig. 10.16(a) supplying a d.c. load, and 10.16(b) supplies an a.c. load. A fully

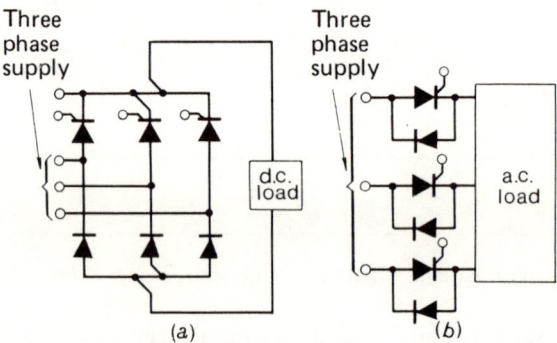

Fig. 10.16   (a) A three-phase half-controlled bridge circuit and (b) a three-phase half-controlled a.c. controller.

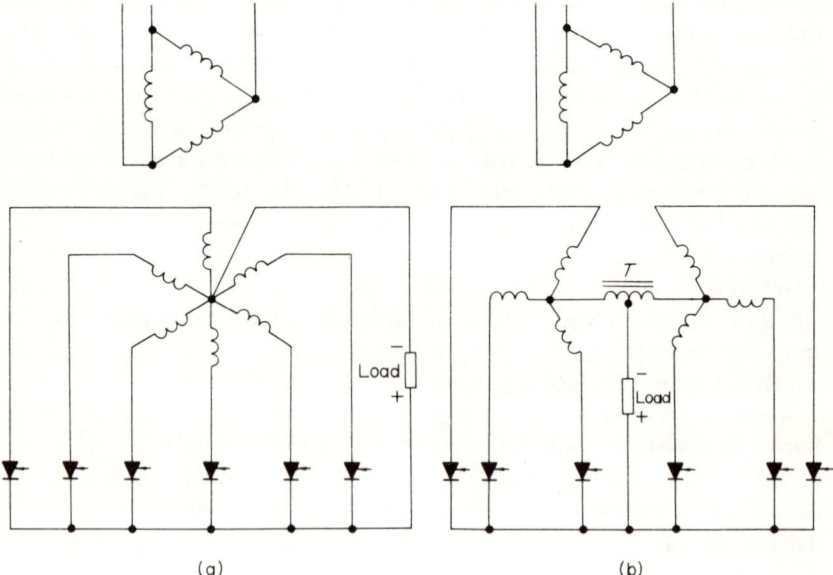

Fig. 10.17   (a) A three-phase centre-tap circuit and (b) a three-phase double-star circuit.

controlled version of the bridge circuit enables it to be used as an invertor in the manner described earlier. In the a.c. controller, the diodes provide return paths for the load current flowing in the other lines. Fully controlled versions of Fig. 10.16(b) are used in cases where a low harmonic content is desirable.

Two versions of three-phase full-wave circuits (or six phase half-wave) are shown in Fig. 10.17. The circuit in Fig. 10.17(a) is not used very often since the thyristors are not efficiently utilized. For example, when the circuit provides its maximum output the thyristors have a conduction angle of only 60 degrees, whereas in Fig. 10.17(b) it is 120 degrees. Moreover, the power factor and transformer utilization of circuit 10.17(b) are improved when compared with circuit 10.17(a). The interphase transformer in circuit 10.17(b) allows the two secondary circuits to operate independently of one another, the potential of the centre-tap being midway between the potentials of the two secondary star points. Since both halves of the circuit carry current simultaneously then, for a given load current, each conducting thyristor carries about half of the total load current, resulting in a better output voltage regulation and lower transformer power loss in circuit 10.17(b) than in circuit 10.17(a).

10.10 Thyristor d.c. controllers

A thyristor is used to control the flow of direct current between a d.c. source and a load by connecting it as shown in Fig. 10.18(a). In this mode, the thyristor is used as a *series chopper* or *switch,* and the average load voltage E_{av} can be controlled in a number of ways, two of which are shown in the figure.

By chopping the supply voltage at a constant frequency so that T is constant, diagram 10.18(b), the output voltage can be varied by altering the

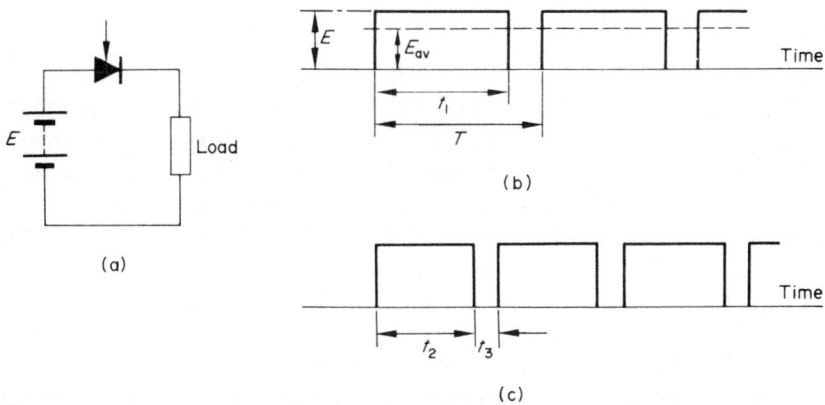

Fig. 10.18 One form of thyristor d.c. controller.

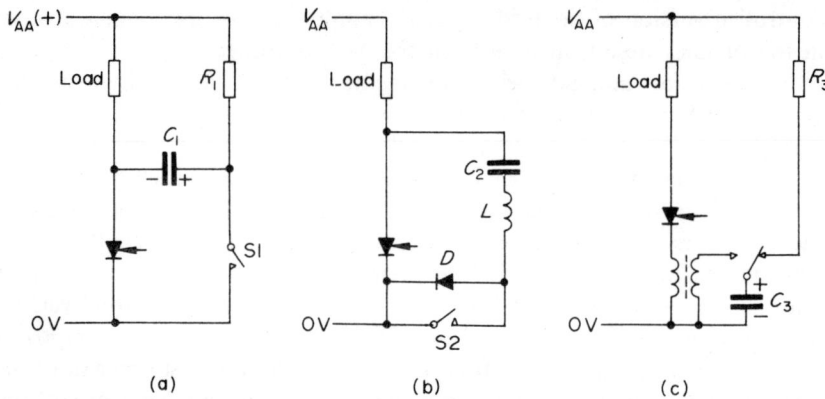

Fig. 10.19 Basic circuits for (a) capacitor commutation, (b) resonance commutation, and (c) pulse commutation.

period t_1 for which the thyristor conducts. This is known as *variable mark-to-space ratio control*. Alternatively, the 'mark' period t_2, Fig. 10.18(c), can be maintained constant and the space period t_3 can be varied. In this way, the periodic time and hence the pulse repetition frequency is adjusted. This is sometimes known as *variable frequency control*.

The three principal methods employed in turning thyristors off in d.c. controllers are illustrated in Fig. 10.19 and are:

(a) Capacitor commutation or parallel capacitor commutation
(b) Resonance commutation (also known as harmonic commutation or parallel capacitor-inductor commutation)
(c) Pulse commutation.

The principle involved in each case is much the same, in that the current through the thyristor is reduced to zero for a time greater than its turn-off time.

In *capacitor commutation*, Fig. 10.19(a), when the thyristor is ON the capacitor C_1 is charged to V_{AA} with the polarity shown via resistor R_1. When S1 is closed, a reverse bias of V_{AA} is applied to the thyristor to turn it off. The average load current is controlled by controlling the instants at which the thyristor is triggered and at which S1 is switched. In practical circuits, S1 is replaced by a thyristor (see also Fig. 10.20).

In the *resonant commutation* circuit, Fig. 10.19(b), capacitor C_2 is charged via inductor L and diode D during the time that the thyristor is cut off. When the thyristor is triggered, the capacitor cannot discharge since diode D is reverse biased. The operation of closing switch S2 completes a resonant discharge path, causing the thyristor to turn off.

The *pulse commutated* circuit in Fig. 10.19(c) uses a transformer to inject a reverse voltage pulse in series with the thyristor to turn it off.

The capacitor commutated circuit and its variations are well suited to a wide range of industrial applications, while resonance commutation circuits are better suited to high voltage, low current applications. The capital cost of pulse commutation circuits is higher than that of other types due to the use of the coupling transformer. This type of circuit is more frequently found in some forms of frequency convertors (see section 10.11).

A version of a capacitor commutated circuit is shown in Fig. 10.20 in which TH2 replaces S1 in Fig. 10.19(a). The waveforms shown commence with TH1 conducting and capacitor C charged to V_{AA}. At the instant that TH2 is triggered, the anode voltage of TH1 momentarily falls to $-V_{AA}$, causing the voltage across load resistor R_{L1} to rise to $2V_{AA}$. As a result, the current in R_{L1} momentarily rises to $2V_{AA}/R_{L1}$. The anode voltage of TH1 then begins to recover towards $+V_{AA}$ along an exponential curve of time constant CR_{L1}.

An approximate value for the capacitance of C can be computed on the basis that the charge CV_{AA} held by the capacitor is equal to the product It_{off}, where I is the *initial* discharge current (which, with a resistive load is $2V_{AA}/R_{L1}$) and t_{off} is the turnoff time of the thyristor. That is

$$CV_{AA} = It_{off}$$

or
$$C = It_{off}/V_{AA}$$

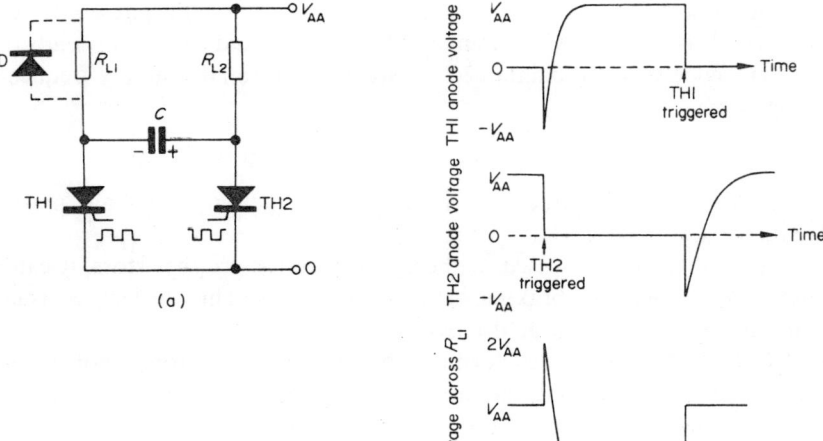

Fig. 10.20 (a) A capacitor commutated d.c. controller and (b) typical waveforms.

With a resistive load this reduces to

$$C = 2t_{off}/R_{L1}$$

A more precise relationship is obtained by considering the equation of the anode voltage of TH1 following the instant that it is turned off. The value of t_{off} must either be less than or equal to the time taken for the anode voltage to rise from $-V_{AA}$ to zero volts. The circuit equation is

$$V_{AA} = -V_{AA} + 2V_{AA}[1 - \exp(-t_{off}/R_{L1}C)]$$

Solving for C yields

$$C = t_{off}/R_{L1} \ln 2 \simeq 1.45 \, t_{off}/R_{L1}$$

which agrees quite well with the approximate result obtained above.

10.11   Frequency convertors

In industrial installations, frequency convertors are used to provide power for variable speed motor drives and other iron-cored loads. Static frequency convertors fall broadly into two categories, which are cycloconvertors and d.c. link convertors.

Cycloconvertors are direct a.c.-to-a.c. convertors, one form of which is shown in Fig. 10.21 and comprises a matrix of thyristors which are fired in a predetermined pattern. In the case considered, the periodic time of the fundamental frequency of the output waveform is $6T$, where T is the periodic time of the supply frequency. Thus, with a 60 Hz supply, the output frequency would be 10 Hz, and with a 50 Hz supply the output would be at 8.33 Hz. It is a frequent requirement of iron-cored and a.c. motor loads that the magnetic flux be maintained constant; whatever the supply frequency. Now since

$$\text{Magnetic flux} \propto \text{Voltage/Frequency}$$

it follows that as the frequency is reduced, so the output voltage must be reduced.

Since all the devices used in the cycloconvertor are thyristors, it can be used for regenerative braking for machine drives. This, in fact, is usually only of value in very large drives.

The output voltage waveform of this type of convertor is not a particularly smooth one, but this does not appear to have any ill effects on induction motor performance. Improved output waveforms can be obtained by the use of six-phase or twelve-phase supplies.

Due to the principle of operation of the cycloconvertor, the maximum output frequency is limited to about 60 per cent of the supply frequency, i.e., about 36 Hz with a 60 Hz supply and 30 Hz with a 50 Hz supply. If the

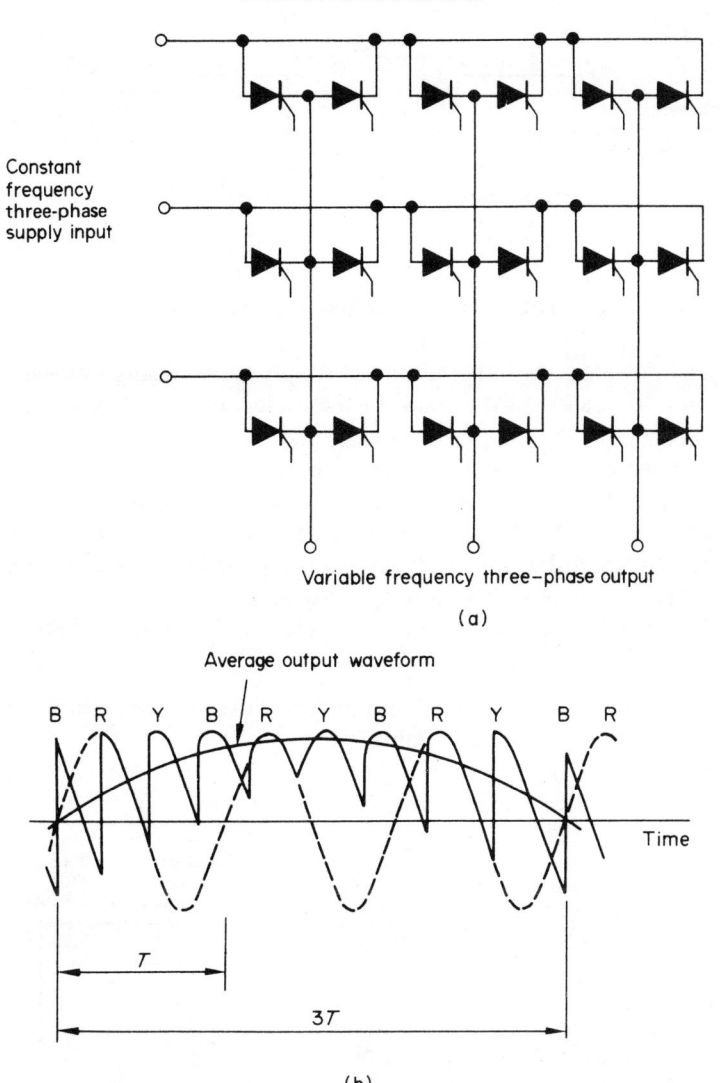

Constant frequency three-phase supply input

Variable frequency three-phase output

(a)

Average output waveform

B   R   Y   B   R   Y   B   R   Y   B   R

Time

T

$3T$

(b)

Fig. 10.21   (a) One form of cycloconvertor and (b) output waveforms.

harmonic content in the output waveform is to be kept to a reasonable level, the maximum output frequency should be limited to about one-third of the supply frequency. Consequently, cycloconvertors are more frequently found in applications which require a low output frequency.

The *d.c. link convertor*, illustrated in Fig. 10.22, uses a d.c. link between the rectifier and invertor sections of the convertor. Link convertors have a number of advantages over cycloconvertors including the following:

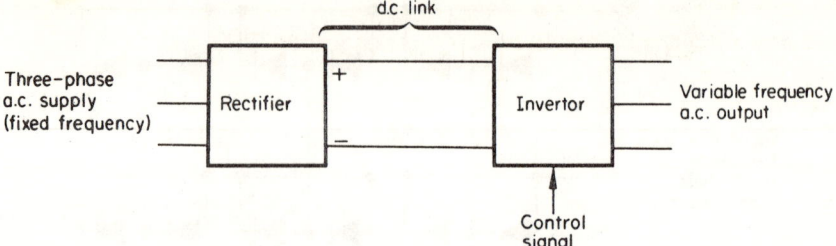

Fig. 10.22   The basis of the d.c. link convertor

(a) The power factor imposed on the supply is practically unity since the load is, in effect, a d.c. load. In the cycloconvertor, the power factor depends on the load power factor and the angle of firing.
(b) In most cases the circuitry associated with d.c. link convertors is less complex than that of otherwise equivalent cycloconvertors.
(c) The output frequency of d.c. link convertors is only limited by the rate at which the thyristors can switch. This gives an upper limit on output frequency of several hundred hertz.
(d) The invertor section can be used independently from a battery as a d.c.-to-a.c. convertor.

The invertor section of the d.c. link convertor usually uses either a bridge circuit or a push-pull type of circuit.

A *bridge circuit* is shown in Fig. 10.23(a), details of the commutation circuit being omitted for simplicity. Flow of current from left to right

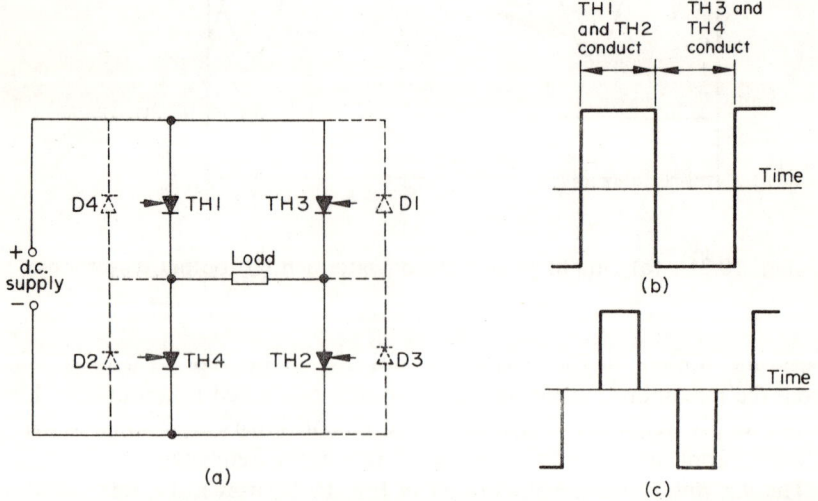

Fig. 10.23   (a) A basic bridge invertor with waveforms for (b) maximum output voltage, and (c) a reduced output voltage.

through the load is brought about by triggering thyristors TH1 and TH2, and flow of current from right to left by triggering TH3 and TH4. With a resistive load, the output voltage waveform has a rectangular shape, the maximum output voltage being obtained with the waveform shown in Fig. 10.23(b). A reduced output voltage is obtained by reducing the width of the positive and negative pulses, as shown in Fig. 10.23(c). At the end of each 'half-cycle' the current through the thyristors is force-commutated to zero by one of the methods described in section 10.10.

With an inductive load, an alternative path for the flow of lagging current is provided by diodes D1 to D4; their operation is as follows. During the positive half-cycle, current flows from left to right through the load and, at the end of this cycle TH1 and TH2 are turned off and gate pulses are applied to TH3 and TH4. However, TH3 and TH4 cannot conduct as the inductive current is flowing in the 'wrong' direction. By shunting TH3 by D1 and TH4 by D2, the energy stored in the inductive load can be returned to the supply and, when the load current finally falls to zero, TH3 and TH4 begin to conduct. This allows current to flow from right to left through the load.

A basic *push-pull invertor* circuit is shown in Fig. 10.24, and it employs a

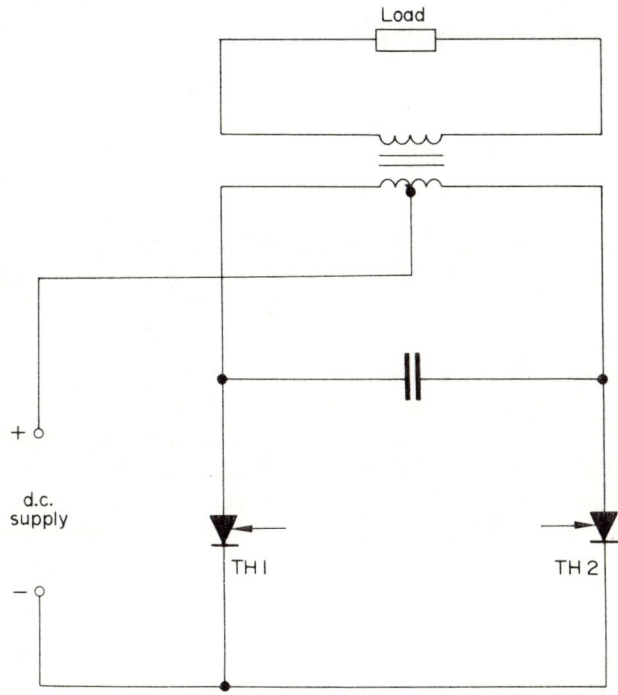

Fig. 10.24   A push-pull invertor.

capacitor commutated circuit similar to the type described in section 10.10. Triggering TH1 in Fig. 10.24 turns TH2 off, and firing TH2 turns TH1 off, thereby reversing the direction of flow of current through the two halves of the primary winding. This results in an alternating e.m.f. being applied to the load. The output frequency is determined by the gate control circuit, and while it is possible to alter the operating frequency of the circuit, it is not easy to alter the output voltage. It might be thought that a satisfactory way of doing this is to reduce the supply voltage, but this has the drawback of reducing the energy stored in the capacitor for commutation purposes. A method sometimes used to control the output voltage is to use two invertor circuits which have a common transformer. By shifting the phase of the trigger pulses in the two circuits, one with respect to the other, an alternating output with control of the type illustrated in Fig. 10.18(b) is achieved.

Of the two types, bridge invertors are the most popular as they do not require the load to be transformer-coupled to the invertor. In a few instances, transformer-coupled circuits do have an advantage as, for example, where the load must be isolated from the supply or where a voltage level change is required.

10.12   Thyristor protection

Semiconductor devices are prone to damage from many causes including overcurrents, overvoltages, and transients. The type of protection incorporated in any equipment depends to a great extent on the application of the equipment. The following is a summary of the more important aspects of protection.

Overcurrent protection is often provided in the form of a current limit circuit which restricts the current flowing in the circuit to a safe value. A schematic diagram of one form of current limit protection is illustrated in Fig. 10.25. In this circuit, the magnitude of the load current is a function of the p.d. across resistor R. This resistance has a low value and, in small systems, the p.d. across it when carrying the maximum current need only be a fraction of a volt. This voltage is increased in magnitude by transformer T, after which the signal is rectified by BR and smoothed by capacitor C. Under normal load conditions the voltage across C is less than the breakdown voltage of the Zener diode Z, so that no current is injected from this source into the pulse generator. As a result, the pulses applied to the gate of the thyristor are completely under the control of the signal applied to the control line.

When the load current exceeds a predetermined value, the voltage across C exceeds the Zener diode breakdown voltage. The current which then flows from the current limit circuit into the pulse generator is arranged to phase back the thyristor gate pulses so as to reduce the load current to a safe value.

Circuit breakers and contactors are used as back-up protection and are also used to isolate the equipment from the power supply. Circuit breakers themselves can, unfortunately, be a source of voltage transients either when energizing a transformer primary winding or when interrupting transformer magnetizing current.

Fast-acting semiconductor fuses are frequently used as a means of protecting thyristors against damage under short-circuit conditions. The energy required to melt a fusible element is related by what is known as its i^2t rating and, for the fuse to protect the thyristor, its i^2t value must be less than that of the thyristor. The fuse must be selected so that its *arcing time* is not too short, otherwise large voltage transients can be generated if the load is inductive. A problem which frequently arises under short-circuit conditions is the large magnitude of the initial rush of current when the short-circuit is applied. This is known as the *asymmetric peak* current, and has a value much greater than the steady short-circuit current. A degree of protection is afforded by including a choke in the input line to limit the fault current. For short-circuit protection, the choke should be air-cored to retain its inductance at high values of current.

It is possible for thyristors to fail due to what is known as di/dt failure, which is explained in the following. Immediately the thyristor has been triggered, the gate current is concentrated into a very small area of the gate region since it has not had sufficient time to 'spread' across the whole region. As a result, the initial rush of anode current is concentrated into a very small section of the total area of the device. If the rate of change of

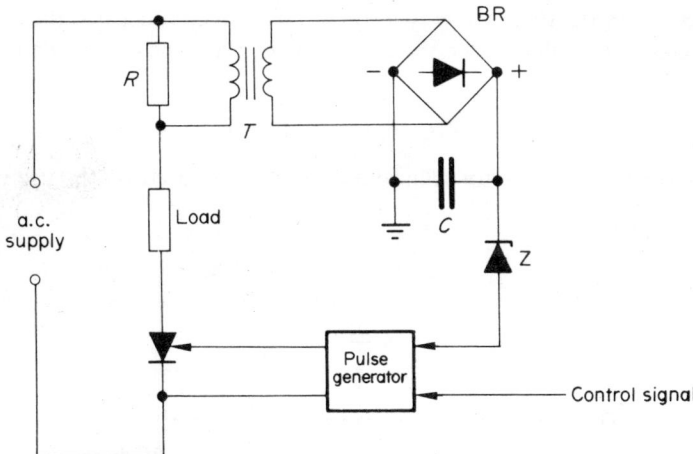

Fig. 10.25 One method of applying current limit protection.

anode current exceeds a critical value, then a local 'hot spot' may develop and the thyristor may fail catastrophically. In most circuits, there is a certain amount of inductance which limits the value of di/dt, so that failure from this cause is not often a problem. The minimum value of circuit inductance L_{min} can be determined approximately by

$$L_{min} = \frac{V_s}{di/dt}$$

where V_s is the supply voltage and di/dt is the rated maximum di/dt of the device, which could be as high as $1000 \, \text{A}/\mu s$.

There are two principal reasons for voltage breakdown. Firstly, the rate of change of anode voltage (dV/dt) can cause the thyristor or triac to be triggered into conduction. Secondly, excessive voltage can result in an increase in leakage current and lead to breakdown.

Failure due to excessive dV/dt can be understood by reference to Fig. 10.26, in which capacitor C represents the anode-gate capacitance of the thyristor. When the device is cut off, it is possible for a high rate of change of anode voltage to cause C to draw a large charging current. If this current exceeds the gate turn-on current, then the thyristor is triggered into conduction. This type of breakdown can occur even if the supply voltage is lower than the rated breakdown voltage of the thyristor.

A common method of reducing the value of dV/dt is to connect an R–C network in parallel with the triac or thyristor, as shown in Fig. 10.27(a). If the supply voltage suddenly increases by V_1, the anode voltage rises along an exponential curve in the manner shown in Fig. 10.17(b). Resistor R has a small value and is included in the circuit to limit the capacitor discharge current when the thyristor is triggered. For maximum effectiveness, R can be shunted by the diode shown in broken line; this means that R is short-circuited when the thyristor is turned off, but is in circuit when the thyristor is triggered. The value of the time constant $R_L C$ can be computed from the relationship

$$R_L C = 0.63 \, V_1 / (dV/dt)$$

Mains-borne and contactor transients can be minimized by the use of RLC

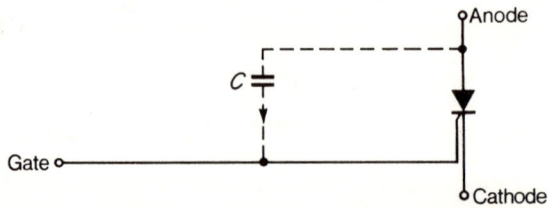

Fig. 10.26 Failure due to excessive dV/dt.

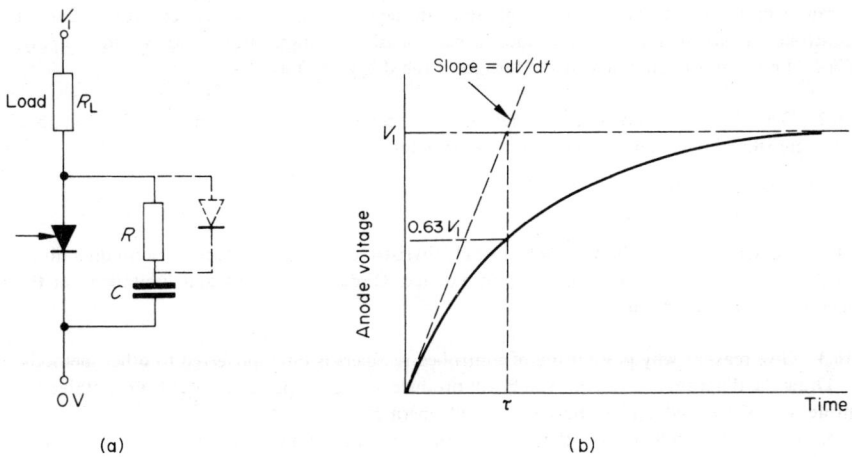

Fig. 10.27 An R-C dV/dt 'snubber' circuit.

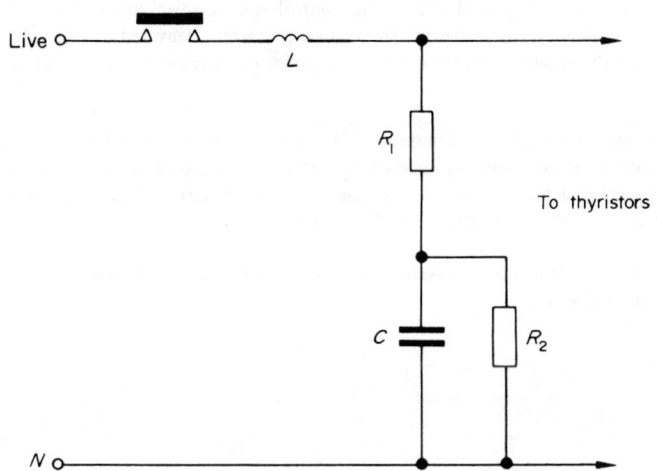

Fig. 10.28 A circuit to reduce the effects of voltage transients.

circuits of the type shown in Fig. 10.28. For dV/dt limitation the inductor can have a ferrite core and be physically small, but if the inductor is also to be used for fault current limitation then it should be air-cored. A compromise between the two is an iron-cored coil.

Problems

10.1 Sketch a typical thyristor characteristic and hence explain the meaning of the terms

(a) forward breakover voltage
(b) reverse breakdown voltage
(c) maintaining current.

Draw a basic circuit diagram for a phase-controlled inverse-parallel a.c. controller. Such a controller is used to supply a resistance furnace of $10\,\Omega$ resistance from a 400 V r.m.s. supply. Calculate the power supplied to the furnace if the delay angle is $90°$.

10.2 Describe using appropriate diagrams the operation of a thyristor. Sketch the anode characteristics of the device and indicate curves for

(a) zero gate current
(b) nominal forward gate current.

Draw a circuit diagram, showing how a single thyristor can be used to control a unidirectional load current. Assume a sinusoidal supply voltage. Discuss the merits and limitations of the method of gate control used.

10.3 Give reasons why pulse firing of controlled rectifiers is often preferred to other methods.

Draw the diagram of a circuit which will produce a delayed pulse or pulse train suitable for firing a thyristor and explain how the circuit operates.

Sketch a curve showing how the average output voltage of a controlled rectifier varies with the firing delay angle. Explain the shape of the curve.

10.4 Describe any one type of solid state controllable rectifier. Show how this could be embodied in a system for the control of the voltage or current delivered to a variable load, and briefly describe the system operation. State what safety features would be incorporated in the system.

10.5 A d.c. motor armature of resistance $2\,\Omega$ is supplied from a 141.4 V r.m.s. single-phase source via a diode. If the armature back e.m.f. is 150 V, calculate (i) the peak armature current, and (ii) the average armature current. Assume that the p.d. across the diode is negligible and that the effects of armature reactance and reaction are negligible.

10.6 Describe how thyristor convertors may be used to control the speed of an induction motor over a wide range.

11. Electronic power supplies

11.1 Types of power supply

Four main types of power supply are in general use:

- (a) d.c. output, a.c. input
- (b) d.c. output, d.c. input
- (c) a.c. output, d.c. input
- (d) a.c. output, a.c. input.

The first type listed above is more frequently used than the other types and forms the basis of the work in this chapter, but a word about the other types is not out of place. The second type listed is sometimes known as a *d.c. line controller* and is used to control the flow of power in d.c. lines. An example of this type was described in section 10.10, in which the flow of direct current in a load was controlled by a thyristor *chopper* circuit. A range of modern d.c.-to-d.c. convertors is discussed in section 11.12. Type (c) above was described in section 10.11 (Frequency Convertors) in which the d.c.-to-a.c. convertor was described as an invertor. Group (d) includes frequency convertors, also described in section 10.11.

The basis many of forms of a.c.-to-d.c. power supplies is shown in Fig. 11.1. The incoming supply voltage is, where appropriate, transformed to the correct voltage level, after which it is rectified and filtered to provide the

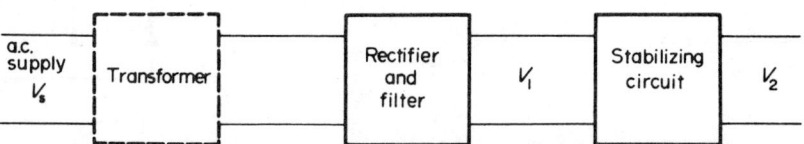

Fig. 11.1   The basis of stabilized power supplies.

unstabilized voltage V_1. The transformer is an expensive and bulky piece of equipment, and where possible it is eliminated from the circuit. However, where a voltage of a different level than the supply voltage is required or where the load must be isolated from the supply, then a transformer must be used. For many applications, the output from the rectifier and filter section is adequately smooth without the need for further signal modification. In other applications, it is necessary to add a stabilizing section to ensure that the output voltage (or current in some cases) is maintained at a constant level.

By increasing the frequency of the signal applied to the transformer, its physical size can be reduced. Where size is an important parameter, the technique of frequency conversion is adopted as follows. The frequency of the supply is increased by applying it to a frequency convertor (see section 10.11), the transformer being part of the convertor. The resulting high frequency, low voltage at the secondary of the transformer is then rectified and smoothed as before. These circuits are known as *switched-mode* supplies (see section 11.12).

11.2   Unstabilized power supplies

The three basic forms of single-phase rectifier circuit are shown in Fig. 11.2. Ideal constants for these circuits are listed in Table 11.1 in which V_2 is the average output voltage, and V_{SM} and V'_{SM} are peak values of supply voltage. The figures listed in the 'percentage ripple' section are computed from the relationship

$$\text{r.m.s. fundamental ripple voltage} \times 100/V_2$$

In circuits of the kind shown, it is prudent to select rectifiers which have peak inverse voltage ratings of at least four times the r.m.s. supply voltage (or transformer secondary voltage). This rule-of-thumb makes some allowance for the normal type of voltage transients which appear on the supply line.

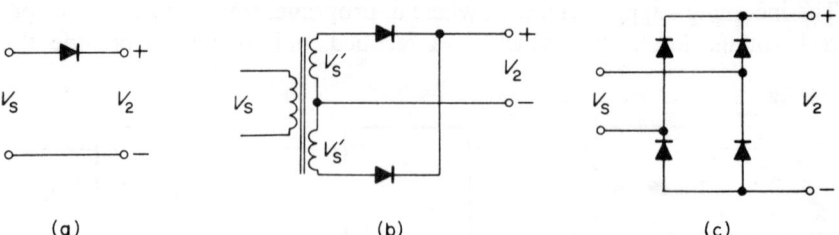

(a)   (b)   (c)

Fig. 11.2   Basic single-phase rectifier circuits; (a) half-wave, (b) full-wave centre-tap, and (c) full-wave bridge.

Table 11.1

Ideal parameters of single-phase rectifier circuits

	Half-wave	Full-wave centre-tap	Full-wave bridge
V_2 in terms of r.m.s., supply voltage	$0.45\,V_S$	$0.9\,V_S$	$0.9\,V_S$
V_2 in terms of peak supply voltage	$0.318\,V_{SM}$	$0.636\,V_{SM}$	$0.636\,V_{SM}$
Fundamental ripple frequency	f	$2f$	$2f$
Percentage ripple	111	47.2	47.2
Peak inverse rectifier voltage in terms of r.m.s. supply voltage	$1.414\,V_S$	$2.828\,V_S$	$2.828\,V_S$

The results set out in Table 11.1 are for ideal systems, and are based on the assumption that the power supply and the diodes have no resistance, and that the leakage reactance of the transformer (where used) is zero. All these factors tend to reduce the output voltage. The output voltage may, in fact, be about 10 to 20 per cent less than the theoretical value as a result of these factors.

11.3 Ripple filters

Basic forms of ripple filter circuit are shown in Fig. 11.3. The capacitor filter in Fig. 11.3(a) is simple and cheap, but has the disadvantage that it draws current from the power source in a series of pulses. The magnitude of the output ripple voltage from Fig. 11.3(a) can be estimated by assuming an output voltage waveform of the type shown in Fig. 11.4. During the discharge period, the output voltage falls at a fairly constant rate, falling from V_{SM} at the rate of V_{SM}/R_LC, hence

$$\delta V_L = V_{SM}T/R_LC$$

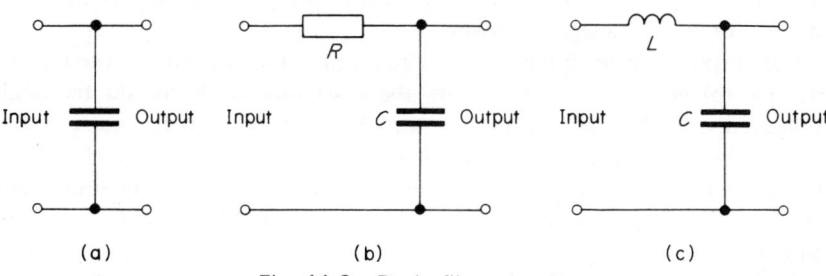

(a)   (b)   (c)

Fig. 11.3 Basic filter circuits.

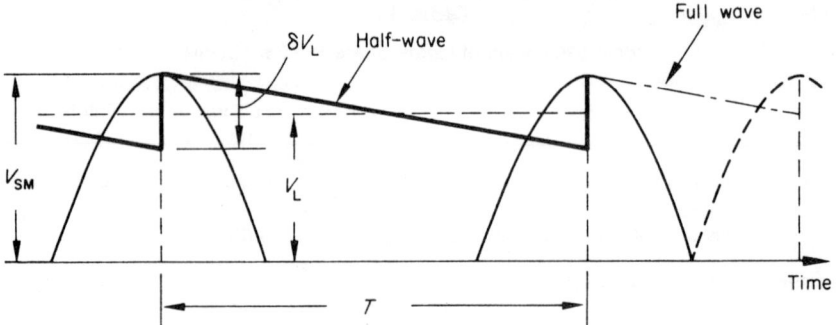

Fig. 11.4   Simplified output voltage waveforms from a capacitor filter.

and the average output voltage of the half-wave circuit is

$$V_L \simeq V_{SM} - \frac{\delta V_L}{2} = V_{SM}\left(1 - \frac{T}{2R_LC}\right) = V_s\left(1 - \frac{T}{2R_LC}\right)\sqrt{2}$$

For the full-wave circuit, the period of discharge is $T/2$, giving an ideal output voltage of $V_s(1 - T/4R_LC)\sqrt{2}$.

The effective output resistance of circuits of this kind can be estimated from a knowledge of the ripple voltage as follows. Under no-load conditions the output voltage is ideally V_{SM}, and when the circuit is providing a load current of V_L/R_L, the mean output voltage is $(V_{SM} - \delta V_L/2)$. That is, the output resistance R_{out} is

$$R_{out} = \frac{\text{(No-load output voltage)} - \text{(Output voltage on load)}}{\text{Load current}}$$

$$= \frac{\delta V_L/2}{V_L/R_L} = \frac{R_L \delta V_L}{2V_L}$$

The above calculations are based on the assumption that the capacitor charges in zero time, indicated by the sudden rise in output voltage in Fig. 11.4 at the peak supply voltage. The calculations also depend on the fact that the voltage decay is linear. These assumptions are generally accurate enough for initial design purposes.

The output voltage ripple is further reduced if a ripple filter of the type in Fig. 11.3(b) is used. In this circuit, the resistance of R should be small compared with the load resistance in order to reduce the p.d. caused by the flow of direct current in it. Also, to ensure that the capacitor effectively shunts the ripple current from the load, the reactance of C at the fundamental ripple frequency should be about one-tenth of the minimum value of load resistance.

Further reduction in output ripple is brought about by means of the L–C

filter in Fig. 11.3(c). The components of the L–C circuit should be chosen so that the fundamental ripple frequency is well above the resonant frequency of the filter. With a 50 or 60 Hz full-wave output, the fundamental ripple frequencies are 100 and 120 Hz, respectively. The filter circuit should be designed so that its resonant frequency is below about 20 Hz. As a simple rule-of-thumb design guide, a filter with a resonant frequency of 10 Hz should have an LC product of about 250×10^{-6}, and a filter with a resonant frequency of 20 Hz should have an LC product of about 65×10^{-6}, where L is in henrys and C is in farads. As in the previous filter, the reactance of capacitor C at the fundamental ripple frequency should be less than about one-tenth of the minimum value of load resistance. Values of inductance in common usage with filter circuits lie in the range 3 to 30 H.

Example 11.1: Design a choke input filter for a full-wave 50 Hz rectifier, the mean output voltage from the filter being 50 V and the maximum average output current being 0.1 A.

Solution: The minimum value of load resistance is

$$50/0.1 = 500 \, \Omega$$

The reactance of the capacitance at $2 \times 50 = 100$ Hz should be less than about 50 Ω, hence

$$C = 1/2\pi \times 100 \times 50 \, F \simeq 32 \, \mu F$$

For a filter with a resonant frequency of 10 Hz, the inductance should have a value of

$$L = 250 \times 10^{-6}/32 \times 10^{-6} = 7.8 \, \text{H}$$

For improved filtering, L should have a value greater than that calculated above.

11.4 The π filter

The performance of the choke input filter is improved by the addition of a capacitor at the input, in the manner shown in Fig. 11.5. The design procedure broadly follows that given above. That is, the reactance of C_2 at the fundamental ripple frequency should be less than about one-tenth of the load resistance, and the resonant frequency of the filter circuit should be about 10 Hz with a 50 Hz or 60 Hz supply frequency. In this case, the effective capacitance for resonance calculations is $C_1 C_2/(C_1 + C_2)$. For the conditions in example 11.1, if the 7.8 H choke is retained, suitable values of capacitance are $C_1 = 64 \, \mu F = C_2$.

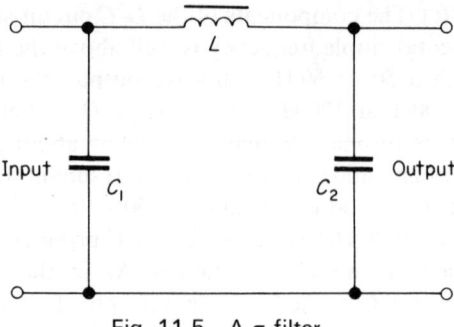

Fig. 11.5   A π filter.

11.5   Principles of stabilized supplies

A block diagram showing the basic elements of a regulated power supply is given in Fig. 11.6. It is, in fact, one form of *series regulator* in which the controlling element (which is either a transistor or a thyristor is connected in series with the output line. The signal which controls the effective resistance of the series controlling device is derived from the difference between a reference signal and a signal proportional to the quantity being controlled. It the quantity being controlled, which may be the output voltage or current, differs from the desired amount then the comparison element adjusts its output; this in turn alters the resistance of the controlling element in order that the quantity being controlled reaches the desired value. In series regulators, the controlling device always has a p.d. across

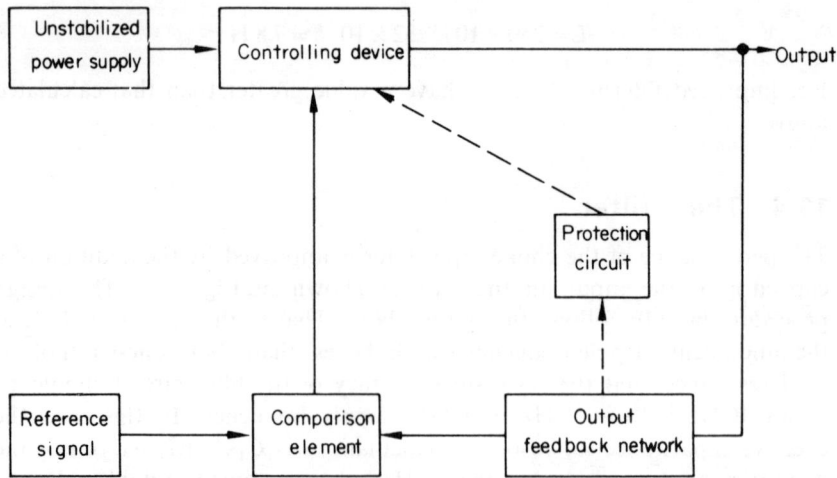

Fig. 11.6   A block diagram of a regulated power supply.

it, and this alters in the manner described above in order to make allowances for fluctuations in load resistance and in supply voltage, etc. From this discussion we see that regulators are part of the negative feedback amplifier family and, in many cases, are versions of the voltage follower circuits described in sections 3.4 and 3.5.

In *shunt regulators* the controlling device is shunted across the load, and its function is to shunt part of the supply current from the load. For a shunt regulator to perform efficiently, the regulating device must pass current all the time it is operating.

11.6 A voltage reference source

A simple voltage reference source is the Zener diode circuit shown in Fig. 11.7(a), in which the output voltage V_0 is approximately equal to the break-

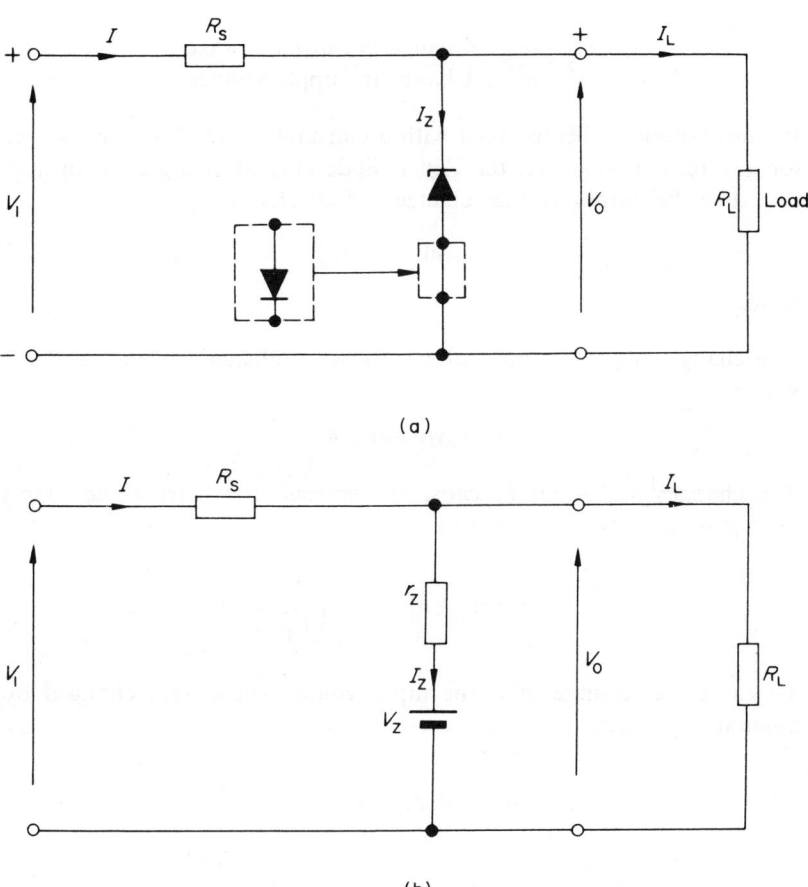

(a)

(b)

Fig. 11.7 A simple voltage reference source.

down voltage of the diode. In fact, V_0 differs from V_Z as a result of flow of current through the internal resistance r_Z of the Zener diode, as illustrated in the equivalent circuit in Fig. 11.7(b).

The output resistance of this type of voltage source can be computed from Fig. 11.7(b) by calculating the effective resistance between the output terminals with R_L disconnected and the supply source replaced by its internal resistance, which we may assume to have a negligible value. The battery V_Z is also disconnected and replaced by a short-circuit. The value of the output resistance R_0 is, therefore

$$R_0 = \frac{r_Z R_S}{r_Z + R_S}$$

A second and important parameter of voltage sources is the *stabilization factor S*, where

$$S = \frac{\delta V_0}{\delta V_1} = \frac{\text{Change in output voltage}}{\text{Change in supply voltage}}$$

the measurements being taken with a constant value of load resistance. If, for any reason whatever, the Zener diode current changes by an amount δI_Z, then the output voltage change in Fig. 11.7 is

$$\delta V_0 = r_Z \delta I_Z$$

hence
$$\delta I_Z = \delta V_0 / r_Z$$

The change δV_0 in output voltage causes a change δI_L in load current, where

$$\delta I_L = \delta V_0 / R_L$$

The changes in I_Z and I_L cause the current drawn from the supply to change by an amount δI, where

$$\delta I = \delta I_L + \delta I_Z = \delta V_0 \left(\frac{1}{R_L} + \frac{1}{r_Z} \right)$$

To cause this change in I, the input voltage must have changed by an amount δV_1, where

$$\delta V_1 = \delta V_0 + R_S \delta I = \delta V_0 \left(1 + \frac{R_S}{R_L} + \frac{R_S}{r_Z} \right)$$

hence
$$S = \frac{\delta V_0}{\delta V_1} = 1 \left/ \left(1 + \frac{R_S}{R_L} + \frac{R_S}{r_Z} \right) \right.$$

In fact, the simple regulator shown in Fig. 11.7 does not provide a stable output voltage due to the effects of the slope resistance of the diode. Any variation in current drawn from the circuit will cause the operating point on the characteristic to shift thereby altering the output voltage by a small amount. Also, the effects of ambient temperature change can cause the breakdown voltage to change. Carefully selected diodes known as *voltage reference diodes*, which have a low slope resistance and low thermal voltage drift, are used in applications where the output voltage is to be maintained at a steady value. Even so, for good voltage stability it is necessary to ensure that the diode operates under conditions of constant current and constant temperature.

Greatly improved performance is obtained if two stages of the type in Fig. 11.7 are cascaded.

11.7 A basic series voltage regulator circuit

A simple form of series regulator is shown in Fig. 11.8. This circuit is basically an emitter follower which uses a Zener diode as the voltage reference source. Due to the emitter follower action, the output voltage is maintained at a value slightly less than the Zener diode breakdown voltage. The output voltage can be reduced below this value by the modification shown in inset (ii) to Fig. 11.8(a). Resistor R_S shown in inset (i) is sometimes included in the circuit to limit the maximum value of fault current.

From the work on the emitter follower in chapter 3, the output resistance of the series regulator is, from eq. (3.21)

$$R_{out} = (r_Z + h_{IE})/h_{FE}$$

The stabilization factor can be determined as follows. Suppose that the supply voltage changes by an amount δV_1, which causes the output voltage to change by δV_0. Since the Zener diode voltage may be regarded as having a constant value, the change δV_0 is applied directly to the emitter junction of the transistor. If the equivalent voltage amplification factor *for the transistor* is A, then the change in collector-emitter voltage is $A\delta V_0$. That is

$$\delta V_1 - \delta V_0 = A\delta V_0$$

hence
$$\delta V_0 = \delta V_1/(1+A) \tag{11.1}$$

therefore
$$S = \delta V_0/\delta V_1 = 1/(1+A) \tag{11.2}$$

From the work in chapter 1 we saw that, for bipolar transistors, $A = h_{FE}/h_{OE}h_{IE}$, so that for the transistor regulator

$$S = \frac{1}{1 + h_{FE}/h_{OE}h_{IE}}$$

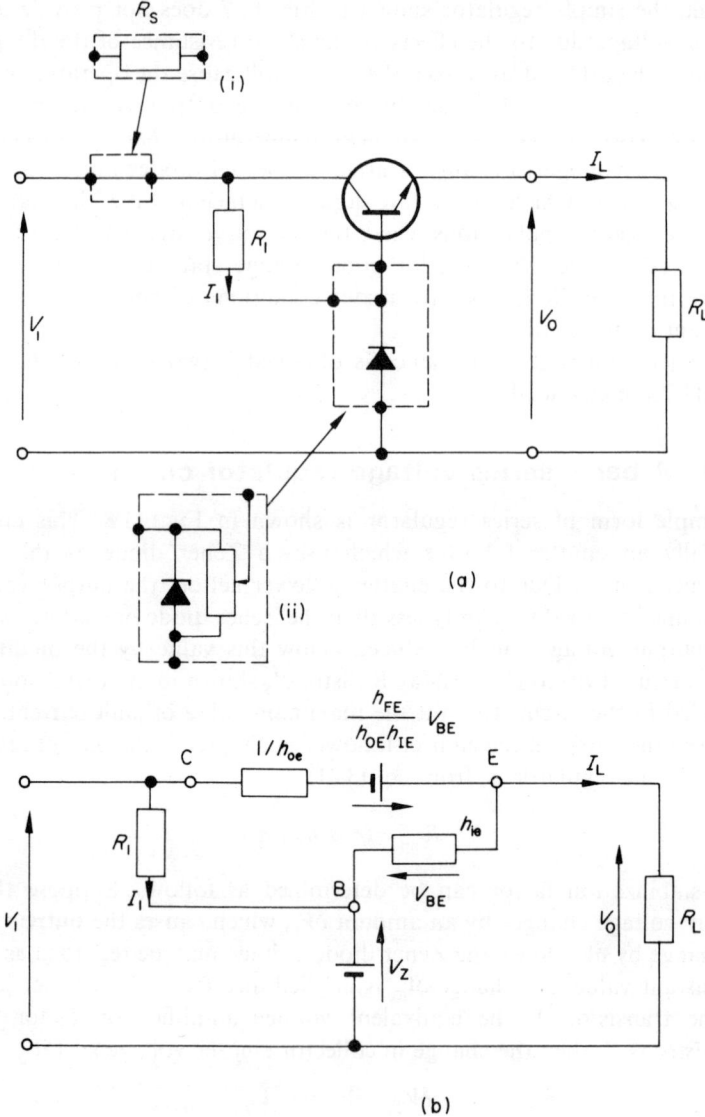

Fig. 11.8 A series voltage regulator.

The stabilization factor can be calculated more accurately from the equation below which is derived from the equivalent circuit, Fig. 11.8(b),

$$V_1 = V_0 - \frac{h_{FE}}{h_{OE} h_{IE}} V_{BE} + \frac{I_L}{h_{OE}}$$

Solving for $\delta V_0/\delta V_1$ yields the result

$$S = 1 \bigg/ \left(1 + \frac{h_{FE}}{h_{OE}h_{IE}} + \frac{1}{h_{OE}R_L} \right)$$

The output current from this type of regulator can be improved by replacing the series transistor by a Darlington-connected pair of transistors. Also, the stabilization factor can be improved by driving the Zener diode by a constant current source. Constant current sources are discussed in section 11.8.

In the event that a short-circuit is applied to the output of the regulator in Fig. 11.8, the transistor has to pass the short-circuit current while supporting the unstabilized supply voltage between its collector and emitter. As a result, the transistor must have a power rating which is in excess of the normal output power of the regulator. It is usual in commercial forms of voltage regulator to limit the flow of fault current to a safe value by electronic means. Protection of power supplies is dealt with in more detail in section 11.11.

11.8   Constant current supply sources

A basic form of constant current regulator is illustrated in Fig. 11.9, in which series negative current feedback is used. The output current is measured by inserting resistance R in series with the load current circuit. The difference in potential between the voltage across R and that across the Zener diode is applied between the base and emitter of the transistor, and this voltage regulates the current flowing in the load. If we use a fairly large value of V_Z, then we may neglect V_{BE} of the transistor, so that

$$IR \simeq V_Z$$

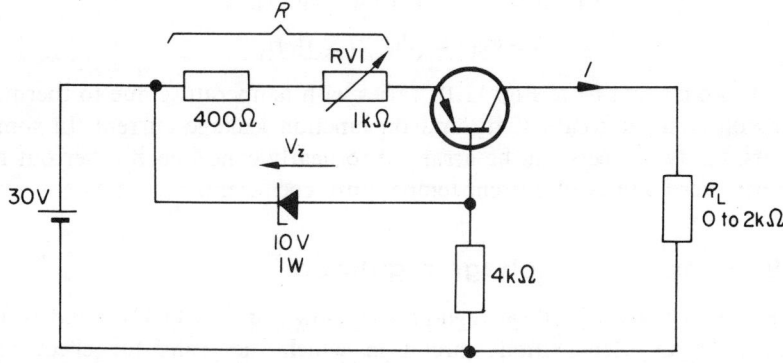

Fig. 11.9   A constant current regulator.

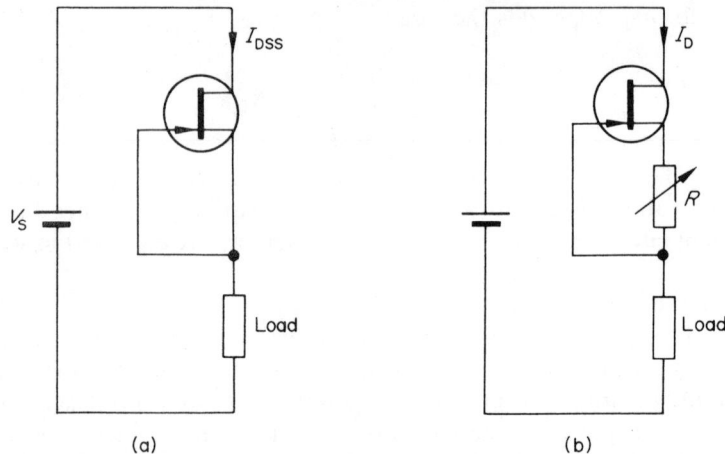

Fig. 11.10   The JUGFET as a constant current source.

or $$I = V_Z/R$$

In Fig. 11.9, if we adjust R to have a total value of 1 kΩ, then I has the value of 10 mA for any load in the range between a short-circuit and, for a supply voltage of 30 V, 2 kΩ. The point at which the circuit ceases to act as a series regulator is, as stated earlier, when the voltage across the transistor falls to zero. This occurs in the case of Fig. 11.9 when $R_L = 2$ kΩ.

The inherent properties of the JUGFET allow it to be used as a constant current source in the manner shown in Fig. 11.10. If the gate and source are linked together so that $V_{GS} = 0$, as shown in Fig. 11.10(a), and the voltage across the FET is greater than the pinch-off voltage V_P of the FET, then a current equal to the FET saturation current I_{DSS} flows in the circuit. The circuit current can be controlled to a value *less than* I_{DSS} by modifying the circuit in the manner in Fig. 11.10(b). From a consideration of the basic equations of the FET, it can be shown that the value of the series resistance R required to limit the drain current to a value I_D is

$$R = V_P[1 - \sqrt{(I_D/I_{DSS})}]/I_D$$

The output current from Fig. 11.10 varies with temperature due to thermal effects on channel conductivity and on junction leakage current. In some circuits, the two effects can be arranged to nearly cancel each other out to give near-zero values of current-temperature coefficient.

11.9   Basic shunt voltage regulators

A rudimentary shunt voltage regulator is shown in Fig. 11.11(a) and is, in effect, a 'super' Zener diode circuit in which the transistor effectively magnifies the Zener diode current. The circuit is capable of handling a

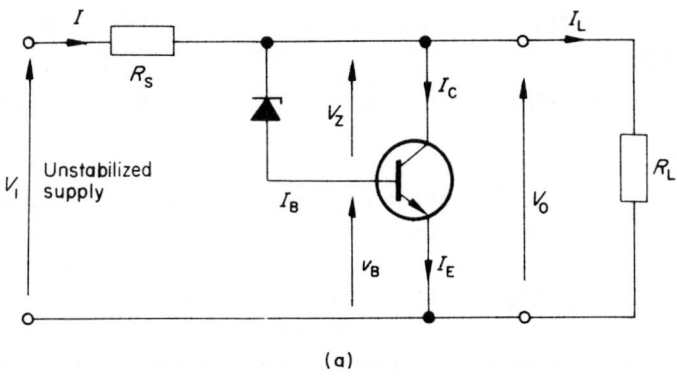

(a)

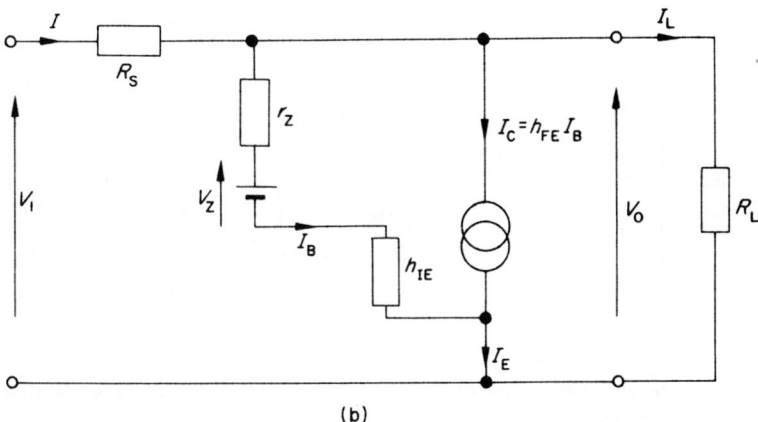

(b)

Fig. 11.11   A basic form of shunt voltage regulator.

current of $(1+h_{FE})I_{Z(max)}$, where h_{FE} is the common-emitter current gain of the transistor and $I_{Z(max)}$ is the maximum current flowing through the Zener diode (i.e., it is the Zener diode current when R_L is disconnected). The output voltage is

$$V_0 = V_B + V_Z$$

and, if $V_B \ll V_Z$, then $V_0 \simeq V_Z$. The equivalent circuit of the regulator is illustrated in Fig. 11.11(b), from which it can be shown that the output resistance R_0 and stabilization factor S are

$$R_0 = R_S \left/ \left\{ \frac{(1+h_{FE})R_S}{r_Z + h_{IE}} + 1 \right\} \right.$$

$$S = 1 \left/ \left\{ \frac{(1+h_{FE})R_S}{r_Z + h_{IE}} + 1 + \frac{R_S}{R_L} \right\} \right.$$

11.10   Series regulators using amplifiers in the feedback path

Further improvements in the output resistance and the stabilization factor when compared with the basic series regulator are brought about by including amplification in the feedback path, in the manner shown in Fig. 11.12. Now, let us suppose that the supply voltage V_1 changes by an amount δV_1 so that the output voltage changes by δV_0. Since the voltage reference source V_R is constant, the change in voltage at the amplifier terminals in the feedback path is $\beta \delta V_0$, and this causes the emitter junction voltage of the series regulator to change by $A_v \beta \delta V_0$, where A_v is the voltage gain of the feedback *amplifier*. If the *voltage amplification factor of the series transistor* is A_1, then the change in voltage across the series transistor is $A_1 A_v \beta \delta V_0$.

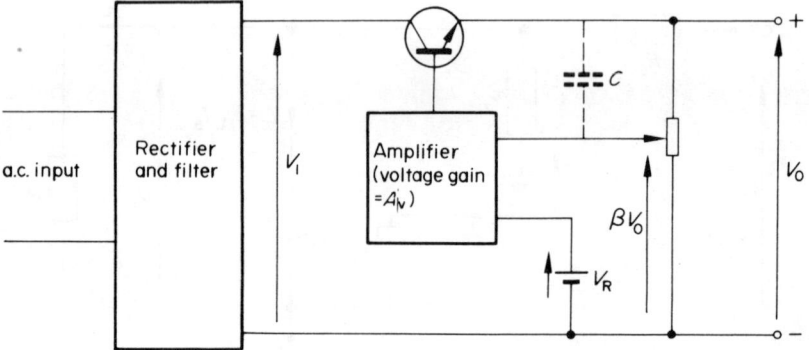

Fig. 11.12   A block diagram of a series voltage regulator using an amplifier in the feedback loop.

Hence

$$\delta V_1 - \delta V_0 = A_1 A_v \beta \delta V_0$$

therefore

$$\delta V_0 = \delta V_1 / (1 + A_1 A_v \beta) \tag{11.3}$$

and

$$S = \delta V_0 / \delta V_1 = 1 / (1 + A_1 A_v \beta) \tag{11.4}$$

Comparing eqs. (11.4) and (11.2), we see that if $A_v \beta > 1$, then the stabilization factor of Fig. 11.12 is improved when compared with that of Fig. 11.8 since the denominator of eq. (11.4) is greater than the denominator of eq. (11.2). The value of A_1 may be taken to be $h_{FE}/h_{OE}h_{IE}$, where the parameters refer to the series regulating transistor.

The basic configuration is still that of an emitter follower, which has the additional gain $A_v \beta$ with the feedback loop. Applying the basic theory of feedback amplifiers to this situation we deduce that the regulator output resistance R'_{out} is

$$R'_{out} = R_{out} / (1 + A_v \beta)$$

where R_{out} is the output resistance of the emitter follower before the additional feedback is applied. If we take R_{out} to have the value h_{IE}/h_{FE}, then for Fig. 11.12

$$R'_{out} = h_{IE}/h_{FE}(1 + A_v\beta)$$

If the gain A_v of the feedback amplifier is fairly large, then the voltage between its input terminals is small, hence $V_R = \beta V_0$ or

$$V_0 = V_R/\beta$$

Capacitor C in Fig. 11.12 is sometimes included to improve the performance of the regulator. It is the function of this capacitor to provide a low reactance path to the flow of ripple current, so that any ripple which appears on the output line is fed back to the amplifier, and corrections are made to reduce its value.

A regulated power supply of the type described above is shown in Fig. 11.13 in which Q1 and R_1 form the error voltage amplifier, and Z_1 is the voltage reference source. The circuit also includes a *pre-regulator* comprising R_2 and Z_2, which stabilizes the supply to Q1; the Darlington-connected

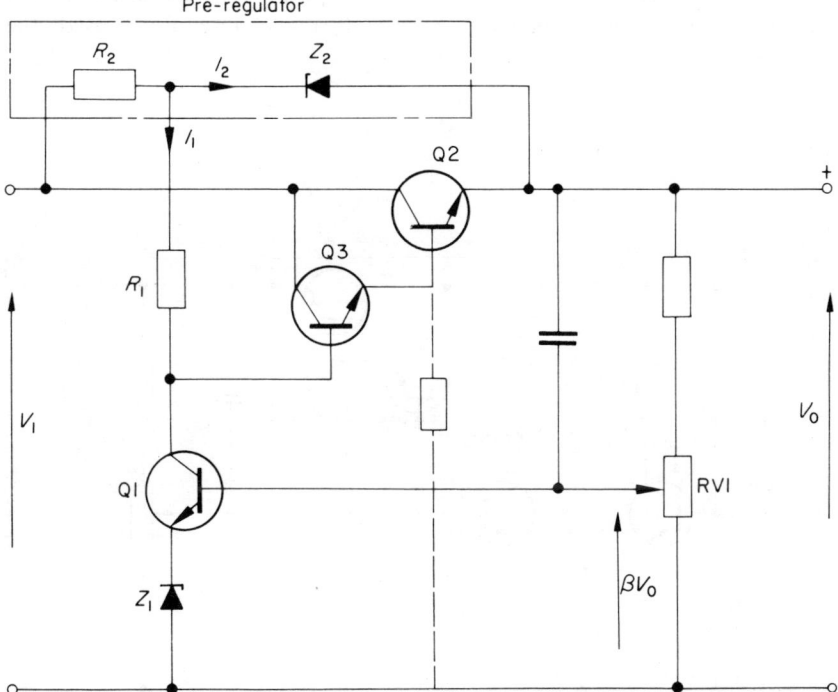

Fig. 11.13   A series regulator incorporating an amplifier in the feedback loop.

transistors Q2 and Q3 enable the regulator to handle a large current while only needing a small base current for control purposes. In some circuits, it is advisable to connect the resistor shown by the dotted connection between the base of Q2 and earth to provide a path for the leakage current of Q2. This connection is one means of preventing the possibility of loss of control over the output voltage at very low values of load current.

The use of a long-tailed pair amplifier in the feedback loop improves the performance of series regulators even further, because of the inherent thermal stability of these amplifiers when compared to simpler designs. A circuit using an amplifier of this type is shown in Fig. 11.14. Many professional types of regulators incorporate a number of long-tailed pair amplifiers in the feedback loop to obtain a very low output resistance together with a good thermal stability and stabilization factor.

It is difficult to provide precise compensation for changes in Zener diode breakdown voltage, and the effects of temperature change are often minimized by maintaining the Zener diode at a constant temperature in an 'oven'.

11.11 Protection of transistor power supplies

The most popular form of protection incorporated in transistor power supplies is *overcurrent protection*. Several forms of circuit are used, the

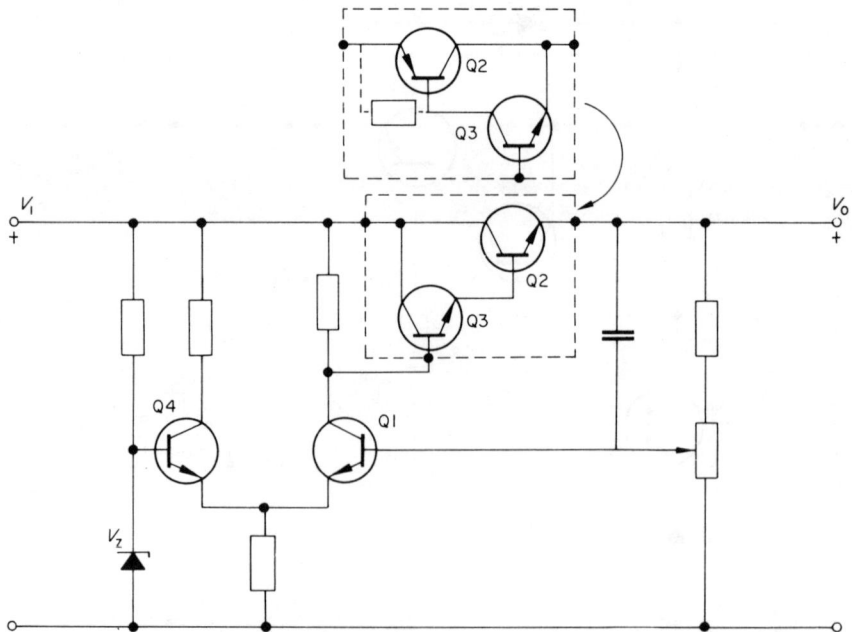

Fig. 11.14 A series regulator which includes a long-tailed pair amplifier.

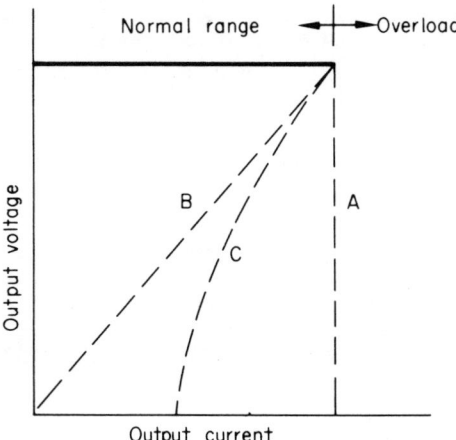

Fig. 11.15 Overcurrent protection characteristics.

principal characteristics being shown in Fig. 11.15. In circuits which have a *constant current limiting* characteristic, the load current is limited to a constant value, shown by curve A. This type of protection is particularly valuable where power supplies are connected in parallel as it prevents one of the supplies 'hogging' all the load current. Current limitation is brought about by adjusting the effective resistance of the regulating device so that the output current is limited to the required value.

Other regulators have some form of *cut-out characteristic* or current trip characteristic, curve B, in which the output voltage is suddenly reduced to zero when the load current exceeds a predetermined value. Protection of this kind is often obtained by allowing the overload current to trigger a flip-flop which reduces the base drive of the series transistor to zero. This type of protection is not as convenient in use as current limitation.

Curve C in Fig. 11.15 is a typical *re-entrant characteristic* or *fold-back characteristic* in which both output voltage and current are reduced under fault conditions. A feature of this type of characteristic is that the fault power dissipated in the regulating device is much less than in the case where a simple constant current type of protection is used. However, in the case of lamp loads (and other loads with a lamp-type characteristic), re-entrant protection can cause the output voltage to fall when the lamp is first switched on, from which it may not recover due to the characteristics of the load and the protection circuit.

A simple form of current limit circuit is shown in Fig. 11.16, in which resistor R and *silicon* diode D are connected across the unregulated supply. The transistor is a *germanium* device. Under low-load conditions, the p.d. across R_2 and the emitter junction of the transistor are much less than that across D, so that the transistor is saturated and has little effect on load

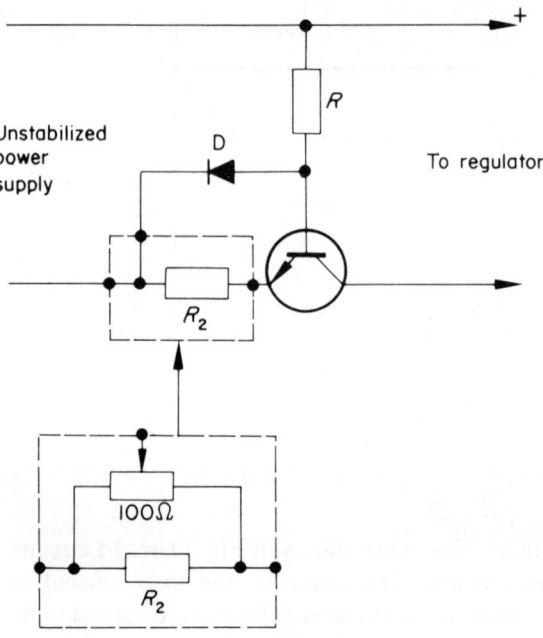

Fig. 11.16   A simple current limit circuit.

current. The p.d. across R_2 increases with load current and, at some value of
load current, the transistor comes out of saturation and the current through
it is limited by its base current. The value of load current at which this
happens depends on the value of R_2. The circuit in the inset of Fig. 11.16

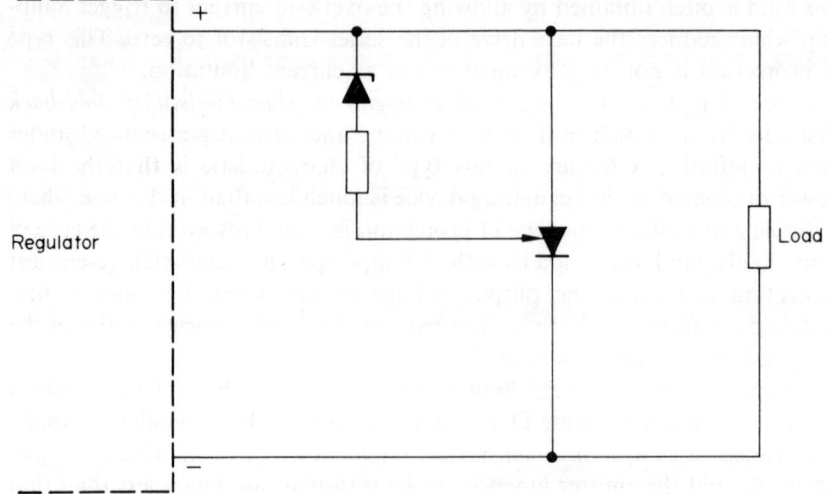

Fig. 11.17   A simple overvoltage protection circuit.

provides a means of controlling the value at which the current is limited. When the potentiometer wiper is at the most left-hand point of its travel, the current is limited to its lowest value.

Overvoltage protection is of great importance in power supplies for integrated circuits, as the application of an overvoltage can damage IC's. A popular method of preventing an overvoltage is by *crowbar protection*, in which the protection circuit applies a short-circuit to the output terminals of the regulator when an overvoltage is detected. Thyristors are frequently used as the 'crowbar' element.

A basic form of overvoltage protection is illustrated in Fig. 11.17. When the output voltage from the regulator exceeds the Zener diode breakdown voltage, the current flowing in the Zener diode triggers the thyristor and applies a short-circuit to the output. The magnitude of the current flowing through the thyristor is restricted by the current limiting circuit of the regulator.

11.12 Switched-mode power supply (SMPS)

The voltage and current regulators which have been described earlier in the chapter have two basic disadvantages which are

(a) They incorporate a large and costly mains-frequency voltage step-down transformer and
(b) The regulating device is in continuous operation, leading to a large amount of power being continuously dissipated.

These disadvantages are largely overcome in switched-mode power supplies, the principle of each of the three main types being outlined below.

11.12.1 General principle of SMPS

The basic block diagram of a SMPS is shown in Fig. 11.18. The a.c. supply is rectified and smoothed before being applied to a d.c.-to-d.c. convertor; it is in this latter section of the convertor where the essential differences lie between the three types of SMPS. The d.c.-to-d.c. convertor contains an

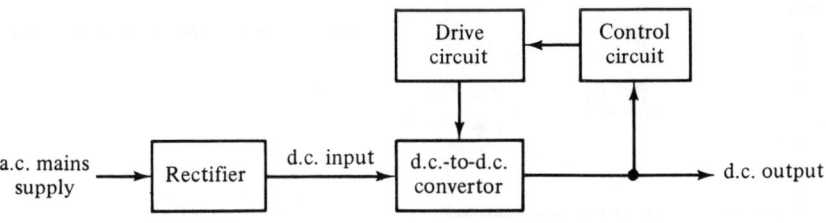

Fig. 11.18 A block diagram of a SMPS.

electronic 'chopper' which operates at a frequency well above the audio range, i.e., 20 kHz–50 kHz. The 'chopper' converts the d.c. input signal to a high frequency alternating signal, which is transformed to the required voltage, after which it is rectified and smoothed to provide the d.c. output from the regulator. The transformer is a necessary part of the convertor circuit as it provides electrical isolation between the high voltage input and the low voltage output; since the transformer operates at a high frequency, it is very small and is highly efficient.

The output voltage is controlled by a negative feedback circuit comprising a control circuit and a drive circuit, the latter 'driving' the chopper element of the d.c.-to-d.c. convertor.

11.12.2   A 'flyback' d.c.-to-d.c. convertor

A simplified circuit of a flyback convertor is shown in Fig. 11.19; this type of convertor utilizes the e.m.f. induced in the secondary winding of transformer T during the 'flyback' period when current is cut-off in transistor TR.

An impulsive waveform is applied to the base of transistor TR by the drive circuit (see also Fig. 11.18), and during this period of time, δt, the transistor TR is turned on. At this time, the collector current in the transistor rises and the magnetic circuit stores energy. The transformer is connected so that diode D1 is reverse biased during this period of time, and the load is supplied from the energy already stored in capacitor C.

When the voltage applied to the base of transistor TR by the drive circuit is suddenly reduced to zero, the collector current is reduced and the e.m.f. induced in the secondary winding during this period (the 'flyback' period) reverses polarity. This voltage forward biases diode D1, and the energy

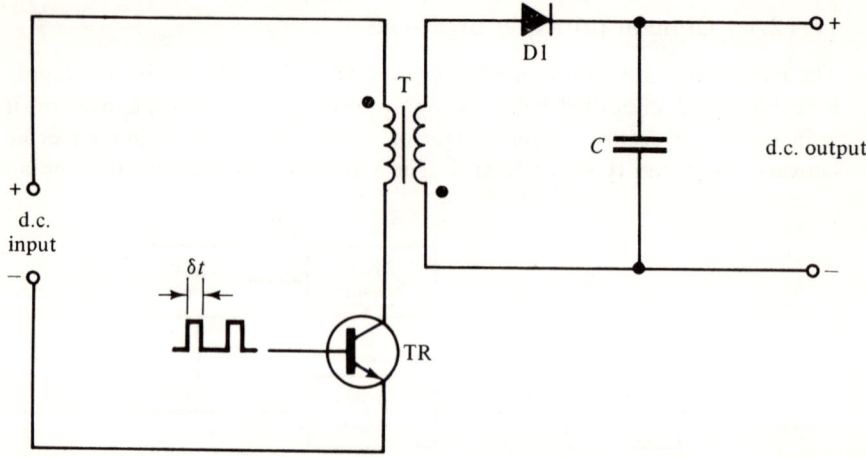

Fig. 11.19   A basic 'flyback' convertor.

stored in the magnetic field is discharged not only into capacitor C but also into the load during the flyback period.

The flyback convertor is the simplest and least expensive of the d.c.-to-d.c. convertors described in this chapter and, since each output needs only one diode and one capacitor, it is widely used in convertors which provide multiple outputs.

11.12.3 A 'forward' d.c.-to-d.c. convertor

The basis of a 'forward' d.c.-to-d.c. convertor is shown in Fig. 11.20. The primary and secondary windings are connected so that when TR is turned on by a positive pulse from the drive circuit, diode D1 is forward biased and energy is transferred to the load via inductor L; at the same time, capacitor C is charged.

When the transistor is turned off by the drive circuit, diode D1 is reverse biased. However, due to the nature of inductance, the current flowing in inductor L cannot suddenly be cut off, and the e.m.f. of self-inductance induced in it forces current to continue flowing through the load; the return path of the current is through the *freewheel diode* or *flywheel diode* D2.

Also, when transistor TR is turned off, the magnetic energy built up in the transformer core during the period when TR was turned on is returned to the d.c. input supply via the demagnetizing winding and diode D3.

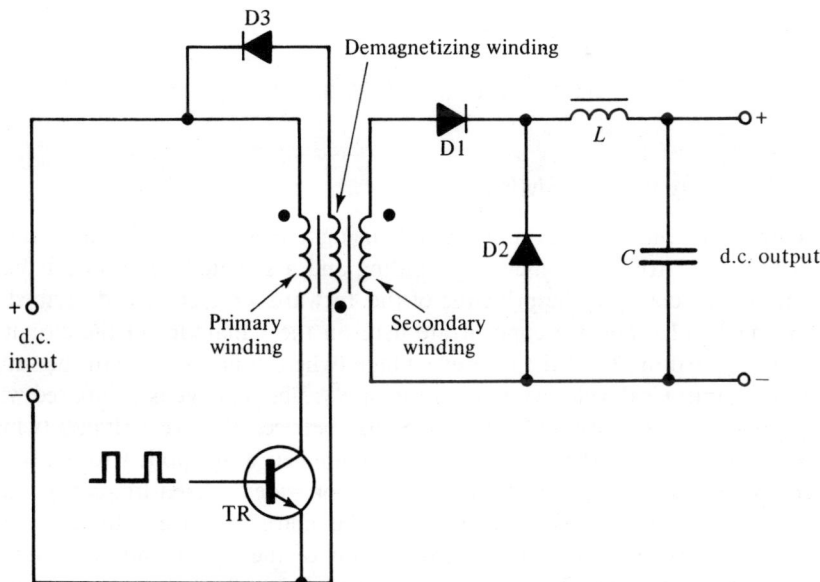

Fig. 11.20 The basis of a 'forward' convertor.

For power levels between about 10 W and 1 kW, the 'forward' convertor provides a smoother d.c. output than does the 'flyback' convertor.

11.12.4   A 'push-pull' convertor

Transistors TR1 and TR2 in Fig. 11.21 are operated in push-pull by the drive circuit, so that when TR1 is on then TR2 is off and vice versa. This induces a bi-phase supply in the secondary winding of the centre-tapped transformer, and diodes D1 and D2 rectify the induced voltage. The resulting unidirectional voltage is smoothed by the LC filter to reduce the ripple at the output. The 'push-pull' convertor is more useful than the other convertor circuits for power levels greater than about 1 kW.

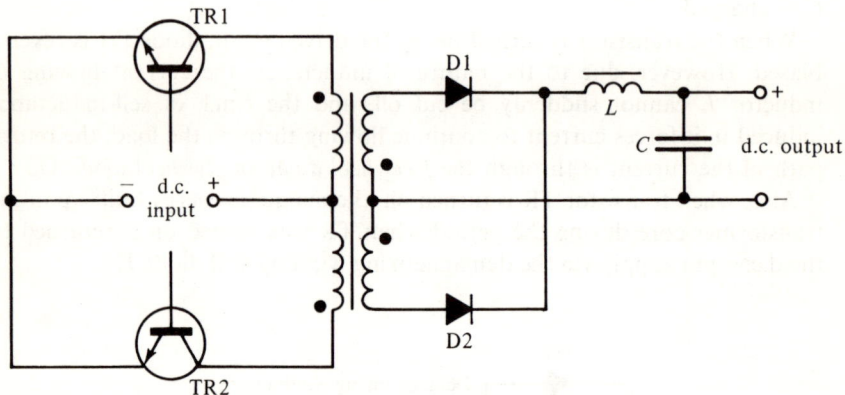

Fig. 11.21   A simplified 'push-pull' convertor.

11.12.5   A typical SMPS

A simplified block diagram of a SMPS using a 'forward' convertor is shown in Fig. 11.22 (the demagnetizing winding and associated circuitry has been omitted for clarity). The principle of the 'forward' convertor is described in section 11.12.3, and we concentrate here on the remainder of the circuit.

A proportion βV_o of the output voltage (where β has a value in the range zero to unity) is developed across resistor R_2. This voltage is compared with a stable reference voltage V_r, the difference between the two voltages (which we shall call the 'error' voltage) being amplified and applied to a *voltage controlled oscillator*, VCO. This oscillator provides a fixed-frequency, variable mark-to-space ratio signal (*note*: the mark-to-space ratio of a rectangular wave signal is the on-to-off ratio of the waveform). The output signal from the VCO is used to control the length of time for which transistor TR is turned on and turned off.

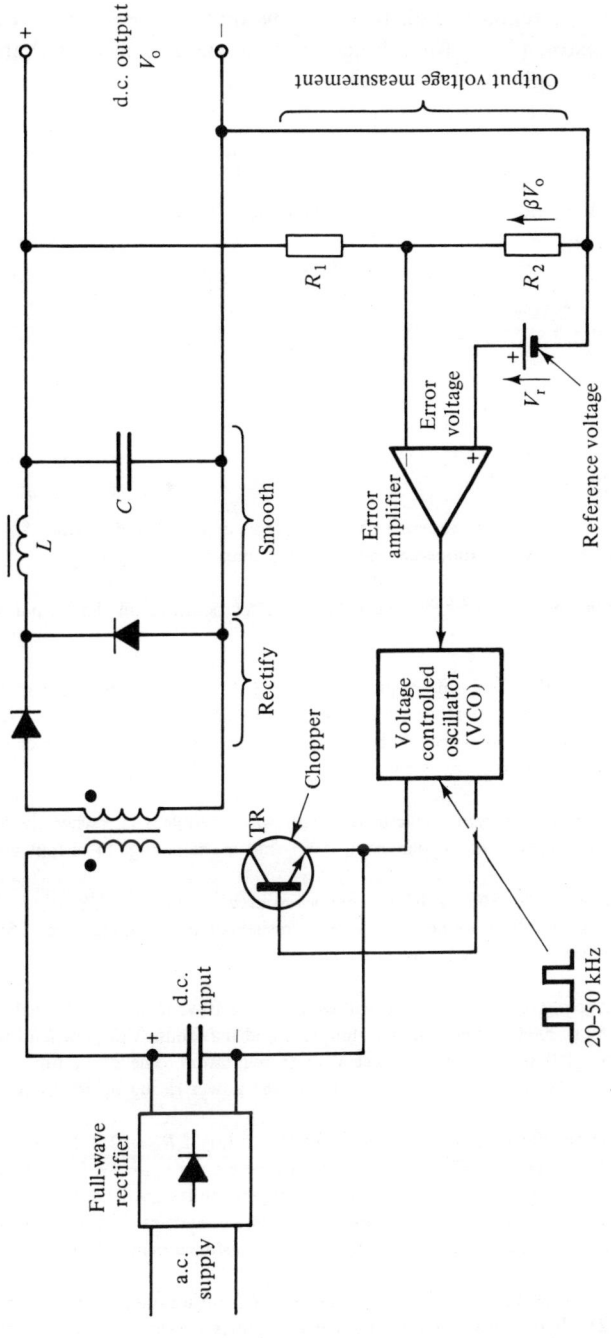

Fig. 11.22 A typical SMPS.

If a large current is drawn from the output terminals of the SMPS, the output voltage tends to fall. In turn, this increases the error voltage which turns transistor TR on for a longer period; the net effect is that the output voltage is brought back to a value close to its original value.

The value of the ratio β is given by

$$\beta = R_2/(R_1 + R_2)$$

and if the loop gain of the system is large, then

$$\beta V_o \simeq V_r$$

or

$$V_o \simeq V_r/\beta$$

Problems

11.1 Draw the static characteristic of a Zener diode, and outline the principle of operation of the device. Show how the diode can be used in a simple shunt regulator.

11.2 Describe the effect of a change in ambient temperature on the output voltage of the regulator in problem 11.1.

11.3 In a stabilizer of the type in problem 11.1, the diode has a breakdown voltage of 10 V and a slope resistance of $10 \, \Omega$, and the series current limiting resistance has a value of $100 \, \Omega$. Calculate the output resistance of the regulator, and hence compute the change in output voltage when the load current is changed by 50 mA.

11.4 Calculate the stabilization factor of the circuit in problem 11.3 when the load resistance is $500 \, \Omega$. If the supply voltage changes by 5 V, what is the change in output voltage?

11.5 Design a Zener diode voltage reference source to supply 0.2 A at 15 V from a 50 V unstabilized d.c. supply. Assume that the slope resistance of the diode is zero. State the rating of the Zener diode and of the series resistance.

11.6 A voltage reference diode has a voltage rating of 10 V. If the unstabilized supply voltage is 20 V and R_S is $200 \, \Omega$, determine the minimum and maximum values of load resistance over which the circuit can maintain correct voltage regulation. The minimum allowable diode current is 1 mA. What is the minimum value of the power rating of the reference diode?

11.7 A shunt regulator uses the circuit in Fig. 11.11(a). If $R_S = 50 \, \Omega$, the Zener diode slope resistance is $10 \, \Omega$, and the relevant transistor parameters are $h_{FE} = 49$, $h_{IE} = 10 \, \Omega$, calculate (a) the output resistance of the regulator, (b) the change in output voltage when the load current changes by 1.5 A, the supply voltage remaining constant, and (c) the change in output voltage when the supply voltage changes by 5 V, the load resistance being $10 \, \Omega$.

11.8 Design a series regulator to supply 2 A at 6 V from a nominal 15 V supply, which may vary by $\pm$ 2V. If it is necessary to assume the values of any parameters, state the values chosen.

11.9 Modify the circuit you have shown for problem 11.8 to include overcurrent and overvoltage protection.

11.10 A circuit similar to that in Fig. 11.8(a) is used to provide a regulated output of 15 V at a variable current from 0.1 to 2 A. The transistor current gain is 100, and its base-emitter voltage is given by the relationship $V_{BE} = 0.3 + 0.01\,I_B$ volts, where I_B is in milliamperes. If V_1 varies between 20 and 25 V, determine (a) a suitable rating for R_1, (b) the voltage rating of the Zener diode, and (c) the ratings of the diode and the transistor.

11.11 A regulating circuit similar to Fig. 11.12 has $\beta = 0.5$, $A_v = -100$, and $A_1 = -20$. What will be the voltage change of the regulated output when the voltage applied to the input of the regulator changes by 10 V?

11.12 If the output resistance of the unregulated power supply in problem 11.11 is 50 Ω, what would be the approximate output resistance when the regulator in problem 10.11 is added to the circuit?

12. Closed-loop control systems

12.1 Closed-loop control

Closed-loop control systems are part of the negative feedback amplifier family and, in many cases, include non-electrical elements with the loop.

A basic *single-loop* system is illustrated in Fig. 12.1 and, as we shall see in this chapter, it is possible to reduce the block diagrams of many apparently complex control systems to this simple form. The elements required in the construction of a control system are:

(a) A *transducer* or *sensor* to measure the output quantity

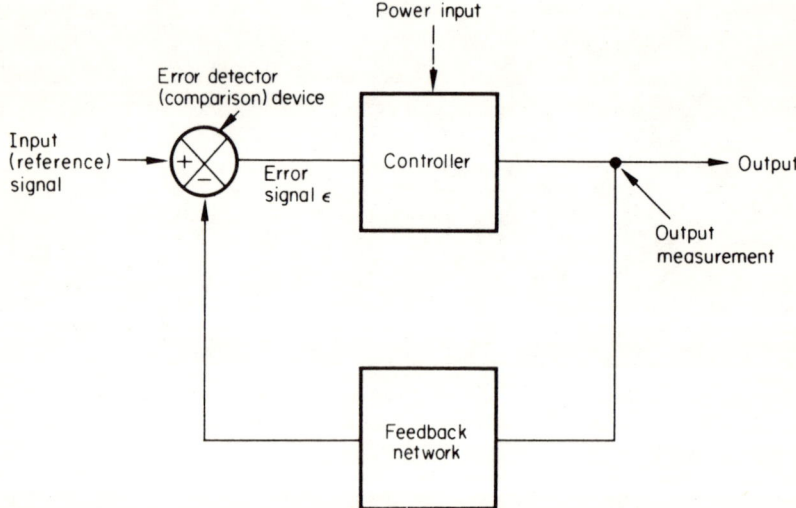

Fig. 12.1   Block diagram of a single-loop control system.

370

(b) A *feedback network* or *β-network*
(c) An input signal or *reference signal*
(d) A *comparison element* or *error detector* which compares the signal fed back with the reference signal
(e) A *controller* which amplifies the error signal and converts it to the required level.

It is true to say that anything that can be measured can be controlled, and transducers using a wide variety of operating principles are employed for the measurement of system variables. In most instances, it is convenient to use an *analogue* form of transducer which provides a signal proportional to the quantity being measured. In other cases, *digital* transducers are used, especially where they offer improved performance over analogue devices.

As is the case in many forms of electronic feedback amplifier, the feedback network may simply be a resistive potential divider. In some applications, it may be desirable to modify the feedback signal in some way, e.g., it may be necessary to integrate or to differentiate the signal; the components necessary to modify the signal can conveniently be incorporated into the feedback network.

The reference signal could be a voltage or current which is derived from a stabilized supply, and is compared in the error detector with the signal from the feedback network. The difference between the two signals is known as the *error signal* or *deviation*.

The controller is an amplifier in which the error signal is used to control the flow of power between the power supply and the system output. When we refer to the *gain* of the controller we, in fact, talk about the ratio of the output to the error; however, it is important to bear in mind that the principal function of the error signal is to *control* the flow of power through the controller.

12.2 Types of system

Closed-loop control systems can be broadly classified into:

(a) Servomechanisms (or servos)
(b) Regulators.

The class *servomechanisms* includes all types of position control systems, such as those found in the machine tool industry. In these systems, the output is in the form of the angular position of a shaft or the position of, say, the bed plate of a machine. When the object being controlled reaches its final (*steady-state*) position, it is (under ideal conditions) in perfect alignment with the required (reference) position. From this we see that under ideal operating conditions, the steady-state error signal in servomechanisms is zero.

Regulators include many forms of industrial control system such as those for controlling speed, temperature, voltage, etc. A significant difference between servos and regulators is that the steady-state error signal is finite in basic forms of regulators, compared with zero in servos. The reason is that, under steady-state operating conditions, regulators are required to provide an output, e.g., speed, temperature, etc. Since practical forms of controller do not have an infinite gain, then an error signal must exist in order to provide the steady-state output.

12.3 Error detection circuits

Two of the more popular circuits in use are the shunt feedback circuit, Fig. 12.2(a), and the series feedback circuit, Fig. 12.2(b).

In order to demonstrate the operation of the shunt feedback circuit we will develop a simplified analysis of the network. The circuit is basically that of the shunt feedback amplifier described in section 3.11 of chapter 3. If we assume that the amplifier input current I is zero (as will soon be shown to be reasonably accurate, even if the input resistance is not very high), then

$$I_1 + I_2 = 0$$

that is

$$\frac{V_1 - \epsilon}{R_1} + \frac{V_2 - \epsilon}{R_2} = 0$$

Solving for ϵ yields

$$\epsilon\left(\frac{1}{R_1} + \frac{1}{R_2}\right) = \frac{V_1}{R_1} + \frac{V_2}{R_2} \tag{12.1}$$

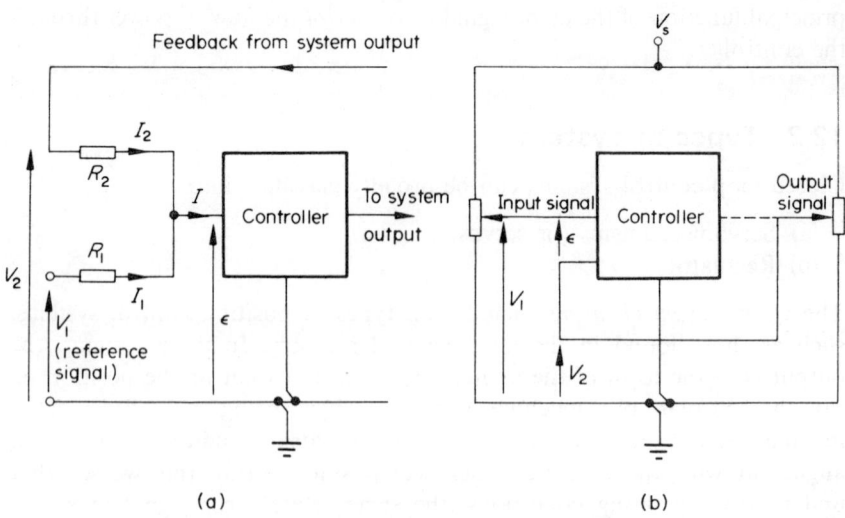

(a) (b)

Fig. 12.2 Voltage comparison circuits.

where V_1 is the reference signal applied to the control system, V_2 is the signal fed back from the output, and ϵ is the error signal. From eq. (12.1) we see that if the error voltage is to be proportional to the *difference* between V_1 and V_2, then V_1 and V_2 must be of opposite polarity to one another, e.g., when V_1 is positive then V_2 is negative, and vice versa. Under ideal conditions, and if the gain is very high, the error voltage is very small so that we may reduce eq. (12.1) to

$$\frac{V_1}{R_1} = -\frac{V_2}{R_2}$$

Thus, if the voltage gain of the amplifier is high, then the error voltage and amplifier input current are small, whatever the value of the input resistance of the amplifier. A feature of the shunt feedback arrangement is that it allows a comparatively simple form of amplifier to be used if so desired.

In the case of the series feedback circuit, Fig. 12.2(b), it is necessary to use a differential input amplifier. In this circuit, a common supply voltage is used so that V_1 and V_2 have the same polarity. These voltages are applied differentially to the first stage of the controller amplifier, and

$$\epsilon = V_1 - V_2$$

The circuit in Fig. 12.2(b) refers to a *remote position control* (r.p.c.) servo-system in which the output voltage is derived from the wiper of a potentiometer which is coupled to the output shaft. The servosystem operates in such a way that, when balanced, $V_1 = V_2$, and $\epsilon = 0$. The outer arms of the system resemble a Wheatstone bridge, and a system of this kind is often referred to as a *self-balancing bridge system*.

A popular form of position sensing transducer for use with a.c. r.p.c. servos is the *synchro*, see Fig. 12.3. The *synchro transmitter* has a fixed part known as the *stator*, which carries three groups of windings distributed in slots around the stator, the axis of the groups being 120 degrees apart. Figure 12.3 shows a simplified version in which concentrated windings are

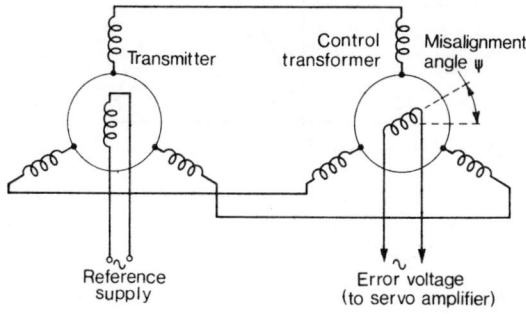

Fig. 12.3 Synchro error detecting system.

used. The *rotor* or rotating member carries a coil which is energized by a reference a.c. source (in the frequency range 50–400 Hz) at a voltage in the range 50–115 V. Depending on the angular position of the rotor, sinusoidal voltages of differing magnitudes are induced in the stator windings. These voltages are *single-phase* in nature, so that the stator voltages are either in phase with or antiphase to the reference voltage. When the rotor is rotated, the relative magnitudes of the stator winding voltages alter, but remain either in phase with or antiphase to one another.

The synchro error detecting device is known as a *control transformer* (CT), and is similar in general construction to that of the transmitter. When stator windings of the transmitter are connected to those of the CT in the manner shown in Fig. 12.3, the current circulating through them causes a pulsating magnetic field to be produced in the magnetic circuit of the CT, the magnetic axis of the field in the CT being in the same relative direction to that in the transmitter. As a result, the e.m.f. induced in the rotor winding of the CT gives an indication of its position relative to that of the rotor of the transmitter. When the two windings are electrically aligned, the r.m.s. value of the e.m.f. induced in the CT winding is at its maximum value. When they are electrically perpendicular to one another, the CT rotor winding voltage is zero.

When synchros are used in control systems, it is arranged that when the input and output shafts are *mechanically* aligned, then the two rotor windings are *electrically* perpendicular and the CT rotor winding induced voltage is zero. A typical characteristic for a synchro error detecting system is shown in Fig. 12.4. The *magnitude* of the misalignment angle is related to the r.m.s. value of the C.T. rotor winding voltage. The *direction* of the misalignment (i.e., clockwise or anticlockwise) is indicated by the phase relationship between the CT rotor voltage and the reference voltage (the transmitter rotor voltage). In Fig. 12.4, clockwise misalignment results in the two voltages being in phase with one another, and anticlockwise

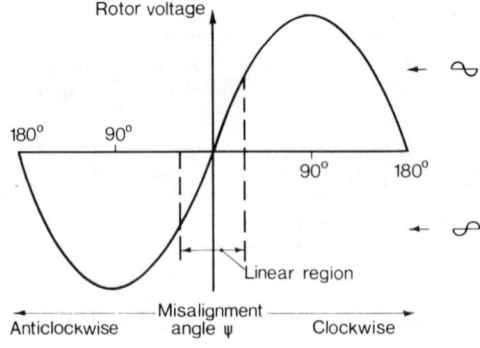

Fig. 12.4   Characteristic of a synchro error detecting system.

misalignment results in the two voltages being antiphase to one another. By measuring both the magnitude and the phase of the CT output voltage, the necessary correcting action can be taken by the system.

A simplified diagram of an a.c. r.p.c. servo is illustrated in Fig. 12.5, in which information about the angular position of the output shaft is fed back by a mechanical link (the belt drive) to the control transformer.

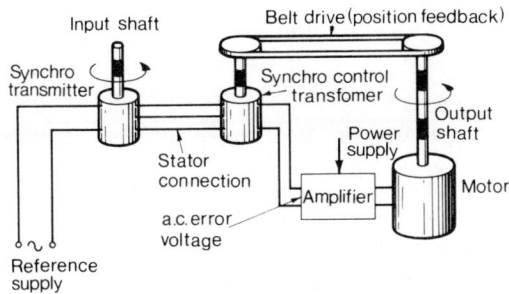

Fig. 12.5 An a.c. servosystem.

12.4 Remote position control systems

Remote position control servos use motors which can vary in rating from a few watts to many thousands of kilowatts. In this section of the book, we will concentrate on the operation of the smaller type of servos.

A block diagram of one form of d.c. r.p.c. system using series feedback is shown in Fig. 12.6. In this circuit, any misalignment between the input and output shafts results in an error voltage being applied to the amplifier. The error signal is amplified and causes the motor to develop a torque to turn the load in a direction to reduce the error. The system shown in Fig. 12.6 is simplified when compared with a practical system since, as we shall see later, it is necessary to incorporate additional circuit elements to minimize oscillations of the output shaft as it settles down to its final position.

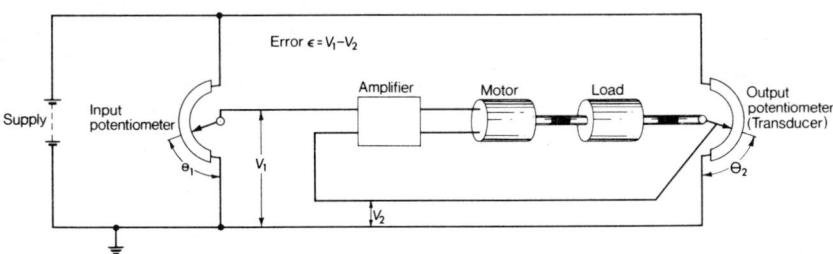

Fig. 12.6 A d.c. servosystem employing series feedback.

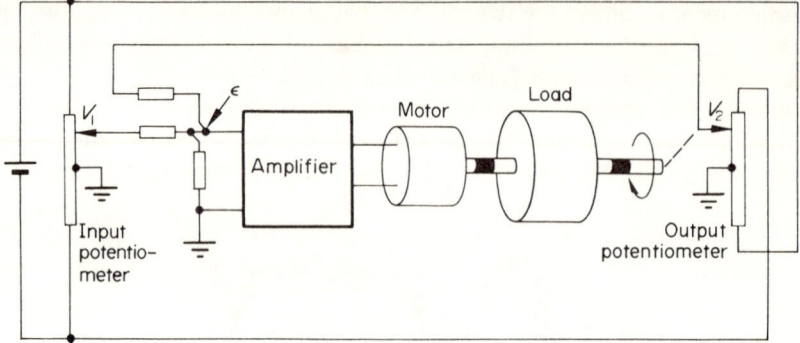

Fig. 12.7   A servosystem employing shunt feedback.

A system employing shunt feedback is shown in Fig. 12.7. Here, the signals V_1 and V_2 are of opposite polarity, so that under steady-state conditions the error voltage is zero.

Most practical forms of control system incorporate a speed reduction gear box between the motor and the load, so that high-speed servo motors can be used. In some large systems the motor and load are directly coupled together, but this is not common practice.

One form of a.c. r.p.c. system has already been illustrated in Fig. 12.5, the a.c. error signal being generated by the synchro system. The superior controllability of d.c. machines when compared with a.c. machines has led to the widespread use of a.c.-d.c. systems, in which the error signal is detected by means of a synchro system and the final drive is provided by a d.c. motor.

A hybrid a.c.-d.c. control system is shown in Fig. 12.8, which incorporates the basic elements of this type of control. The box marked p.s.r. contains a

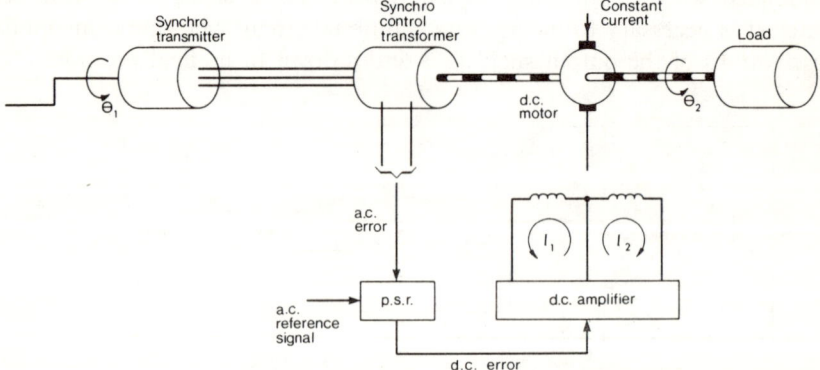

Fig. 12.8   A hybrid a.c. – d.c. r.p.c. system.

phase-sensitive rectifier which gives a d.c. error signal which is proportional to the a.c. error signal; the polarity of the d.c. error is dependent on the phase relationship existing between the a.c. error and the a.c. reference signal. If the two a.c. signals are in phase with one another, then the polarity of the d.c. error signal is, say, positive. If they are antiphase, then the error voltage is negative.

The d.c. work motor used in Fig. 12.8 is of the *split-field* type, that is the field winding is centre-tapped, the current in the two halves of the winding flowing in opposite directions. A basic d.c. machine equation is

$$T \propto \Phi I_a$$

where T is the shaft torque developed, Φ is the net magnetic flux in the field system, and I_a is the armature current. In the system considered, I_a is maintained at a constant value by means of a constant current circuit. As a result, the torque equation reduces to

$$T \propto \Phi$$
$$\propto (I_1 - I_2)$$

When the error signal is zero, the currents in the two halves of the motor field winding are equal to one another and the torque developed is zero, and the armature is stationary. The application of an error signal results in an imbalance between I_1 and I_2, so that the motor armature develops a torque to drive the output shaft and the control transformer rotor in a direction to reduce the error to zero. This action finally reduces the imbalance between I_1 and I_2 to zero, by which time the load shaft has reached angular alignment with the position of the input shaft.

12.5 Regulator systems

A popular form of speed controller is shown in Fig. 12.9 in which the motor is powered from the output of BRI via a thyristor. The thyristor is included in a shunt feedback loop, and the firing angle is controlled by the difference in potential between the speed reference signal and the speed feedback signal. The equation relating armature speed N to its armature 'back' e.m.f. E is

$$E \propto \Phi N$$

where Φ is the field flux. In the case considered the flux is constant, so that in the case of Fig. 12.9 the equation reduces to

$$E \propto N$$

That is, the motor speed is proportional to the back e.m.f. of the machine. Unfortunately, we do not have direct access to the back e.m.f., but we can

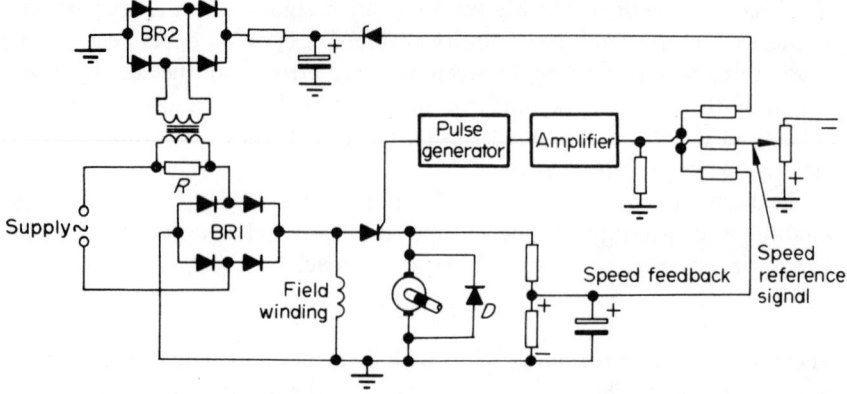

Fig. 12.9   A speed control system using a thyristor.

measure the armature supply voltage either directly or by means of the resistive potential divider network connected across the armature. Now

$$V = E + I_a R_a$$

where $I_a R_a$ is the p.d. in the armature resistance. If $I_a R_a$ has a negligible value when compared with E then, to a first approximation

$$V \simeq E$$

If this assumption is valid, we can use the armature voltage as a measure of output speed in the manner shown in Fig. 12.9. Since the output voltage from the thyristor is in the form of a series of pulses, it is smoothed by the capacitor which shunts the β-network output before being fed back to the controller. In order to satisfy the conditions of shunt feedback, the polarity of the signal fed back is of opposite polarity to the reference signal.

When the speed reference signal is increased, the pulses applied to the thyristor are phased forward to deliver more current to the armature, causing the speed to increase. Steady-state conditions are reached when the error voltage falls to a level which just maintains the new speed.

The circuit in Fig. 12.9 incorporates current limit protection, and operates in the same manner as that described in section 10.12 of chapter 10.

In the above we assumed that the $I_a R_a$ drop was insignificant. In many instances this is not the case, and the application of a load to the machine shaft causes the output speed to fall or *droop*. With armature voltage feedback, the speed droop is less than in the case when no feedback is applied but none the less it reduces the accuracy with which the speed is controlled. One method of overcoming this defect is to introduce a small amount of positive current feedback into the loop. This is known as *armature voltage drop compensation*, the principle being outlined in Fig. 12.10.

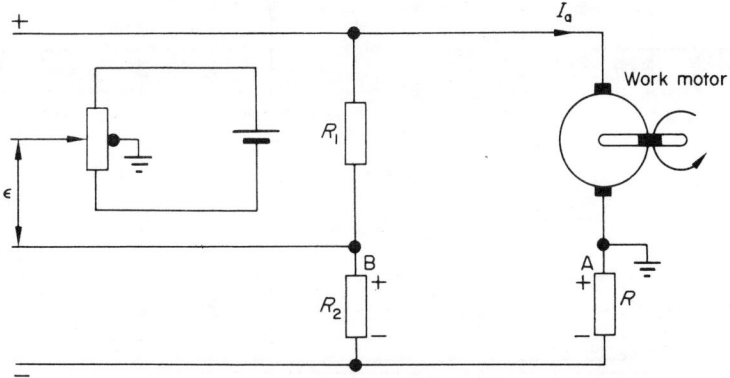

Fig. 12.10   Armature voltage drop compensation.

Here, the resistors R_1 and R_2 act as a β-network for the armature voltage signal. The voltage across R_2 is βV, where $\beta = R_2/(R_1 + R_2)$ and V is the armature voltage. The voltage at point B relative to A is

$$V_{BA} = \beta V - I_a R \tag{12.2}$$

Now, the equation for armature voltage is

$$E = V - I_a R_a \tag{12.3}$$

and, multiplying eq. (12.3) by the factor β, gives

$$\beta E = \beta V - \beta I_a R_a \tag{12.4}$$

If V_{BA} in eq. (12.2) can be made to have the same value as βE, then V_{BA} will be proportional to the back e.m.f. of rotation, i.e., V_{BA} would be proportional to the shaft speed. For this relationship to hold good, the coefficients of the terms in eqs. (12.2) and (12.4) must be equal to one another. That is

$$R = \beta R_a$$

With this value of resistance in circuit the speed droop is effectively eliminated.

Yet another type of regulator, one for regulating the temperature of an oven, is shown in Fig. 12.11. The temperature sensing element is a thermistor which is included in the bridge circuit on the left-hand side of the figure. In this circuit, the load is on the a.c. side of the rectifier, while the thyristor is on the d.c. side. When the oven temperature is low, the thermistor resistance is high, and this causes the emitter junction of transistor TR to be forward biased. As a result, TR is saturated and the maximum voltage is applied to the R–C network of the pulse generator. Consequently, the capacitor charges at a high rate, and the pulse generator generates a train of

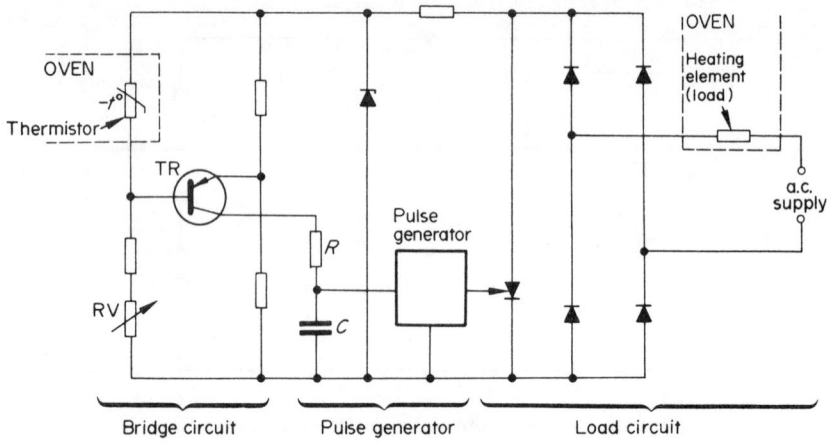

Fig. 12.11   One form of oven temperature control.

pulses with a high repetition rate. In turn, this causes the thyristor to trigger at an early point in the cycle, and the oven heats up quickly. As the oven temperature rises, so the thermistor resistance reduces and causes the transistor current to diminish, and the thyristor gate pulses to be phased back. Steady-state operation is achieved when the pulses are phased back sufficiently to maintain a constant temperature in the oven. The oven temperature setting is controlled by adjusting the value of variable resistor *RV*.

12.6 Stabilizing techniques

Control systems contain energy storage elements within the systems in the form of capacitors, inductors, and, in electro-mechanical systems, inertia of moving parts, etc. Once the system has been excited by an input signal these elements begin to store energy; when the error voltage falls to zero the output from the controller should fall to zero and the system output should stabilize at its steady-state value. But, under certain circumstances, the stored energy may be released to cause the output to continue to change. In this way, the output *overshoots* the desired level before the rate of change of the system output begins to decrease. Various types of system response are possible, several of which are illustrated in Fig. 12.12; these curves correspond to the case where the input is suddenly changed from one level to another. Such a change is known as a *step change* of input signal.

The type of response which is generally desired for this type of input signal disturbance is the one shown as an *underdamped* curve in Fig. 12.12. In this case, the output overshoots the desired level by about 5 to 10 per cent before settling down to the correct level. This type of response usually

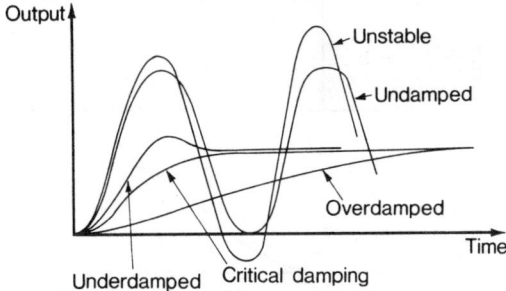

Fig. 12.12 Possible forms of response of a control system to a step input change.

provides a short settling time together with good steady-state accuracy. The actual amount of *damping* applied to the system is defined by the system *damping factor*. If this factor has a value less than unity but greater than zero, then the system is said to be underdamped and the output executes one or more oscillations before settling down. For example, a damping factor of 0.5 results in the output executing about one complete oscillation with a peak overshoot of about 16 per cent. If the damping factor is 0.2, then the output completes about three oscillations with a first peak overshoot of about 54 per cent.

When the system damping is reduced to zero (damping factor = 0), the system is said to be *undamped* and the output oscillates continuously, the peak-to-peak amplitude of the output swing remaining constant from one cycle to the next. A system is said to be *unstable* if the amplitude of the oscillations continues to increase. Ultimately, in an unstable system a component or sub-system will fail. Fortunately, in most cases, a feature of the circuit such as amplifier saturation limits the magnitude of the output oscillations.

If the damping factor is unity, the system is *critically damped*, and the output reaches its final value without overshoot. Should the damping factor be greater than unity, the system is *overdamped* and has a sluggish response.

Many systems are basically unstable, and it is necessary to employ some form of *damping system* in order to achieve the slightly underdamped response of the type in Fig. 12.12. In order to understand the problem more fully, let us consider the waveforms which occur in a position control system which is overdamped. A series of idealized waveforms is illustrated in Fig. 12.13.

The step change in the input signal is illustrated in curve (a) and the output waveform in curve (b). In the time interval between points A and B, curve (b), the required input R applied to the amplifier must be positive for the motor to provide a driving torque to the load. During the time interval between B and C, signal R must be negative in order to retard the load to

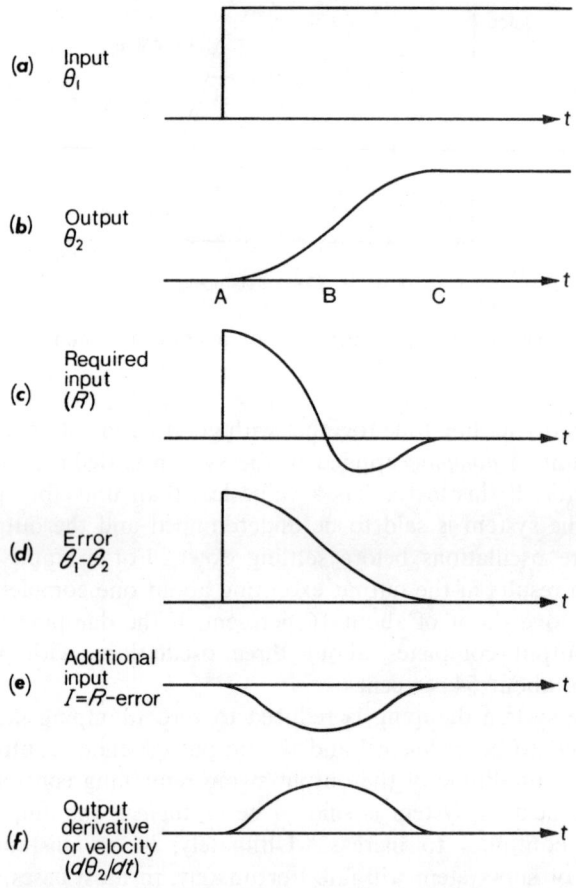

Fig. 12.13   Waveforms in an overdamped r.p.c. servo.

prevent the energy stored in the load causing the output shaft to overshoot the final position. Thus, the net input to the amplifier must first of all be positive in order to accelerate the load and, finally, must be negative to retard the load.

In a basic system, the only signal present is the error signal ($\epsilon = \theta_1 - \theta_2$), and is shown in curve (d). Quite clearly, this curve does not have the same shape as the required input R, curve (c), and it is necessary to combine the error signal with some other signal in order to derive the required response. That is

$$\text{Required input } R = \text{Error} + \text{Additional input } (I)$$

or $$I = R - \text{Error}$$

This subtraction having been performed, we are left with curve (e) in Fig.

12.13. It simply remains to generate a waveshape of the kind in curve (e) in order to give the system an overdamped response of the type in curve (a). In the following, we will consider ways in which this can be done.

12.6.1   Output derivative damping

If we differentiate the output signal, curve (b) in Fig. 12.13, the resulting wave-shape is as shown in curve (f) in the same figure. The general shape of curve (f) is similar to that of the additional input signal I, curve (e), but is of the opposite polarity. Clearly, if we *subtract* the output derivative signal (or *output velocity* or *output rate* signal) from the error signal, then the required form of input signal is generated.

A block diagram of a control system employing *output derivative stabilization* or *output velocity stabilization* is shown in Fig. 12.14, in which the output derivative signal is subtracted from the error signal to provide stabilization.

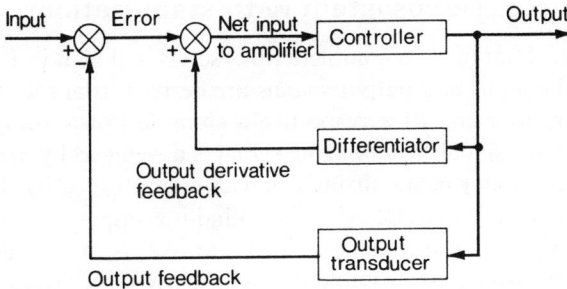

Fig. 12.14   A block diagram of a control system employing output derivative (output velocity) stabilization.

12.6.2   Error-rate damping

An alternative source of stabilizing signal is the error signal itself. If the error signal, curve (d) in Fig. 12.13, is differentiated, the resulting curve is similar to the additional input, curve (e). The required input waveform, curve (c), is therefore generated by a circuit which provides a signal which is the *sum* of the error and the rate of change of error, i.e., a signal which is defined by the expression

$$k_1 \epsilon + k_2 \, dc/dt \qquad (12.5)$$

where k_1 and k_2 are constants.

One such circuit is shown in Fig. 12.15. To a first approximation, the current flowing in R_1 is proportional to error, and the current flowing in C is proportional to the rate of change of error. Since the sum of the two currents flows in R_2, the equation describing the voltage across R_2 has the

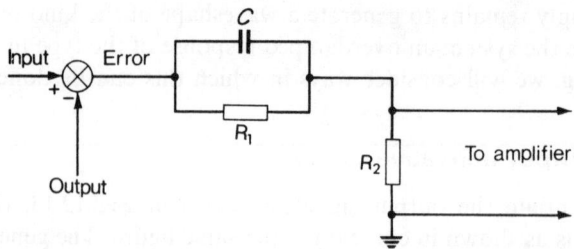

Fig. 12.15 An *R–C* circuit which provides an output proportional to error plus rate of change of error.

same general form as eq. (12.5). The voltage across R_2 can then be used as the input to the controller.

12.7 An r.p.c. servosystem with stabilization

The schematic diagram of a complete r.p.c. servo is shown in Fig. 12.16. In this system, the input and output signals are derived from the wipers of the potentiometers *RV*1 and *RV*2, respectively, shunt feedback being applied to generate the error signal. A stabilizing signal is developed by the tachogenerator, and the amount of stabilizing feedback is controlled by the setting of *RV*3. In commissioning equipment of this kind it is important to ensure that the connections of the tachogenerator are correct as, if they are reversed, then the system would be unstable if a sufficiently large signal is fed back.

The system shown incorporates a speed reduction gear box between the

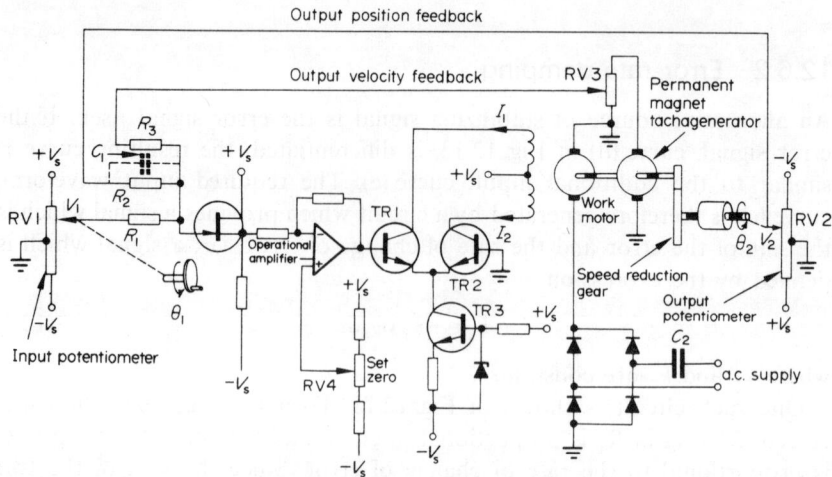

Fig. 12.16 A stabilized r.p.c. servo.

motor/tacho element and the load to allow a high-speed motor to be employed.

It is possible to eliminate the need for the tachogenerator and its associated circuitry by deriving the $(-d\theta_2/dt)$ signal by shunting resistor R_2 by capacitor C_1. The source follower acts as a high input impedance buffer element, the resulting signal from the source follower being amplified by the operational amplifier and is applied to the power output stage.

The output stage comprises an emitter-coupled amplifier containing transistors TR1 and TR2, together with the constant current source TR3. The work motor is a split-field machine with the two halves of the field winding being excited by the currents from TR1 and TR2. With the input and output shafts aligned, the system is initially balanced by adjusting the set zero control RV_4, so that $I_1 = I_2$. A constant current circuit maintains the armature current at a constant value, simply by including capacitor C_2 in series with the armature rectifier, the reactance of C_2 being sufficiently great to ensure constant current operation.

A change in input shaft position which causes V_1 to become positive results in the operational amplifier output voltage swinging in a negative direction. In turn, this reduces I_1 and increases I_2, and causes the motor armature to rotate to drive the output potentiometer wiper. When the magnitudes of V_1 and V_2 are equal to one another, the error voltage becomes zero once more and the torque developed by the motor falls to zero. If the system is either critically damped or overdamped, then the output shaft remains in the new position. If underdamped, the inertia of the system causes the load to continue to rotate for a little time, and the polarity of the error voltage reverses. This causes the torque developed by the motor to reverse so that the direction of rotation of the output shaft is quickly reversed to bring the two shafts into alignment once more. This process continues until the system finally settles down with the two shafts aligned.

12.8 Transient performance of control systems

The transient response of a control system to a step input signal provides several important parameters. These parameters may be determined from the transient response curve in Fig. 12.17.

The amount of *first overshoot* is expressed in terms of the per-unit value of the final steady-state output, and gives an indication of the stresses to which the system is subjected. The per-unit first overshoot can be shown to be given by the equation

$$\text{p.u. overshoot} = 1/\{\exp[\zeta\pi/\sqrt{(1-\zeta^2)}]\} \qquad (12.6)$$

where ζ is the system damping factor.

The *rise time* t_r of the system is the time taken for the output to change

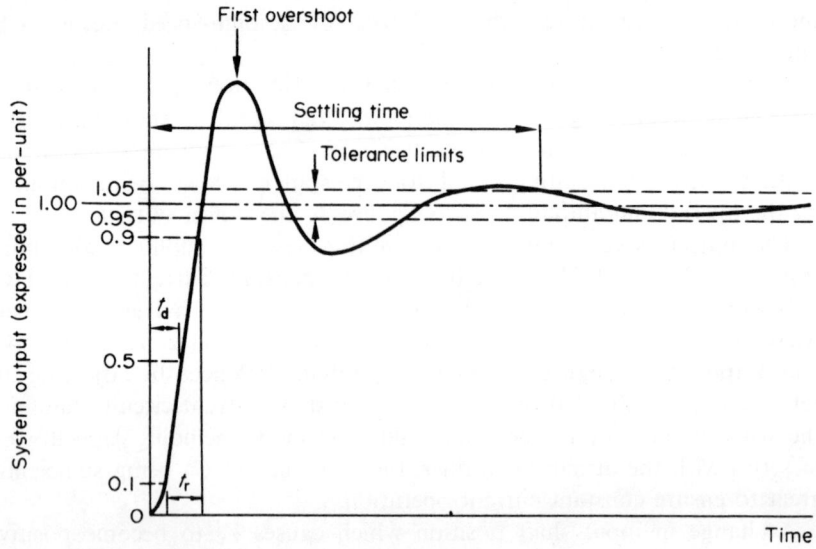

Fig. 12.17   Underdamped response to a step input signal.

from 10 to 90 per cent of the final steady-state value. The *delay time* t_d is defined as the time required for the response to a step input to reach 50 per cent of the final value. The *settling time* t_s is an important parameter, since it gives the time taken for the response to *enter and remain within the system tolerance limits. The tolerance limits frequently used are* ± 2 *and* ± 5 per cent of the total change, and in Fig. 12.17 the later limit is chosen for clarity.

12.9   Analysis of an r.p.c. servosystem

For the purposes of analysis, it is convenient to reduce the block diagram of a control system of the type in Fig. 12.16 to that in Fig. 12.18. To simplify the analysis, we will initially ignore the stabilizing feedback loop associated with the tachogenerator in Fig. 12.16.

Commencing with the error signal ϵ radians, we see that the potentiometer system develops a voltage $K_p\epsilon$, where K_p is the potentiometer *transfer function* having the dimensions of volts per radian. The amplifier and motor can be considered as a single unit having a transfer function K, with dimensions of newton-metres/volt. In fact, the transfer function of the amplifier section is usually fairly complex; a detailed treatment can be found in specialized books on the subject of control systems. Since the gearbox reduces the motor speed by a factor of N, it also magnifies the torque by the same factor, and the torque at the output shaft is

$$K_p KN\epsilon = K_p KN(\theta_1 - \theta_2) \text{ newton-metre}$$

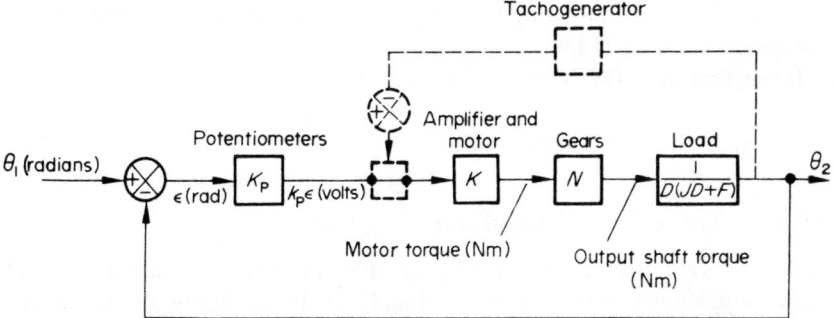

Fig. 12.18 A block diagram of a closed-loop control system.

The torque at the output shaft must not only be adequate to overcome the inertia torque, $J\,(d^2\theta_2/dt^2)$, but also the friction torque, $F(d\theta_2/dt)$, where J is the total system inertia in kg-m^2 referred to the load shaft and F is the viscous friction coefficient in Nm/(rad/s) referred to the load shaft. Thus,

$$J\frac{d^2\theta_2}{dt^2}+F\frac{d\theta_2}{dt}=K_{\mathrm P}KN(\theta_1-\theta_2)$$

If $K_{\mathrm S}=K_{\mathrm P}KN$, where $K_{\mathrm S}$ is a system constant having dimensions Nm/rad, then the equation simplifies to

$$\frac{d^2\theta_2}{dt^2}+\frac{F}{J}\frac{d\theta_2}{dt}+\frac{K_{\mathrm S}}{J}\theta_2=\frac{K_{\mathrm S}}{J}\theta_1 \tag{12.7}$$

Equation (12.7) may be rewritten in the following standard form.

$$\frac{d^2\theta_2}{dt^2}+2\zeta\omega_{\mathrm n}\frac{d\theta_2}{dt}+\omega_{\mathrm n}{}^2\theta_2=\omega_{\mathrm n}{}^2\theta_1 \tag{12.8}$$

where $\qquad \omega_{\mathrm n}=\sqrt{(K_{\mathrm S}/J)}=$ Undamped natural frequency

and the damping factor or damping ratio is

$$\zeta=\frac{F}{2\sqrt{(K_{\mathrm S}J)}}$$

$$=\frac{\text{Actual damping}}{\text{Damping required for critical response}}$$

With a step input signal change, the system response is evaluated by solving eq. (12.8). For an underdamped system, its solution is

$$\theta_2=\theta_1\left\{1-\exp(-\omega_{\mathrm n}t)(\cos\omega t+\frac{\zeta}{\sqrt{(1-\zeta^2)}}\sin\omega t)\right\} \tag{12.9}$$

From this equation we can compute the shape of the output response, one form having been illustrated in Fig. 12.17. Equation (12.6) was, in fact, computed from eq. (12.9).

Inspecting eqs. (12.7) and (12.8), we see that when the output shaft is stationary, i.e., when $d^2\theta_2/dt^2 = 0 = d\theta_2/dt$, then $\theta_2 = \theta_1$. That is, the input and output shafts are aligned and the error is zero.

12.9.1 The effect of the stabilizing loop

The system equation is modified by the introduction of the stabilizing feedback loop, shown in broken line in Fig. 12.18. In the figure, the tachogenerator is assumed to be mounted on the output shaft but, in fact, it is mounted on the motor shaft in the manner of Fig. 12.16 in order to take advantage of the higher speed at the motor shaft. The tachogenerator itself has a transfer function of K_T volts/(rad/s) and, since it is mounted on the motor shaft and not the output shaft, its output voltage for a load shaft speed of $d\theta_2/dt$ is $NK_T(d\theta_2/dt)$. With this modification the amplifier input signal is given by the expression

$$K_P(\theta_1 - \theta_2) - NK_T(d\theta_2/dt)$$

The resulting equation of the system is

$$KN\{K_P(\theta_1 - \theta_2) - NK_T(d\theta_2/dt)\} = Jd^2\theta_2/dt^2 + Fd\theta_2/dt$$

This reduces to the form

$$J\frac{d^2\theta_2}{dt^2} + \frac{d\theta_2}{dt}(F + KK_TN^2) + K_S\theta_2 = K_S\theta_1 \tag{12.10}$$

where $K_S = KK_PN$, as before. Rewriting the equation in standard form, the equation becomes

$$\frac{d^2\theta_2}{dt^2} + 2\zeta'\omega_n\frac{d\theta_2}{dt} + \omega_n^2\theta_2 = \omega_n^2\theta_1 \tag{12.11}$$

We see that the undamped natural frequency is $\sqrt{(K_S/J)}$, and is unchanged when compared with the system before the additional damping was added. However, the damping factor ζ' of the modified system with output velocity damping is increased when compared with that without damping. The new damping factor is

$$\zeta' = \frac{F + KK_TN^2}{2\sqrt{(K_S/J)}}$$

12.10 Stability margins of control systems

The degree of stability that a control system has is determined from the open-loop frequency response diagram. The method of applying the technique was described in section 3.16, and results in a frequency response diagram known as a *Nyquist diagram*. There is, however, a significant difference between the type of Nyquist diagram normally used in connection with a control system and that used for a feedback amplifier. The difference lies in the fact that in control systems the signal fed back is assumed to be subtracted from the input signal, whereas in feedback amplifier theory the signal fed back is assumed to be added to the input signal. Thus, for stability in a control system, *the point* $(-\mathbf{1}, \mathbf{0})$ *on the Nyquist diagram must lie on the left of the frequency response curve as it is traversed in the direction of increasing frequency.* [In the case of feedback amplifiers, the critical point was shown to be $(+1, 0)$ (see chapter 3).]

To indicate the degree of stability of the system, certain margins are specified at well defined points on the Nyquist diagram. These are the *phase margin* and the *gain margin*, and are defined in terms of frequencies ω_1 and ω_2, respectively, on Fig. 12.19. Frequency ω_1 is known as the *gain crossover frequency*, and is the frequency at which the gain is unity. Frequency ω_2 is the *phase crossover frequency*, and is the frequency at which the phase shift is 180 degrees. The phase margin of the system is the angle in degrees by which the phase lag can be increased (at the gain crossover frequency) without change in gain in order that the phase shift shall be 180 degrees. The gain margin is the amount by which the loop gain can be increased without change in phase shift (at the phase crossover frequency) in order that the gain shall be unity.

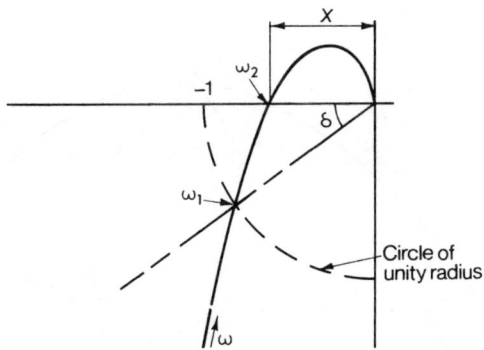

Fig. 12.19   Gain and phase margins of a control system.

The numerical value of the gain margin in the case of Fig. 12.19 is $1/X$, where X is the loop gain at frequency ω_2. This margin is usually expressed in decibels and is

$$20 \lg(1/X)$$

Acceptable minimum values of gain margin lie in the range 10–20 dB. The phase margin is equal to δ in Fig. 12.19, and its value often lies in the range 30 to 60 degrees.

A significant difference between completely electronic systems and electro-mechanical control systems is in the range of frequencies over which the tests are made. In all electronic systems, a popular range of test frequencies is from a few Hz to several MHz. In electromechanical systems, the range of frequencies commonly employed is from about 0.001 Hz to about 30 Hz, although this will depend on the system.

In systems which have complex feedback paths, the amount of labour involved in plotting the Nyquist diagram can be reduced by using the *inverse Nyquist diagram* (sometimes known as the Whiteley diagram), and is shown in Fig. 12.20. This diagram is a plot of the inverse of the loop gain with frequency, the gain and phase crossover frequencies being shown as ω_1 and ω_2, respectively, in Fig. 12.20. Once again, for stability, the $(-1, 0)$ point must lie on the left of the curve as it is traversed in the direction of increasing frequency. The gain and phase margins are defined for the Nyquist diagram.

Inverse Nyquist diagrams are also useful when used in connection with *describing function* studies of non-linear systems (see section 12.11).

Another type of diagram on which the gain (in decibels) and phase shift

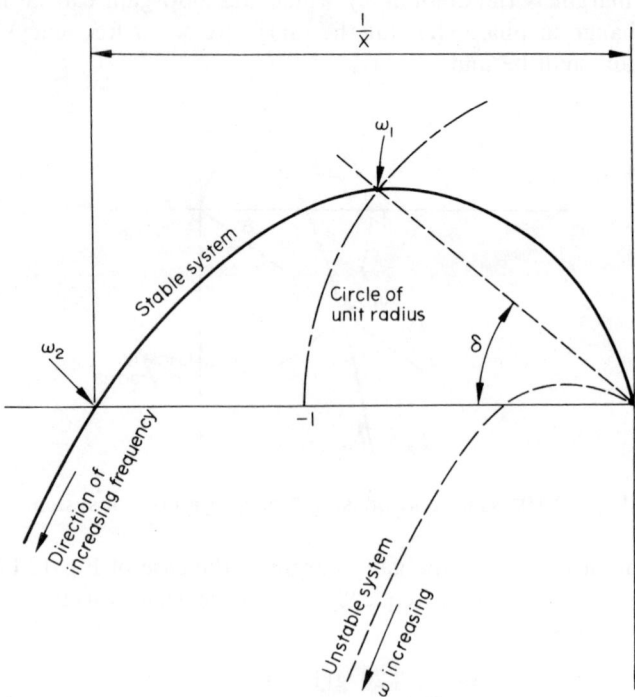

Fig. 12.20.   An inverse Nyquist diagram.

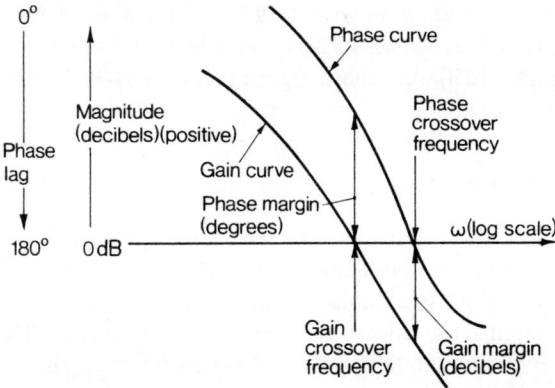

Fig. 12.21   Bode diagram for a stable system.

(in degrees) curves obtained from an open-loop test are plotted individually to a common base of frequency is known as a *Bode diagram*. A Bode diagram of a stable system is illustrated in Fig. 12.21, the gain and phase margins being measured in the manner shown.

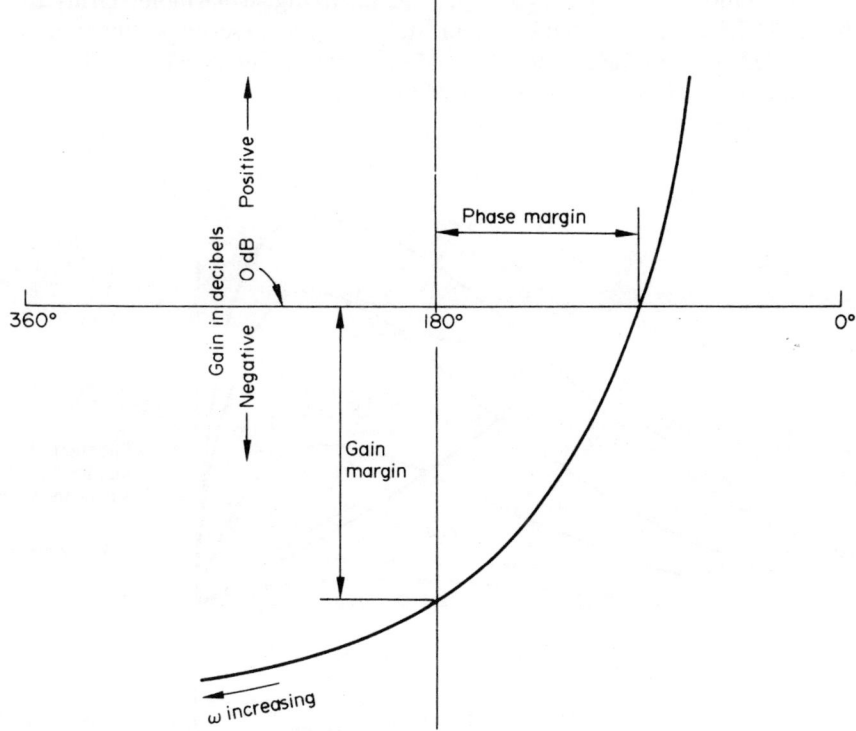

Fig. 12.22   A Nichol's chart.

Yet another method of determining the gain and phase margins of a control system is the Nichol's chart, on which the open-loop response is plotted using gain (dB) and phase lag (degrees) as axes. A Nichol's chart for a stable system is shown in Fig. 12.22.

12.11 Non-linear systems

To some extent, all systems exhibit non-linear effects. For example, amplifiers of all types have saturation limits, transducers have limited performance, mechanical parts exhibit non-linear friction characteristics, etc.

In many systems the effects of non-linearities are small enough to be ignored, and it is possible to carry out design studies on these systems using linear theory. Where this is not the case, special techniques have been developed, a powerful method being the *describing function* technique. The principle is described in the following.

If we energize a non-linear element with a sine wave, the output signal can be analysed into a fundamental frequency term and its constituent harmonic components. The describing function of the element is the ratio of the *fundamental* component of the output from the non-linear element to the magnitude of the sinusoidal input signal. All higher harmonic terms are ignored. In order to study system stability, the describing function is multiplied by a coefficient of (-1) and is plotted together with the inverse Nyquist diagram of the linear part of the system, as shown in Fig. 12.23.

If G_D is the describing function of the non-linear element, then in a very

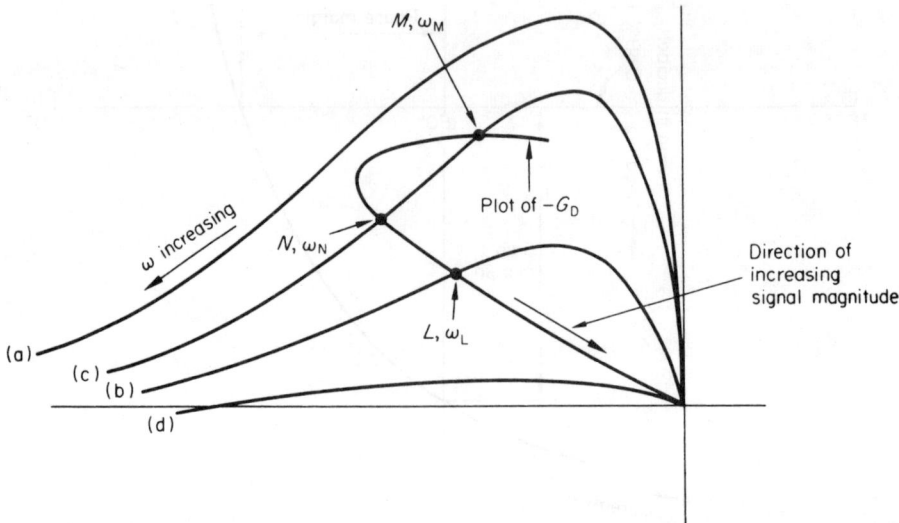

Fig. 12.23 Stability of a system containing a non-linear element.

broad sense we can regard the $-G_D$ locus as being the equivalent of the $(-1, 0)$ point on the inverse Nyquist diagram.

Thus, the system is completely stable if the whole of the $-G_D$ locus lies on the left of the inverse Nyquist diagram as the inverse Nyquist diagram is traversed in the direction of increasing frequency. The system is completely unstable if the $-G_D$ locus lies entirely to the right of the inverse Nyquist diagram.

Figure 12.23 shows one form of $-G_D$ locus together with a number of inverse Nyquist diagrams. The $-G_D$ curve shown is typical of a relay with 'hysteresis' due to differential pull-in and drop-out voltages. If the remainder of the system has an inverse Nyquist diagram similar to that shown in curve (a), then the system is stable. If the inverse locus is of type (b), which cuts the $-G_D$ locus at L, then the output oscillates at frequency ω_L corresponding to point L on the inverse Nyquist diagram with an amplitude corresponding to point L on the $-G_D$ locus. This is known as *limit cycling*, and is a special case of instability.

If the inverse locus is of the type in curve (c), then the situation is more complex. For magnitudes of excitation applied to the non-linear element which are less than M, the system is stable. When the excitation signal increases to a value equal to M the system becomes unstable. When the loop is closed, oscillations commence at frequency ω_M and build up to amplitude N at frequency ω_N. However, when the excitation voltage is increased beyond N, the system becomes stable once more, so that a stable limit cycle occurs at frequency ω_N with amplitude N.

In the case of curve (d), the whole of the $-G_D$ locus lies on the right of the inverse Nyquist diagram, so that if the loop is closed the system would be unstable.

Problems

12.1 Discuss, with the aid of appropriate diagrams, TWO of the following

(a) a synchro pair with a phase-sensitive rectifier
(b) a d.c. servo amplifier
(c) a d.c. split-field motor with a d.c. tachogenerator
(d) a d.c. constant-current armature supply for a servo motor.

12.2 (a) What factors govern (i) the linearity, and (ii) the resolution of a typical rotary potentiometer as used in control systems?

(b) A linear $100\,k\Omega$ potentiometer with an arc length of $344°$ is connected to a $15\,V$ stabilized d.c. supply. Calculate (i) the potentiometer constant in volts/radian, and (ii) the output voltage when a $100\,k\Omega$ resistor is connected to the slider, which is set at one-half the arc length. Discuss the way in which the error in output voltage will vary as the slider moves along the arc.

12.3   Sketch circuit diagrams of resistance-capacitance networks which will provide an output which approximates to

(a) proportional plus derivative of input
(b) proportional plus integral of input.

Show how each of these networks could be incorporated in a closed-loop control system to improve its performance.

Discuss for each case the effects on the transient and steady-state response of the control system obtained by including the network.

12.4   With the aid of block diagrams, explain the difference between open- and closed-loop control systems.

Draw a block diagram of a closed-loop speed control system for a d.c. motor when the system is fed from an a.c. supply. Quote typical examples of equipment that may be used in each section of the block diagram. State how

(a) the accuracy of speed control
(b) the motor power output

will affect the choice of motor type and motor power supply equipment.

12.5   Explain what is meant by the following terms as applied to control engineering

(a) open-loop system gain
(b) undamped natural frequency
(c) velocity error.

A control system has an open-loop gain K (N-m/rad), a coefficient of viscous friction F (N-m.sec), and a load inertia J (kg-m^2). Discuss fully what effect each of the factors has on (i) the undamped natural frequency, and on (ii) the steady-state velocity error, assuming a ramp input to the system.

12.6   Draw a block diagram for a d.c. position control servo system incorporating tachogenerator damping. Show how the amount of such damping could be varied. Give a brief description of the equipment in the system.

Sketch graphs showing the change of the system output θ_0 with time, in response to a step input change θ_i:

(a) when the system is critically damped,
(b) when the system is underdamped, and
(c) when no damping is present.

What is the steady value of θ_0 for each of these conditions?

12.7   The terminal voltage of a d.c. generator is to be accurately maintained by a closed-loop control system. Draw the circuit diagram for an appropriate control system. Explain clearly the manner in which the error-actuating signal is derived. Describe the operation of the system when the generator voltage changes to a value (a) above, and (b) below the desired value.

Describe one method for compensating the effect of the IR drop in the generator armature circuit.

12.8   Draw a circuit diagram and explain the operation of a photo-electric furnace temperature indicator and controller. Discuss the required characteristics of the photo-sensitive device and show how the desired value of temperature could be adjusted.

12.9 The following results were obtained for an open-loop frequency response test on a control system.

Frequency (rad/sec)	0.1	0.5	1.0	5.0	10	50	100
Amplitude (dB)	37	24	17	−5	−19	−60	−78
Phase-lag (degrees)	90	105	124	186	214	256	266

Plot the BODE diagram and (a) determine the gain and phase margins, and (b) comment fully on the practical implications of these results.

12.10 Repeat problem 12.9, but use a NYQUIST diagram.

12.11 The open-loop transfer function of a control system is given by the expression

$$\frac{10}{j\omega(1+0.1j\omega)(1+0.01j\omega)}$$

Draw the relevant section of the Bode diagram to determine the gain and phase margins. Explain what is meant by the gain crossover frequency and the phase crossover frequency, and determine their values in this problem.

12.12 Repeat problem 12.11 using a Nyquist diagram.

12.13 An amplifier has a transfer function given by the expression

$$\frac{40}{(1+0.1\,D)(1+0.05\,D)}$$

where operator $D = d/dt$. The amplifier is to be used in a closed-loop control system in which 10 per cent of the output is fed back to the input.

Derive an expression for the time variation in output voltage if the reference voltage is suddenly changed from zero to 20 V. What is the steady-state output voltage and the peak overshoot (expressed as a percentage)?

Solutions to numerical problems

Chapter 1

1.4 $h_{ie} = 1200\,\Omega$; $h_{re} = 6 \times 10^{-4}$; $h_{fe} = 82$; $h_{oe} = 128\,\mu S$.

1.5 The capacitor voltage reduces linearly at the rate of 1 V/s.

1.8 $y_{fs} = 3.6\,\text{mS}$; $y_{os} = 0.1\,\text{mS}$.

1.13 13 W.

Chapter 2

2.1 $R_L = 500\,\Omega$; $R_B = 104\,\text{k}\Omega$; feedback resistor $= 46\,\text{k}\Omega$.

2.2 1; 0.442.

2.3 $200\,\text{k}\Omega$; -56.25.

2.4 (a) $0.99\,\text{k}\Omega$; (b) -58.4.

2.6 (a) 90; (b) $11.6\,\text{k}\Omega$.

2.7 40; 63.5 degrees.

2.8 (a) 985; (b) $39.4\,\Omega$.

2.9 (a) $1\,\text{k}\Omega$, 50; (b) -1665; 41.6×10^3.

2.10 $1.62\,\mu F$ (minimum value).

2.11 (b) (i) $1.125\,\text{kHz}$; (ii) -217; (iii) $79.5\,\text{Hz}$.

2.12 (a) $1.6\,\text{ms}$; (b) $4.6\,\text{ms}$.

2.13 (b) (i) -2.25; (ii) $1\,\text{M}\Omega$.

2.17 (a) (i) $15.5\,\text{MHz}$, (ii) 3.22; (b) (i) $19.6\,\text{MHz}$, (ii) 2.55; (c) (i) $31\,\text{MHz}$, (ii) 1.61.

Chapter 3

3.1 30.4; 33.4; 27.6.

3.5 (a) 953 to 1015; (b) $2\,\text{M}\Omega$; (c) $200\,\text{Hz}$.

3.6 $A/(1 + A\beta)$; 0.02; (i) 20 per cent, (ii) 11.2 per cent.

3.7 (a) $1510\,\Omega$; (b) $510\,\Omega$.

3.8 0.184.

3.9 $0.1\,\text{V}$.

3.10 $20\,\text{dB}$.

3.11 (a) $79\,\text{k}\Omega$; (b) $12.1\,\Omega$.

3.12 (i) $44.8\,\text{mV}$; (ii) $431\,\text{mV}$.

3.13 $100\,\text{k}\Omega$.

3.14 $100\,\text{k}\Omega$; $200\,\text{k}\Omega$.

3.15 (a) 0.001; (b) $50.1\,\text{M}\Omega$; (c) $0.2\,\Omega$.

Chapter 4

4.1 2.28 MHz; 0.025.
4.2 191 kHz; 0.25.
4.3 318.3 kHz.
4.11 412 kHz; 413 kHz; 2215.
4.12 (a) increase; (b) no change.
4.13 250 Hz.

Chapter 5

5.3 1 V; 0.607 V.
5.4 0.2 V.
5.5 5 pF.
5.6 5.27 pF.

Chapter 6

6.4 1663 pulses/sec; 48.4 μs.
6.5 4.55 kHz.
6.6 $R_1 = R_2 = 2$ kΩ; $R_3 = R_4 = 40$ kΩ; $C_1 = 24$ pF, $C_2 = 12$ pF. (Note: See Fig. 6.3.)
6.7 (i) 0.7 ms; (ii) 0.103 ms.
6.8 $V_{C1} = 10$ V; $V_{B1} = -2.5$ V; $I_{C1} = 0$; $I_{B1} = 0$; $V_{C2} = 0$; $V_{B2} = 0.6$ V; $I_{C2} = 2.13$ mA; $I_{B2} = 94$ μA.
6.9 13.8 μs.
6.11 1442 pulses/sec.
6.12 1442 pulses/sec.
6.13 0.1 V/μs.
6.14 95 μs.
6.15 5 V.

Chapter 7

7.1 (a) $\bar{A} \vee B$; (b) $X \vee Y$; (c) $\bar{X}.\bar{Y}$; (d) 1.
7.2 (a) $\overline{A.\bar{B}.C} + A.B.\bar{C} + A.B.C$; (b) $A.(B+C)$.
7.4 (a) $A \vee B \vee C = \bar{A}.\bar{B}.\bar{C}$; (b) $A \vee B$.
7.5 $B + \bar{A}.\bar{C}$.
7.6 (a) $\bar{A}.C \vee \bar{B}.C.E \vee A.\bar{C}.\bar{D}.\bar{E}$.
7.7 $B.\bar{C} \vee \bar{B}.C$; A is redundant.
7.8 (c) $A.\bar{B} + \bar{A}.B$.
7.10 $\bar{A}.C \vee B.\bar{C}$.
7.11 (b) $A.B$.
7.12 (b) $X = A.B + A.C + B.C$; $Y = A + B.C$; $Z = B + A.\bar{C}$.

Chapter 9

9.8 LDA 2000, CMA, STA 2001, INR A, STA 2002, HLT; $201B_{16}$; 10_{16}; 11_{16}.
9.9 21, 00, 20, 7E, 23, 86, 23, 77, 76; 2018_{16}; OC_{16}.
9.10 MVI A, 00, OUT 20, IN 23, MOV C, A, IN 22, MOV B, A, IN 20, HLT

Chapter 10

10.1 810 W.
10.5 (i) 25 A; (ii) 3.8 A.

Chapter 11

11.3 9.1 Ω; 0.455 V.
11.4 0.0892; 0.446 V.

11.5 $R_S = 175\,\Omega$, 7 W; 15 V, 3 W diode.

11.6 $204\,\Omega$ to $\infty\,\Omega$; 0.5 W.

11.7 (a) $0.397\,\Omega$; (b) 0.595 V; (c) 0.038 V.

11.10 (a) $225\,\Omega$ (maximum value); (b) 15.5 V; (c) 0.31 W (minimum rating), 20 W.

11.11 0.01.

1.12 $0.05\,\Omega$.

Chapter 12

12.2 (i) 2.5 V/rad; (ii) 6 V.

12.9 (a) Gain margin approximately 3.5 dB; phase margin about 8 degrees.

12.11 Phase margin 47 degrees; gain margin 20 dB; gain crossover frequency 7.7 rad/s; phase crossover frequency 31 rad/s.

12.13 $160\,[1 - 1.135\exp(-15t)\sin(27.8t + 1.0769\,\text{rad})]$ volts; 160 V; 18.35 per cent.

Appendix

8085 Instruction listing by function

DATA TRANSFER GROUP – No flags affected

MOVE

MOV A, B	78		MOV E, B	58		
MOV A, C	79		MOV E, C	59		
MOV A, D	7A		MOV E, D	5A		
MOV A, E	7B		MOV E, E	5B		
MOV A, H	7C		MOV E, H	5C		
MOV A, L	7D		MOV E, L	5D		
MOV A, M	7E		MOV E, M	5E		
MOV A, A	7F		MOV E, A	5F		
MOV B, B	40		MOV H, B	60		
MOV B, C	41		MOV H, C	61		
MOV B, D	42		MOV H, D	62		
MOV B, E	43		MOV H, E	63		
MOV B, H	44		MOV H, H	64		
MOV B, L	45		MOV H, L	65		
MOV B, M	46		MOV H, M	66		
MOV B, A	47		MOV H, A	67		
MOV C, B	48		MOV L, B	68		
MOV C, C	49		MOV L, C	69		
MOV C, D	4A		MOV L, D	6A		
MOV C, E	4B		MOV L, E	6B		
MOV C, H	4C		MOV L, H	6C		
MOV C, L	4D		MOV L, L	6D		
MOV C, M	4E		MOV L, M	6E		
MOV C, A	4F		MOV L, A	6F		
MOV D, B	50		MOV M, B	70		
MOV D, C	51		MOV M, C	71		
MOV D, D	52		MOV M, D	72		
MOV D, E	53		MOV M, E	73		
MOV D, H	54		MOV M, H	74		
MOV D, L	55		MOV M, L	75		
MOV D, M	56		—	–		
MOV D, A	57		MOV_ M, A	77		

MOVE IMMEDIATE

MVI B	06
MVI C	0E
MVI D	16
MVI E	1E
MVI H	26
MVI L	2E
MVI M	36
MVI A	3E

LOAD (Reg pair) IMMEDIATE

LXI B	01
LXI D	11
LXI H	21
LXI SP	31

STORE/LOAD

SHLD	22
STA	32
LHLD	2A
LDA	3A

LOAD/STORE A INDIRECT

STAX B	02
STAX D	12
LDAX B	0A
LDAX D	1A

EXCHANGE HL/DE

XCHG	EB

DATA MANIPULATION GROUP – ARITHMETIC

ADD*

ADD B	80
ADD C	81
ADD D	82
ADD E	83
ADD H	84
ADD L	85
ADD M	86
ADD A	87
ADC B	88
ADC C	89
ADC D	8A
ADC E	8B
ADC H	8C
ADC L	8D
ADC M	8E
ADC A	8F
SUB B	90
SUB C	91
SUB D	92
SUB E	93
SUB H	94
SUB L	95
SUB M	96
SUB A	97
SBB B	98
SBB C	99
SBB D	9A
SBB E	9B
SBB H	9C
SBB L	9D
SBB M	9E
SBB A	9F

ADD/SUBTRACT* IMMEDIATE

ADI	C6
ACI	CE
SUI	D6
SBI	DE

INCREMENT**

INR B	04
INR C	0C
INR D	14
INR E	1C
INR H	24
INR L	2C
INR M	34
INR A	3C

DECREMENT**

DCR B	05
DCR C	0D
DCR D	15
DCR E	1D
DCR H	25
DCR L	2D
DCR M	35
DCR A	3D

REGISTER PAIR†† INCR/DECR

INX B	03
INX D	13
INX H	23
INX SP	33
DCX B	0B
DCX D	1B
DCX H	2B
DCX SP	3B

DOUBLE ADD†

DAD B	09
DAD D	19
DAD H	29
DAD SP	39

SET/COMPLEMENT CARRY FLAG†

STC	37
CMC	3F

DECIMAL ADJUST A*

DAA	27

COMPLEMENT A††

CMA	2F

* All flags affected
** All flags except carry affected
† Only carry flag affected
†† No flags affected

8085 Instruction listing by function (cont.)

DATA MANIPULATION GROUP – LOGICAL

AND*

ANA	B	A0
ANA	C	A1
ANA	D	A2
ANA	E	A3
ANA	H	A4
ANA	L	A5
ANA	M	A6
ANA	A	A7

EXCLUSIVE – OR*

XRA	B	A8
XRA	C	A9
XRA	D	AA
XRA	E	AB
XRA	H	AC
XRA	L	AD
XRA	M	AE
XRA	A	AF

OR*

ORA	B	B0
ORA	C	B1
ORA	D	B2
ORA	E	B3
ORA	H	B4
ORA	L	B5
ORA	M	B6
ORA	A	B7

COMPARE*

CMP	B	B8
CMP	C	B9
CMP	D	BA
CMP	E	BB
CMP	H	BC
CMP	L	BD
CMP	M	BE
CMP	A	BF

IMMEDIATE*

ANI	E6
XRI	EE
ORI	F6
CPI	FE

ROTATE†

RLC	07
RRC	0F
RAL	17
RAR	1F

NZ	(Z=0)	} Zero
Z	(Z=1)	
NC	(C=0)	} Carry
C	(C=1)	
PO	(P=0)	} Parity
PE	(P=1)	
P	(S=0)	} Sign
M	(S=1)	

Note: P=0 for odd parity
S=0 for plus

TRANSFER OF CONTROL GROUP

JUMP

JMP	C3
JZ	CA
JNZ	C2
JC	DA
JNC	D2
JPE	EA
JPO	E2
JM	FA
JP	F2

CALL

CALL	CD
CZ	CC
CNZ	C4
CC	DC
CNC	D4
CPE	EC
CPO	E4
CM	FC
CP	F4

RETURN

RET	C9
RZ	C8
RNZ	C0
RC	D8
RNC	D0
RPE	E8
RPO	E0
RM	F8
RP	F0

JUMP INDIRECT

PCHL	E9

I/O GROUP

IN	D8
OUT	D3

MACHINE CONTROL GROUP

STACK

POP	B	C1
POP	D	D1
POP	H	E1
POP	PSW	F1
PUSH	B	C5
PUSH	D	D5
PUSH	H	E5
PUSH	PSW	F5
XTHL		E3
SPHL		F9

$(L) \leftrightarrow ((SP))$
$(H) \leftrightarrow ((SP)+1)$
$(SP) \leftarrow (H)(L)$

RESTART

RST	0	C7
RST	1	CF
RST	2	D7
RST	3	DF
RST	4	E7
RST	5	EF
RST	6	F7
RST	7	FF

CONTROL

NOP	00
HLT	76
EI	FB
DI	F3
RIM	20
SIM	30

8085 Instructions by description format

DATA TRANSFER GROUP

Instruction		Result
MOV	r1, r2	$(r1)\leftarrow(r2)$
MOV	r, M	$(r)\leftarrow((rp))$
MOV	M, r	$((H)(L))\leftarrow(r)$
MVI	r, data	$(r)\leftarrow(\text{byte }2)$
MVI	M. data	$((H)(L))\leftarrow(\text{byte }2)$
LXI	rp. data 16	$(rl)\leftarrow(\text{byte }2)$
		$(rh)\leftarrow(\text{byte }3)$
LDA	addr	$(A)\leftarrow((\text{byte }3)(\text{byte }2))$
STA	addr	$((\text{byte }3)(\text{byte }2))\leftarrow(A)$
LHLD	addr	$\left\{\begin{array}{l}(L)\leftarrow((\text{byte }3)(\text{byte }2))\\(H)\leftarrow((\text{byte }3)(\text{byte }2)+1)\end{array}\right.$
SHLD	addr	$\left\{\begin{array}{l}((\text{byte }3)(\text{byte }2))\leftarrow(L)\\((\text{byte }3)(\text{byte }2)+1)\leftarrow(H)\end{array}\right.$
LDAX	rp	$(A)\leftarrow((rp))$
STAX	rp	$((rp))\leftarrow(A)$
XCHG		$\left\{\begin{array}{l}(H)\leftrightarrow(D)\\(L)\leftrightarrow(E)\end{array}\right.$

DATA MANIPULATION GROUP – ARITHMETIC

Instruction		Result
ADD	r	$(A)\leftarrow(A)+(r)$
ADD	M	$(A)\leftarrow(A)+((H)(L))$
ADI	data	$(A)\leftarrow(A)+(\text{byte }2)$
ADC	r	$(A)\leftarrow(A)+(r)+(CY)$
ADC	M	$(A)\leftarrow(A)+((H)(L))+(CY)$
ACI	data	$(A)\leftarrow(A)+(\text{byte }2)+(CY)$
SUB	r	$(A)\leftarrow(A)-(r)$
SUB	M	$(A)\leftarrow(A)-((H)(L))$
SUI	data	$(A)\leftarrow(A)-(\text{byte }2)$
SBB	r	$(A)\leftarrow(A)-(r)-(CY)$
SBB	M	$(A)\leftarrow(A)-((H)(L))-(CY)$
SBI	data	$(A)\leftarrow(A)-(\text{byte }2)-(CY)$
INR	r	$(r)\leftarrow(r)+1$
INR	M	$((H)(L))\leftarrow((H)(L))+1$
DCR	r	$(r)\leftarrow(r)-1$
DCR	M	$((H)(L))\leftarrow((H)(L))-1$
INX	rp	$(rh)(rl)\leftarrow(rh)(rl)+1$
DCX	rp	$(rh)(rl)\leftarrow(rh)(rl)-1$
DAD	rp	$(H)(L)\leftarrow(H)(L)+(rh)(rl)$
STC		$(CY)\leftarrow1$
CMC		$(CY)\leftarrow\overline{(CY)}$
DAA		decimal adjust (A)
CMA		$(A)\leftarrow(\bar{A})$

DATA MANIPULATION GROUP – LOGICAL

Instruction		Result
ANA	r	$(A)\leftarrow(A)\wedge(r)$
ANA	M	$(A)\leftarrow(A)\wedge((H)(L))$
ANI	data	$(A)\leftarrow(A)\wedge(\text{byte }2)$
XRA	r	$(A)\leftarrow(A)\veebar(r)$
XRA	M	$(A)\leftarrow(A)\veebar((H)(L))$
XRI	data	$(A)\leftarrow(A)\veebar(\text{byte }2)$
ORA	r	$(A)\leftarrow(A)\vee(r)$
ORA	M	$(A)\leftarrow(A)\vee((H)(L))$
ORI	data	$(A)\leftarrow(A)\vee(\text{byte }2)$
CMP	r	$(A)-(r)$
CMP	M	$(A)-((H)(L))$
CMI	data	$(A)-(\text{byte }2)$
RLC		$\left\{\begin{array}{l}(b_{n+1})\leftarrow(b_n)\\(b_0)\leftarrow(b_7)\\(CY)\leftarrow(b_7)\end{array}\right.$
RRC		$\left\{\begin{array}{l}(b_n)\leftarrow(b_{n+1})\\(b_7)\leftarrow(b_0)\\(CY)\leftarrow(b_0)\end{array}\right.$
RAL		$\left\{\begin{array}{l}(b_{n+1})\leftarrow(b_n)\\(b_0)\leftarrow(CY)\\(CY)\leftarrow(b_7)\end{array}\right.$
RAR		$\left\{\begin{array}{l}(b_n)\leftarrow(b_{n+1})\\(b_7)\leftarrow(CY)\\(CY)\leftarrow(b_0)\end{array}\right.$

( )	"the contents of"
←	"is transferred to"
↔	"is exchanged with"
∧	AND
∨	OR
⊻	EXCLUSIVE-OR

A, B, C, D, E, H, L	registers
addr	16-bit address
data	8-bit data
data 16	16-bit data
byte 2	2nd byte of instruction
port	addr of 8-bit port
r, r1, r2	register
rp	register pair

8085 Instructions by description format (cont.)

TRANSFER OF CONTROL GROUP

Instruction	Result
JMP addr	$(PC)\leftarrow$(byte 3)(byte 2)
Jcondition addr	$(PC)\leftarrow$(byte 3)(byte 2)
CALL addr	$((SP)-1)\leftarrow(PCH)$
	$((SP)-2)\leftarrow(PCL)$
	$(SP)\leftarrow(SP)-2$
	$(PC)\leftarrow$(byte 3)(byte 2)
Ccondition addr	$((SP)-1)\leftarrow(PCH)$
	$((SP)-2)\leftarrow(PCL)$
	$(SP)\leftarrow(SP)-2$
	$(PC)\leftarrow$(byte 3)(byte 2)
RET	$(PCL)\leftarrow((SP))$
	$(PCH)\leftarrow((SP)+1)$
	$(SP)\leftarrow(SP)+2$
Rcondition	$(PCL)\leftarrow((SP))$
	$(PCH)\leftarrow((SP)+1)$
	$(SP)\leftarrow(SP)+2$
PCHL	$(PCH)\leftarrow(H)$
	$(PCL)\leftarrow(L)$

I/O GROUP

Instruction	Result
IN port	$(A)\leftarrow$(data)
OUT port	(data)$\leftarrow(A)$

MACHINE CONTROL GROUP

Instruction	Result
POP	$(rl)\leftarrow((SP))$
	$(rh)\leftarrow((SP)+1)$
	$(SP)\leftarrow(SP)+2$
POP PSW	$(CY)\leftarrow((SP))_0$
	$(P)\leftarrow((SP))_2$
	$(AC)\leftarrow((SP))_4$
	$(Z)\leftarrow((SP))_6$
	$(S)\leftarrow((SP))_7$
	$(A)\leftarrow((SP)+1)$
	$(SP)\leftarrow((SP)+2)$
PUSH rp	$((SP)-1)\leftarrow(rh)$
	$((SP)-2)\leftarrow(rl)$
	$(SP)\leftarrow(SP)-2$
PUSH PSW	$((SP)-1\leftarrow(A)$
	$((SP)-2)_0\leftarrow(CY)$
	$((SP)-2)_2\leftarrow(P)$
	$((SP)-2)_4\leftarrow(AC)$
	$((SP)-2)_6\leftarrow(Z)$
	$((SP)-2)_7\leftarrow(S)$
	$(SP)\leftarrow(SP)-2$
XTHL	$(L)\leftrightarrow((SP))$
	$(H)\leftrightarrow((SP)+1)$
SPHL	$(SP)\leftarrow(H)(L)$

Instruction	Result
RST n	$((SP)-1)\leftarrow(PCH)$
	$((SP)-2)\leftarrow(PCL)$
	$(SP)\leftarrow(SP)-2$
	$(PC)\leftarrow8*(NNN)$
NOP	no operation
HLT	halt-processor stopped
EI	enable interrupts
DI	disable interrupts
RIM	read interrupt mask
SIM	set interrupt mask
SP	stock pointer register
SPH	high byte of stock pointer
SPL	low byte of stock pointer
PC	program counter
PCH	high byte of PC
PCL	low byte of PC
rl	low register of rp
rh	high register of rp
PSW	processor status word
*	multiply
NNN	binary representation of 'restart' number $0\rightarrow7$

Index